DISPOSAL OF HAZARDOUS WASTES IN INDUSTRIAL BOILERS AND FURNACES

DISPOSAL OF HAZARDOUS WASTES IN INDUSTRIAL BOILERS AND FURNACES

by

C. Castaldini, H.K. Willard
C.D. Wolbach, L.R. Waterland
Acurex Corporation
Mountain View, California

F.D. Hall, W.F. Kemner, G. Annamraju
R. Krishnan, M.A. Taft-Frank, D.N. Albrinck
PEDCo Environmental, Inc.
Cincinnati, Ohio

np
NOYES PUBLICATIONS
Park Ridge, New Jersey, U.S.A.

Library of Congress Catalog Card Number 85-25847
ISBN: 0-8155-1067-5
ISSN: 0090-516X
Printed in the United States

Published in the United States of America by
Noyes Publications
Mill Road, Park Ridge, New Jersey 07656

10 9 8 7 6 5 4 3 2 1

Library of Congress Cataloging-in-Publication Data
Main entry under title:

Disposal of hazardous wastes in industrial boilers
and furnaces.

(Pollution technology review, ISSN 0090-516X ;
no. 129)
Bibliography: p.
Includes index.
1. Hazardous wastes. 2. Incinerators. 3. Power-plants. 4. Furnaces. I. Castaldini, Carlo. II. Series.
TD811.5.D57 1986 628.4'457 85-25847
ISBN 0-8155-1067-5

Foreword

The disposal of hazardous wastes in industrial boilers and furnaces is described in this book. Processes which can convert wastes to energy continue to be of interest in energy intensive industries, and the disposal of hazardous wastes is a primary concern in today's environment-conscious society.

In the search for disposal alternatives, the USEPA has evaluated the potential use of high-temperature processes for the incineration of hazardous wastes. The feasibility of incineration in industrial boilers and furnaces is covered in Parts I and II of the book, respectively.

Part I documents a feasibility study to determine whether industrial boilers could potentially be used for safely destroying hazardous wastes. The study compares the theoretical conditions needed to destroy certain hazardous materials with conditions found in typical industrial boilers. The results suggest that boilers can achieve requisite levels of hazardous waste destruction and that further work in this area could be fruitful.

Part II evaluates the potential use of high-temperature processes for the incineration of hazardous wastes. Many kinds of waste have already been disposed of in boilers and cement kilns; this study considers the many other potential processes, such as metallurgical furnaces, brick and lime kilns, glass furnaces, and sewage sludge combustors. Each process is examined against such criteria as time/temperature profile, geographical location, product quality considerations, institutional factors, and environmental impacts. Promising alternatives are identified.

Industrial boilers and furnaces may become an important and acceptable method of hazardous waste disposal.

The information in the book is from:

A Technical Overview of the Concept of Disposing of Hazardous Wastes in Industrial Boilers, prepared by C. Castaldini, H.K. Willard, C.D. Wolbach, and L.R. Waterland of Acurex Corporation for the U.S. Environmental Protection Agency, June 1984.

Evaluation of the Feasibility of Incinerating Hazardous Waste in High-Temperature Industrial Processes, prepared by F.D. Hall, W.F. Kemner, G. Annamraju, R. Krishnan, M.A. Taft-Frank, and D.N. Albrinck of PEDCo Environmental, Inc. for the U.S. Environmental Protection Agency, April 1983.

The table of contents is organized in such a way as to serve as a subject index and provides easy access to the information contained in the book.

Advanced composition and production methods developed by Noyes Publications are employed to bring this durably bound book to you in a minimum of time. Special techniques are used to close the gap between "manuscript" and "completed book." In order to keep the price of the book to a reasonable level, it has been partially reproduced by photo-offset directly from the original reports and the cost saving passed on to the reader. Due to this method of publishing, certain portions of the book may be less legible than desired.

NOTICE

Contents and Subject Index

PART II
FURNACES, KILNS, AND COMBUSTORS

PART I

INDUSTRIAL BOILERS

The information in Part I is from *A Technical Overview of the Concept of Disposing of Hazardous Wastes in Industrial Boilers,* prepared by C. Castaldini, H.K. Willard, C.D. Wolbach, and L.R. Waterland of Acurex Corporation for the U.S. Environmental Protection Agency, June 1984.

1. Introduction

1.1 BACKGROUND

Approximately 40 million metric tons of hazardous wastes are currently generated annually in the U.S. On May 18, 1980, the Environmental Protection Agency (EPA) promulgated Phase I rules and interim status standards regulating the generation, transporation, treatment, storage, and disposal of these wastes under authority of the Resource Conservation and Recovery Act (RCRA). These and subsequently announced regulations place stringent restrictions on generators of wastes, including requirements for manifesting wastes and chain-of-custody liability. These restrictions may ultimately impose additional costs on the waste generator for offsite disposal. Thus, economic incentives for disposing of wastes onsite may result.

One method of disposing of wastes with high organic content is by thermal destruction. This can be done by high-temperature incineration, which is regulated by RCRA. However, another means of thermal destruction is by cocombustion of the waste, or use of the waste as a fuel (if its heating value is sufficiently high), in devices whose primary purpose is energy production, such as boilers and process heaters.

Current RCRA regulations specifically exempt facilities that burn wastes in energy-producing operations from complying with RCRA incineration mandates. In fact, it is estimated that about 20 million metric tons of wastes that might be classified as hazardous, owing to their characteristics, are currently burned as fuel, or cocombusted with fuel, in energy-producing operations.

The purpose of this study is to provide a basis for answering several key questions:

- What is the present and projected magnitude of hazardous waste combustion or cocombustion in boilers?
- Which waste materials and boiler types are the most compatible?
- Are there potential mechanical difficulties or environmental hazards associated with waste combustion in boilers?

The opportunity for combustion of hazardous wastes in industrial boilers is greater than currently practiced. There are over 40,000 industrial boilers in the U.S. with capacities greater than 10 million Btu/hr. These units offer great potential for hazardous waste combustion.

Regulations published on January 23, 1981, for incineration of hazardous wastes require 99.99-percent destruction and removal efficiencies (DRE) for the principal hazardous organic constituents in the waste incinerated. If the temperature-time profile of a given boiler type is similar to the time and temperature required to destroy a given waste, then the disposal of the hazardous waste in the boiler can become an option open for more extensive evaluation. This evaluation would consider the compatibility of the primary fuel to the waste, potential impacts on the design conditions of the boiler, retrofit requirements, and fugitive and secondary emissions. This report presents information on each of these issues.

1.2 SCOPE OF THIS STUDY

As previously noted, the objective of this study is to assess the problems and potential of destroying hazardous wastes in industrial boilers. The major elements of this overview include:

- Surveying hazardous waste generation and disposal practice
  - -- Document present generation and disposal inventories
  - -- Establish the present extent of waste combustion as fuel
  - -- Project the future extent of waste combustion as fuel
- Assembling an inventory of industrial boiler capacity capable of hazardous waste combustion
  - -- Inventory by size, design type, and fuel fired
  - -- Provide regional inventories
  - -- Select a set of model boilers typifying the installed population
- Characterizing the combustion characteristics of typical boilers
  - -- Establish boiler operating and design characteristics
  - -- Establish boiler temperature-time profiles as a function of boiler characteristics
- Characterizing hazardous wastes suitable for combustion
  - -- Specify the requirements (temperature and time) for waste constituent destruction
  - -- Define the composition of candidate wastes and rank the listed waste streams for potential use as fuel
- Identifying the matrix of wastes/boiler types suitable for hazardous waste combustion
  - -- Combine boiler combustion characteristics with waste constituent destruction requirements
  - -- Specify boilers/wastes achieving 99.99-percent destruction
- Provide an overview of considerations for further evaluating the waste/boiler destruction option

Results and findings associated with each of these elements as they address the objectives of the study are discussed respectively in Sections 3 through 8 of this report. Conclusions and recommendations of the entire study are summarized in Section 2.

2. Conclusions and Recommendations

As discussed in the introduction, the primary purpose of this study was to determine whether hazardous wastes could potentially be safely destroyed under the conditions found in a number of industrial boilers. The study determined that the thermal destruction requirements of most wastes evaluated are satisfied by the conditions found in a large percentage of industrial boilers currently in use. However, because of the nature of such a theoretical evaluation, it can only be concluded that boilers appear to have the potential for safe waste destruction. Further experimental work is necessary to confirm this conclusion. In this respect this study represents only a starting point for more detailed technical evaluation of industrial boilers as vehicles for the safe disposal of hazardous wastes. This study did establish that further work is warranted and could result in significant benefits in both expansion of safe, hazardous waste disposal capacity and the utilization of the energy content that wastes represent.

2.1 CONCLUSIONS

The conclusions developed from this theoretical analysis are:

- Cofiring of hazardous wastes at a small percentage of the base fuel (about 5 to 10 percent) appears to be a viable method of disposing of most hazardous organic material.
- Cofiring many wastes, or firing some high Btu content wastes entirely may produce lower levels of criteria pollutant and trace element emissions than either traditional coal or oil combustion.
- The conditions found in many types of watertube boilers appear to be sufficient to achieve at least 99.99 percent destruction of most hazardous organic compounds.
- The conditions found in firetube boilers do not appear to be sufficient to destroy all hazardous organic materials. There is too great a likelihood that cold tube-wall quenching of the waste degradation reactions is possible before complete destruction can occur.
- Watertube boilers whose furnace exit temperatures are greater than 1,100K (1,500°F) and whose furnace mean residence times are greater than 1 second appear to be best candidates for the destruction of simple hazardous organic waste streams.
- Complex organic waste streams are likely to require approximately 200K (360°F) higher temperatures.

- Insufficient data are available to predict whether hazardous products of combustion could be released from a boiler. Laboratory data should, however, be able to provide some conservative estimates of the likelihood of the release of such materials. The formation of hazardous products of combustion must be addressed by further research.
- In the absence of other data, the autoignition temperature appears to be a possible predictor of the relative ease of a compound's thermal destruction. This needs further confirmation.

2.2 RECOMMENDATIONS

This study showed that certain boilers can theoretically destroy hazardous organic materials. This work suggests the following areas of further research which need to be addressed:

- The available waste destruction kinetic data are based on results obtained from thermal destruction analytical system (TDAS) experiments. While such work is very useful, it does not model the destruction phenomena that occur in a flame. This type of information must now be obtained through a flame-mode destruction apparatus.
- The current TDAS work (as well as future flame-mode work) needs to be broadened to include destruction characteristics of the wastes under the pyrolytic (no oxygen) as well as under high oxygen concentration conditions. Further information on the formation and destruction of hazardous products of combustion is also needed.
- The information that will be obtained by EPA under the hazardous waste manifest system needs to be evaluated to determine other wastes which are potential candidates for destruction in boilers.
- A field program needs to be initiated to verify the preliminary results of this study. The program should include testing of the full boiler system, not just the boiler's emissions. The tests should include:
  - -- Full input/output analysis of all streams.
  - -- Boiler temperatures at a number of points in the combustion chamber
  - -- Boiler residence time variations.
  - -- Testing for the principal organic hazardous components in the waste and for hazardous products of combustion. TDAS data can be used to determine which products of combustion could occur and hence should be tested for.

3. Survey of Hazardous Waste Generation and Disposal

This section presents data on the rate of generation of hazardous wastes, the current practice of disposal in industrial boilers, and the future of this practice as impacted by regulatory and consumer activities.

3.1 HAZARDOUS WASTE GENERATION

Estimates of the quantities of hazardous wastes generated in the U.S. have been developed over recent years and documented in a wide variety of available literature. Studies ranging from in-depth industrywide surveys to studies presenting well-defined and specific segments of an industry provide valuable data on the type of hazardous wastes generated and the location of the generators. Information developed in recent reports, (Ref. 1-4) were the most pertinent data sources for quantifying hazardous waste inventories.

A 1980 report (Ref. 1) presented a brief overview of inventory data available from a variety of federal and state studies completed over the past several years. Although many of these efforts have been quite extensive, the ultimate disposal fate, and knowledge of quantities of hazardous wastes generated are still limited to estimation. The data compiled from the state surveys varied widely from state to state in level of detail and reliability, but many states have completed extensive surveys to determine quantities of wastes generated, waste types, and disposal practices of generators.

With the recent implementation of RCRA, each state must re-evaluate its methods of estimating the volume and determining the disposal practices of the hazardous waste generators. Even then, the basis of quantities will not be consistent since some states may have more stringent regulations than those required by RCRA. As an example, federal legislation contains a small generator exclusion. California has no small generator exclusion, thereby increasing the number of generators contributing to the hazardous waste inventory.

The definition of hazardous waste is the single most important factor in determining waste quantities and generators. Recent changes in EPA's list of hazardous components and hazardous waste definitions have impacted the values used in this report and those from the most recent Agency inventories (Ref. 5).

The data available from federal studies ranges from overview information, such as the assessment of hazardous waste practices in various industries (Ref. 6-8), which were conducted in the mid-1970's by EPA's Office of Solid Waste (OSW), to more specialized studies done on specific topics related to hazardous waste assessment and disposal.

3.1.1 Summary of Hazardous Waste Generated by EPA Region

Table 1 presents estimates of total quantities of hazardous waste generated by manufacturing industries in the U.S. by EPA region. These values were obtained using the best method available to estimate the total quantities of hazardous waste generated. They were based on the assumption that the ratio of the amount of hazardous waste generated by an industry to the number of employees in that industry is approximately constant for each plant in each industry (as grouped at the SIC code level). The table gives estimates of the quantities of hazardous waste generated within 17 of the 19 standard industrial classification (SIC) codes comprising the manufacturing industry, and one nonmanufacturing code. The stated values have limited reliability because of a lack of consistent and comprehensive site-specific data.

Since the RCRA regulations do not differentiate between wastes amenable to thermal destruction and those not amenable, quantities given in Table 1 include both. Therefore, to determine the quantities of waste potentially available for thermal disposal, the wastes which are primarily inorganic must be factored out. The remaining, primarily organic, wastes are called class A here. Of these wastes, the fraction containing high Btu material is called class B.

For class A (organic) hazardous wastes, the quantity of organic waste generated was determined by multiplying an estimate for the fraction of organic material contained in a given industry's waste (characterized by SIC code) by the generation rate of that industry's waste for each geographical area. These fractions were determined from quantities of waste reported in state inventories (Ref. 9-12) and an EPA waste report (Ref. 13). When quantitative values were not available, organic contents listed as high, large, low, or negligible were assigned values of 80, 50, 30 or 5 weight percent, respectively. The organic fractions used in subsequent calculations by SIC code are:

TABLE 1. SUMMARY OF HAZARDOUS WASTE GENERATED BY EPA REGION (1,000 METRIC TONS PER YEAR) (REF. 2)

SIC code		EPA Region I	II	III	IV	V	VI	VII	VIII	IX	X	Total	Percent of total
22	Textile mill products	27	24	4	132	10	6	0	0	0	0	203	0.5
24	Lumber and wood products	4	3	7	20	10	8	3	2	9	21	87	0.2
25	Furniture and fixtures	2	4	4	9	7	2	2	0	5	1	36	0.1
26	Paper and allied products	130	272	130	156	324	77	52	13	117	25	1,296	3.1
27	Printing and publishing	11	25	19	19	38	10	11	4	13	3	153	0.4
28	Chemicals and allied products	185	1,497	2,515	8,763	1,870	8,577	553	1	1,426	123	25,510	62.0
29	Petroleum and coal products	0	122	165	90	384	878	99	61	258	61	2,118	5.1
30	Rubber and miscellaneous plastics products	22	28	22	39	83	17	11	2	22	2	248	0.6
31	Leather and leather products	156	57	24	14	137	0	38	5	38	5	474	1.1
32	Stone, clay, and glass products	1	2	1	3	4	2	1	1	2	1	18	0.1
33	Primary metal industries	2	185	791	434	1,348	452	74	109	101	565	4,061	9.8
34	Fabricated metal products	204	212	172	172	781	137	96	28	164	31	1,997	4.8
35	Machinery, except electrical	26	31	28	29	127	24	20	6	28	4	323	0.8
36	Electric and electronic equipment	67	156	183	82	334	65	28	5	145	27	1,092	2.6
37	Transportation equipment	86	73	80	106	464	80	80	13	199	60	1,241	3.0
38	Instruments and related products	14	21	8	6	21	3	3	2	11	1	90	0.2
39	Miscellaneous manufacturing industries	48	66	24	42	72	18	12	7	24	6	319	0.8
40	Nonmanufacturing industries	119	335	177	237	414	177	118	59	276	59	1,971	4.8
	TOTAL[a]	1,104	3,113	4,354	10,353	6,428	10,533	1,201	318	2,838	995	41,237	100.0
	Percent of total	2.7	7.5	10.6	25.1	15.6	25.5	2.9	0.8	6.9	2.4	100	

[a]Totals may not balance because of rounding of numbers.

SIC code	Organic content (weight percent)	SIC code	Organic content (weight percent)	SIC code	Organic content (weight percent)
20	80	27	30	33	30
22	80	28	70	34	50
23	30	29	40	35	50
24	80	30	80	36	30
25	50	31	80	37	30
26	50	32	50	38	50
				39	30

The class A hazardous wastes listed in Table 2, along with quantities generated, represent the best candidates for thermal disposal. The values were obtained by multiplying the waste quantities for each SIC code (Table 1) by the assigned organic content, and totaling.

Class B (highly combustible) hazardous wastes, a subset of the class A wastes presented in Table 2, represent wastes with the highest heat content. The heat content of these is sufficiently high to support combustion. A study (Ref. 14) provided the basic industry data for compiling the quantity of class B wastes generated. The class B waste quantities identified in SIC codes 28 and 29, (by far the largest contributors to the waste class) were distributed among the region based on the regional generation of waste within that code. For example, from Table 1, 185,000 metric tons of SIC 28 waste were generated in Region I. This represents 0.73 percent of the SIC code 28 waste. Therefore, 0.73 percent of the SIC 28 generated class B total was assigned to Region I. The distribution of the wastes other than for the SIC codes 28 and 29 was determined using the total waste percentages shown at the bottom of Table 1.

The OSW generator notification information, as required by RCRA, and now available only on a limited basis, is expected to provide better estimates of waste generation rates. The number of industries that have to date notified EPA that they generate a hazardous waste is shown in Table 3. Further information on the two types and quantities of hazardous waste generated is unavailable at this time.

3.1.2 Options For Disposal of Hazardous Waste Based on State Inventory Data

Data on the amount of hazardous waste disposed of by the various available methods are presently unavailble. However, a comprehensive and recent (1979) industrial hazardous waste generation and disposal inventory is available for the State of New York (Ref. 9). Using this data base, representative values and approaches to quantifying the hazardous waste volumes that could be burned in boilers were developed. Such values are

TABLE 2. SUMMARY OF HAZARDOUS WASTE GENERATED BY EPA REGION (1,000 METRIC TONS PER YEAR)

	EPA Region										
	I	II	III	IV	V	VI	VII	VIII	IX	X	Total
Total hazardous waste EPA inventory[a] (Table 1)	1,104	3,113	4,354	10,353	6,428	10,533	1,201	318	2,838	995	41,237
Organic wastes class A[b]	550	1,621	2,390	6,735	3,002	6,778	627	105	1,490	369	23,668
Highly combustible wastes class B[c]	19	109	173	535	176	570	45	7	112	19	1,765

[a]Summary of hazardous waste is the OSW estimated values developed by multiplying regional employment figures for each SIC code times waste generation factors.

[b]Class A: Organic quantities were determined by averaging the organic fraction of hazardous waste for each two-digit SIC code and multiplying by the OSW estimated total hazardous waste for that industry segment in each region. These product values were added for each region.

[c]Class B: Highly combustible quantities were determined by selecting a limited set of industrial wastes (those useful for steam generation and having unit quantities) determined by Acurex based partly on Reference 14 and distributed to various regions on a basis of specific industry waste generation (SIC codes 28, 29, or all segments).

TABLE 3. NUMBER OF HAZARDOUS WASTE GENERATORS PER EPA REGION[5]

EPA Region										
I	II	III	IV	V	VI	VII	VIII	IX	X	Total
3,370	5,260	5,080	6,920	9,800	8,270	2,270	1,570	4,315	1,500	48,355

presented as an example since nationwide values could only be determined by having such data from all states.

Of the 1.12 million metric tons of hazardous wastes inventoried in New York, 0.56 million metric tons were disposed onsite by the following methods:

Disposal method	Metric tons per yr
Distillation	48,610
Incineration	94,040
Reclaimed for fuel or material value	148,780
Other (landfill, lagooning, storage, neutralization)	
Total	560,600

The organic fraction of materials in the "other" category (mostly solvents, oils, and still bottoms) was 15,800 metric tons. In addition, wastewater treatment sludge, mixed waste, and lagoon bottoms also contained combustible organic materials. In total, the organic waste categories constitute about one-half of the total hazardous waste material.

Table 4 lists the major hazardous waste generating industries in New York, including those generating nonorganic type wastes. The total quantity of waste tabulated in Table 4 represents greater than 86 percent of the total (all industry) quantity of hazardous wastes generated in New York.

The New York survey also categorized wastes by type as shown in Table 5. Of these wastes, types that can be incinerated include halogenated solvents, other solvents, waste oils, still bottoms, leather tanning waste, and some wastewater treatment sludges. By reviewing the industries contributing to the wastewater sludge category, it was estimated that the organic content of

TABLE 4. MAJOR HAZARDOUS WASTE GENERATING INDUSTRIES IN THE STATE OF NEW YORK (REF. 11)

Type	Approximate hazardous waste generation (metric tons)	Percent of total from survey
Inorganic chemicals	61,250	5.5
Organic chemicals, pesticides, and explosives	290,314	25.9
Miscellaneous chemicals	12,095	1.1
Primary metals	200,012	17.8
Pharmaceuticals	105,900	9.4
Plastics and rubber	105,601	9.4
Leather tanning and finishing	54,438	4.9
Motor vehicles and equipment	44,983	4.0
Photographic chemicals and equipment	40,207	3.6
Machine manufacturing	38,109	3.4
Electroplating and metal finishing	13,534	1.2
Total	966,443	86.2

TABLE 5. VOLUMES OF HAZARDOUS WASTES BY TYPE IN THE STATE OF NEW YORK (REF. 11)

Type	Approximate hazardous waste generation (million liters)[a]	Percent of total from survey
Acids	72.0	7.7
Alkalies	9.0	1.0
Pickle liquor	55.0	5.9
Etching solution	4.0	0.5
Halogenated solvents	4.0	0.4
Other solvents	192.0	20.6
Heavy metal sludges	55.0	5.8
Wastewater treatment sludges	198.0	21.3
Other sludges	63.0	6.8
Waste oils	57.0	6.4
Still bottoms	12.0	1.4
Photographic chemicals	7.0	0.8
Rinse water and other wastewaters	45.0	9.3
	773.0	
Subtotal	982,270 metric tons	
Leather tanning wastes	54,445 metric tons	4.9
Total	1,036,720 metric tons	92.8

[a]Use 1.12 kg/liter to convert to metric tons.

wastewater sludge was 42 volume percent. When this quantity was added to the organic content of the other components, the combustible waste fraction totalled 477,000 metric tons or 43 percent of the total waste quantity generated in the state. The total quantity of these waste types represents over 92 percent of the total hazardous waste listed in the New York survey.

The most significant factors illustrated by this case study are the large quantity of combustible hazardous waste (over 43 percent of the total), the large number of generators that dispose of waste onsite (over half), and the prominence of industries that usually have onsite boilers. Such a study which provides site-specific data on wastes and boilers represents the only method that can be used to quantify the amount of material that could be boiler destroyed if plant operators so choose. If the inventory had separately reported the use of reclaimed material for fuel, the information would be more helpful. However, this inventory provides more exact estimates of hazardous waste that could be used for fuel in boilers than the current EPA RCRA background data (Ref. 2,3).

Such an estimation would utilize the various classes of waste given, stream size, and industry type, and correlate these values with characteristics provided in the literature (Ref. 7,8,15) and boiler availability. While an estimate of the extent of hazardous waste destruction in boilers for the state of New York is not carried out here, this site-specific data show that the quantities of waste which are potentially boiler-disposable are considerably higher than estimates based on general industrywide waste estimates would predict. The primary reason that such inventory data are more useful is the inclusion of quantities of combustible types of wastes. In the case of the data from New York, the extent of solvent waste disposal is much higher than most available data would indicate. Thus, at least some state data indicate that the potential for boiler destruction could be higher than generally expected.

3.2 CURRENT PRACTICE OF DISPOSAL OF HAZARDOUS WASTE IN INDUSTRIAL BOILERS

Though the destruction of hazardous materials in steam boilers and space heaters has occurred randomly at various locations, only limited cases are reported in the literature. In those industries where the practice has gained widespread acceptance, the wastes are generally classified as fuels and not hazardous wastes. Therefore, this survey includes destruction in industrial boilers of hazardous wastes as defined in 40 CFR Part 261 of RCRA, as well as those hazardous wastes exempt from RCRA because of their fuel value. For example, waste oil burned for fuel is exempt from RCRA though it would potentially be classified a hazardous waste if it were disposed of. Waste material fuels not included in this survey include waste gas and biomass (wood, bark, bagasse, food processing products, other undegraded cellulosic materials).

Several studies have been conducted on the reuse or disposal of waste materials in combustion processes. Recent assessments (Ref. 1,14) are the most pertinent to this effort. Also, another recent study (Ref. 4) listed waste materials burned for energy for several industries. Other literature

(Ref. 16,17) amplify data on air emissions or cofiring conditions for similar waste destruction.

Quantification of hazardous waste disposal in boilers includes several groups of waste categories described below. They include: (1) waste-oil firing, (2) pulping-waste firing (3) petroleum refining fuels, and known individual plant burning of other hazardous wastes.

3.2.1 Waste Oil

Excess spent oils contaminated by metal, dirt, water, or organic solids and liquids represent a major hazardous waste material. An estimated 9,458 million liters of oil are used yearly in the U.S. (see Table 6). Since 1960, the quantity of waste that has been re-refined has decreased from a peak of 1,136 million liters to an estimated 189 million liters in 1977. Less severe processing practices than re-refining (see Table 7) are estimated to condition over 379 million liters/yr, generating some 40,000 metric tons of sludge and tank bottoms. Anticipated new re-refining equipment may reverse the decreasing trend and produce higher re-refined oil sales in the 1980's (Ref. 15). Simultaneously, more sludge and bottom deposited materials will be generated.

Currently waste oil boiler feed is estimated as follows:

	Million liters/yr
Direct use as fuel	1,360 (high percent to boilers)
Processing plant fuel	185 (high percent to boilers)
Sold for fuel	2,350 (high percent to boilers)
Total	3,895
Reclaimable waste oil:	
Other uses of reclaimable oil	
Re-refined oil	314
Process waste	273
	587
Fuel	3,895
Total	4,478

Comparable other study values were 4,200 million and 4,700 million liters/yr.

TABLE 6. WASTE OIL INVENTORY (MILLION LITERS PER YEAR) (REF. 18)

Waste oil generation	Million liters
Automotive lubricating oil	2,332
Industrial and aviation lubricating oil	1,491
Other industrial oil	329
U.S. government lubricating oil	68
Marine oil, transportation losses, oil spills	791
Other oil losses in production and transportation	4,375
Subtotal	9,386
Oil and fat industry losses	49
Coke oven oil losses	23
Total generated	9,458
Ultimate disposal	
Directly to environment	3,702
Processing waste to environment	273
Road oil dust control	1,064
Subtotal environment	5,039
Directly for use as fuels	1,355
Used for processing plant fuels	185
Fuels sold by oil processors	2,350
Subtotal fuels	3,890
Road oil and dust control, lost by biodegradation	144
Re-refined oil sales	314
Subtotal other	458
Unknown	72
Total disposal	9,458

TABLE 7. WASTE LUBE OIL RECYCLING METHODS

Reclaiming	
Removes	Water, solids
Process	Heat 100°0 to 180°0F Separatory screening Demulsification Gravity settling or centrifugation
Products	Fuel Parting agents, e.g., concrete forms
Reprocessing	
Removes	Water, solids Additives blended with original oil Compounds formed during use
Process	Heat 150°0 to 200°0F Chemical treatment, acid, caustic, sodium silicate Surfactants (demulsification, solids settling) Gravity settling or centrifugation Separatory screening Filtration
Products	Fuel Noncritical lubricants Parting agents
Re-refining	
Removes	Water, solids Original additives Compounds formed during use Light ends (low boiling fractions) Color bodies and soluble high boiling compounds
Process	Chemical pretreatment (similar to reprocessing) Distillation 550°0 to 650°0F Clay contact Filtration
Products	Distillate fuel Lubricating oils Motor oils Hydraulic oils Machining oils Process oils

Since the above values are typical of 1971 oil usage, a growth factor has been applied to obtain 1980 values. During 1965 to 1975, oil usage increased 11 percent for segments with high reclaimable oil, an annual growth rate of 1.1 percent. Using this growth rate from 1971 to 1980 gives a 9.9 percent increase. The 1971 value of 3,891 million liters yields a 1980 estimate of 4,300 million liters/yr or 4.1 million metric tons.

3.2.2 Waste Chemical Pulping Liquors

Chemical pulping of wood is generally carried out under highly acidic or alkaline conditions and usually produces both a pulp product and spent liquor byproduct. Spent liquors from the alkaline sodium process (Kraft pulping) are evaporated and burned for pulping chemical (and energy) recovery. Sulfite (acid) spent liquors traditionally were disposed by lagooning or wastewater discharge until regulation in the 1950's and 1960's stopped this practice. Most of these latter wastes are now concentrated, burned, or neutralized and sold as concrete additives, oil drilling mud components, and other byproducts. Table 8 presents the pertinent yield factors for various pulping wastes. An estimated 5.3 million metric tons of solids are currently burned (at 65 percent solids) in boilers as the primary fuel with supplemental oil-firing mechanisms.

3.2.3 Petroleum Refining Onsite Waste Burning

Current petroleum refinery production has been estimated to be 16 million bbl/day nationally (Ref. 20). A highly variable fraction of this process material is burned as fuel (frequently fired in boilers) or discharged as process waste. Details of the various wastes are given by Rosenberg et al.(Ref. 6). Although the report is very comprehensive in its coverage of industry practices, no use of materials for energy generation were identified (by definition fuel was exempt). To determine the amount of material burned in boilers, the quantity of fuel was estimated with exclusions for gases and crude oil firing. Utilizing a net liquid fuel use factor of 0.875 percent, the total tonnage burned in boilers is estimated at 7.64 million metric tons/yr nationwide.

3.2.4 Industry Waste Solvent Burning for Energy

Waste solvents generated by the organic chemicals industry are routinely reclaimed as solvent, distilled (also for reclaimed solvent), used for fuel, or incinerated. Other disposal methods such as landfill with residues or storage account for a smaller fraction. Harrison et al. (Ref. 1) has noted the high energy recovery potential of these wastes but found that only a small fraction of the solvents were reclaimed. Several state inventories have included burning of solvents as a specific disposal method, thereby quantifying the disposal in boilers and heaters (Ref. 9,10).

End uses of solvents from degreasing operations include reclamation of 45 percent, while 15 percent is mixed with waste oil (most of which is used as fuel) (Ref. 21). Similar percentages were reported by the State of Massachussets (Ref. 10). These values are higher than an earlier reported

TABLE 8. FACTORS DETERMINING PULPING HAZARDOUS WASTES DESTROYED IN BOILERS

Pulp type	Yield	Waste per unit of fiber	Fiber production (1,000 tons) 1972[a]	1980 Projection	Waste solids 1980	Fraction burned as waste	Waste quantity burned 1,000 tons/yr	1,000 metric tons/yr
Dissolving	0.36	1.78	1,677	1,720	3,062	0.95	2,909	2,645
Sulfite	0.48	1.08	2,130	1,980	2,138	0.80	1,710	1,555
Neutral sulfite semichem	0.72	0.39	3,800	4,200	1,638	0.75	1,228	1,116
							5,847	5,315

[a]Reference 19.

2 percent (Ref. 22). In New York, detailed solvent disposal inventories indicated that most waste solvents were generated by SIC code 28 industries and about 26 percent of these were burned for fuel. Most state inventories do not identify material burned for fuel, thus, the New York data may be the best available. The solvent component in SIC 28 hazardous wastes ranged from 1 to 50 percent in state inventories (Ref. 9,12,23-25). A value of 20 percent of all SIC code 28 industry hazardous wastes was assumed to represent the quantity of solvent. Using 26 percent as the amount of solvent used for fuel, a factor for total hazardous wastes burning as fuel of 5 percent was calculated (0.20 x 0.26 = 0.052). Although this factor may be low for many areas, it amounts to over 1.4 million metric tons of the estimated 28.6 million metric tons of SIC code 28 industry waste generated nationally.

3.2.5 Miscellaneous Waste Firing Examples

Although not quantifiable, a variety of other industries are known to dispose of small amounts of organic waste in boilers. These include plywood (condensed hydrocarbon residues), wood processing (wastewater sludge), textile manufacturing (liquid waste), and petrochemical plants (liquids). Specific examples are shown in Table 9.

3.2.6 Summary of Current Practice

From the above waste generator categories, approximately 19 million metric tons/yr of organic waste materials are estimated to be fired in industrial boilers as fuel. This breaks down as follows:

Fuel source	Quantity (million metric tons)	Estimated accuracy
Waste oils	4.1	±30%
Industry solvents	1.4	Inaccurate
Pulping waste liquors	5.3	±5%
Petroleum refining	7.6	Unknown
Textile/petrochemical liquids	0.4	Unknown
Miscellaneous plants	Undetermined	--
Total	18.8	

Acknowledging that many of the above fuels are commonly accepted as traditional fuels rather than wastes, they are included as defined earlier. In all cases except petroleum refining, these materials were wasted to the environment before environmental pressure or handling costs in addition to

TABLE 9. LIQUID ORGANIC WASTE STREAMS CURRENTLY DISPOSED IN BOILERS (REF. 4)

Composition	Source	Producer	Location	Process	Product	Flowrate (kg/hr)
ND	Low-boiling overheads	Tennessee, Eastman	Kingsport, TN	Ketone	Acetic anhydride	ND
Organics	Heavies removal column	Union Carbide	Brownsville, TX	n-Butane oxidation	Acetic acid, formic acid, methyl ethyl ketone, ethyl acetate, acetic anhydride	ND
Organics	Propionic column	Union Carbide	Brownsville, TX	n-Butane oxidation	Acetic acid, formic acid, methyl ethyl ketone, ethyl acetate, acetic anhydride	ND
Organics	Butyl removal column	Union Carbide	Brownsville, TX	n-Butane oxidation	Acetic acid, formic acid, methyl ethyl ketone, ethyl acetate, acetic anhydride	ND
Organics	Ethyl acetate refining column	Union Carbide	Brownsville, TX	n-Butane oxidation	Acetic acid, formic acid, methyl ethyl ketone, ethyl acetate, acetic anhydride	ND
Light ends	Light ends refining column	Union Carbide	Brownsville, TX	n-Butane oxidation	Acetic acid, formic acid, methyl ethyl ketone, ethyl acetate, acetic anhydride	ND

(continued)

ND = No data

TABLE 9. (continued)

Composition	Source	Producer	Location	Process	Product	Flowrate (kg/hr)
Acetaldehyde volatiles	Distillation	Union Carbide	Taft, LA	Oxidation	Acrolein	129
Water, oxygen, carbon, hydrogen, ash	Waste oil stream from refining	DuPont	Orange, TX	Cyclohexane	Adipic acid	8,600
Water, oxygen, carbon, hydrogen, ash	Waste oil stream from refining	DuPont	Victoria, TX	Cyclohexane	Adipic acid	8,600
Cyclohexanone, 0.5%, cyclohexane, 32% phenol, 8.5% heavies, 59%	Heavy ends	Union Carbide	Taft, LA	Phenol	Cyclohexanol, cyclohexanone	88
High boiling organics	Adiponitrile purification	Monsanto	Decatur, AL	Adiponitrile	Hexamethylene diamine	ND
Low boiling organics	Adiponitrile purification	Monsanto	Decatur, AL	Adiponitrile	Hexamethylene diamine	ND
High boiling organics	Hexamethylenediamine purification	Monsanto	Decatur, AL	Adiponitrile	Hexamethylene diamine	ND
ND	Residual tars	Union Carbide	Institute, WV	Chlorination	Linear alkylbenzene	143
Methanol, 70%; water, 8%; other organics, 22%	Extraction and refining	Tenneco	Houston, TX	High pressure	Methanol	1,224
Water, 25% organics, 70% methanol, 5%	Spent catalyst	Union Carbide	Institute, WV	Methanol hydro-chlorination	Methyl chloride	0.05
Organic polymers	Secondary butyl alcohol purification and water extraction	ARCO Chemical	Channel View, TX	Secondary butyl alcohol	Methyl ethyl ketone	253

ND = No data

increasing fuel costs, encouraged their use as fuel. In all cases, spills or improper alternative disposal could cause adverse environmental damage.

3.3 PROJECTED TRENDS IN HAZARDOUS WASTE DISPOSAL IN INDUSTRIAL BOILERS DURING THE PERIOD 1980 TO 1985

Many factors will influence the future use of industrial boilers for the disposal of hazardous wastes. These include the compatibility of the waste material and boiler type (see Sections 6 and 7), emission and hazardous waste disposal regulations, byproduct recovery potential, and other economic and technical considerations (see Section 8).

The sensitivity of projected degrees of waste disposal are dependent on each of the above factors. Using the waste categories discussed in the previous section, these same materials are evaluated to establish the projected change in quantities disposed of in industrial boilers.

Boiler disposal of groups of waste is expected to become more prevalent during 1980 to 1985. These groups are waste oils, spent solvents, selected hazardous wastes listed in 40 CFR 261.31, and other high energy content organic waste materials being generated and currently disposed of at sites containing industrial boilers. Changes in the practice of boiler disposal for all groups will be influenced by recent and upcoming rules under RCRA. The three most pertinent RCRA related factors are the phase I (45FR33215) exclusion of burning a waste in an energy recovery process from subtitle C rules, phase II rules on incineration and possible future rules for boilers, and waste hauling and disposal costs or limitations as effected by the manifest and landfill rules.

Industries that are using an established disposal practice may change that practice because of changes in costs, onsite restrictions, new hauling requirements, or other factors. New requirements for onsite disposal as well as environmental controls on existing systems will increase the paperwork. Similar requirements for haulers and offsite disposal will increase costs to generators. These factors will be impacted by specific site costs that preclude projecting general trends in hazardous waste management.

Generators and haulers of hazardous wastes are anticipating changes in disposal practice because of significant upcoming rules. For example, if EPA extends the phase II requirement of 99.99-percent destruction from incinerators to include boilers as originally planned (Ref. 26), the practice of boiler destruction will certainly be affected. If the current boiler disposal exclusion is extended, then the conditions under which the exclusion is allowed will become very important.

Since EPA has funded other reviews of current waste management practices and the probable changes in practices that may occur, several findings of these efforts are presented. Other options for handling waste will directly compete with boiler disposal. Costs for competing optional management practices are covered in Table 10 while existing capacity of those options is presented in Table 11. Although capacity is apparently available for the

TABLE 10. REPORTED PRICES FOR HAZARDOUS WASTE MANAGEMENT SERVICES, 1980 (REF. 5)

Waste management option	Dollars per metric ton	Cents per gallon
Landfill[a]		
Wastes that are not acutely hazardous, including sludges	20-90	8-34
Highly toxic, explosives or reactives	100-400	38-152
Land treatment	5-25	2-9
Incineration		
High Btu value, no acute hazard	50-300	19-114
Highly toxic, heavy metals	300-1,000	114-378
Chemical treatment		
Acids, alkalines	15-80	6-30
Cyanides, heavy metals, highly toxic	100-500	38-200
Resource recovery	50-200	19-80
Deep-well injection		
Oily wastewaters	15-40	6-15
Dilute toxic rinse waters	50-100	19-38

[a]By comparison, disposal in sanitary landfills generally costs between $5 and $10 per ton.

TABLE 11. 1980 TOTAL CAPACITY FOR WASTE MANAGEMENT OPTIONS[5]

Waste management option	Capacity (thousand wet metric tons)	Current utilization (percent of capacity)
Land treatment/solar evaporation	2,437	22
Incineration	670	59
Chemical treatment[a]	3,921	54[b]
Resource recovery	1,069	40
Deep-well injection[c]	4,657	17
Subtotal	12,754	
Landfill (lifetime)	27,604	10.2 years
Total, all waste management options	40,358	

[a]Since there is an estimated 10-percent overlap between capacity to chemically treat and landfill material, the total capacity reported here represents capacity less overlap.

[b]Chemical treatment utilization computed on gross capacity of 4,357 wet metric tons.

[c]Although size of reservoir is many times larger, other factors such as unloading, incoming trucks, filtration and pumping dilution water may make it difficult to exceed 50 percent of the pumping rate according to facility operators. Capacity is based on 50 percent of pumping capacity or current utilization rate, whichever is highest.

NOTE

Accuracy of capacity estimates is judged to be ±25 percent.

near term, high costs of some options may suggest savings could be attained by changing disposal to boiler firing if this proves less expensive. (See Table 12 for 1980 to 1982 projected capacity.) A complete discussion of the way RCRA rules will impact various waste treatment options is presented in Reference 2.

3.3.1 Waste Oils

A large class of wastes that may increasingly be disposed in industrial boilers is waste oils -- emulsified, nonemulsified, or mixed. Contaminants are generally metals or organic liquids, but water, dirt, and other inert contaminants are common. As crude oil prices have increased over the past 8 years, the value of spent or contaminated oil has increased primarily because of its fuel value. Most of this oil is not re-refined for use as lubricants.

Several categories of waste oil have been increasingly reclaimed as a fuel additive. As stated earlier, re-refined oil sales volume decreased during the early 1970's and then slowly increased in the past 2 years. Competition came mainly from the fuel usage component of waste oil, which caused the unit price to increase rapidly (Ref. 15). This same factor is expected to decrease the amount of oil wasted directly to the environment and the waste from processing residual oily sludge. Essentially all oil disposed of onsite is currently burned or incinerated. Increased boiler disposal may occur for oils now lagooned, stored onsite, or land disposed. State inventory data (Ref. 9,11,23) and Harrison et al. (Ref. 1) indicate that only 20 to 45 percent of waste oil is currently disposed of onsite. Thus, the remaining oil wastes may be expected to constitute most of the heavily contaminated material available for burning. Impetus for destruction in boilers may increase disposal and displaced fuel as costs increase.

The conversion of oil disposal to boilers has been determined for two assumed rates, an increase of 10 and an increase of 40 percent over current levels, over the next 5 years. Such a conversion rate is applied only to the current disposal practices of lagooning, storage, landfilling, or incineration. An estimated 136 million liters are disposed of by incineration. The other disposal methods handle 371 million liters. Fifty-five percent of the additional waste oil generated in the future is assumed to be burned, a rate 10 percent higher than the current incidence of 45 percent. The anticipated growth of waste oil generation is projected to be 0.9 percent per year (Ref. 23). This translates to 433 million liters of additional generation in 1985. Fifty-five percent of this quantity is projected to be boiler disposed, or 238 million liters of this new waste oil. A summary of increased firing of waste oil (excluding oily sludge) for the assumed use rates is:

TABLE 12. HAZARDOUS WASTE MANAGEMENT INDUSTRY CAPACITY FORECASTS BY WASTE MANAGEMENT OPTION (1,000 WET METRIC TONS) (REF. 5)

	1980			1981			1982		
Waste management option	Lower bound	Baseline	Upper bound	Lower bound	Baseline	Upper bound	Lower bound	Baseline	Upper bound
Lifetime:									
Landfill[a]	20,979	27,604	34,229	23,048	30,334	37,619	19,094	25,124	31,157
Annual									
Land treatment	1,852	2,437	3,022	2,245	2,972	3,695	2,293	3,034	3,774
Incineration	509	670	831	779	1,037	1,295	1,094	1,490	1,886
Chemical treatment	2,980	3,921	4,862	3,546	4,787	6,028	4,450	5,910	7,370
Resource recovery	812	1,069	1,325	888	1,184	1,478	972	1,287	1,602
Deep-well injection	3,539	4,657	5,775	3,539	4,657	5,775	3,539	4,657	5,775

[a]Assumes regional utilization at 1980 rate or 5-year lifetime, whichever is highest, for 1981 and 1982 to determine lifetime capacity.

Oil source currently disposed	Increase use as fuel 1980 to 1985 (million liters)	
Incineration	13.6 10% to fuel	54.9 40% to fuel
Current disposal by means other than incineration or burning	37.1 55% to fuel	149.5 55% to fuel
Amount of increased new waste oil to fuel	238.1	238.1
Total	289.2 (289,640 metric tons)	442.5 (443,180 metric tons)

An OSW study of the burning of waste oil in industrial boilers to be released in the near future may clarify the current practice of waste oil disposal.

3.3.2 Solvent Firing for Fuel

Increased firing of spent solvents is anticipated because of hazardous waste control regulations coupled with their generally high energy content and good firing characteristics (Ref. 1,14). Mixing of spent solvents with waste oils or oily material is common (Ref. 1). Since contradictions in the data on quantities of these wastes disposed by various means preclude a mass balance approach to estimate the extent of solvent firing, only specific solvent usage was reviewed.

Increases in firing solvents for fuel are projected as a result of increased onsite burning of waste solvents from increased solvent usage, increased burning of contaminated spent solvents previously landfilled and deep-well injected, and increased burning of solvent mixed with oily and organic liquid wastes. Boiler disposal of halogenated solvents or solvents currently incinerated is considered to remain negligible.

Currently, landfills, deep-well injection, and storage account for 212,730 metric tons per year of spent solvent disposal. This quantity was determined by multiplying the percent of spent solvents disposed of by these methods (24 percent (Ref. 1)) by the total quantity of spent solvents generated (875,000 metric tons) (Ref. 22,27). Solvent production is anticipated to increase at 1.2 percent per year, or by 57,270 metric tons over 5 years. A solvent content factor was determined from state inventory data (Ref. 9,10) for solvent in organic chemical still bottoms, liquids, and waste oils in the chemical industry. This factor (6.7 percent) times the SIC code 28 hazardous waste

Solvent source	Increased use as fuel, 1980 to 1985 (metric tons per yr)	
	10% to fuel	40% to fuel
Currently directly disposed of in landfills, deep-well injection, or storage (now 212,730 metric tons per yr)	21,273	85,090
Increased production (at 1.2 percent per yr -- 57,270 metric tons)	5,727	22,910
Amount in still bottoms, liquids, or oils codisposed in boilers increase in practice	17,180	68,730
Total	44,180	176,730

generation (25.5 million metric tons) gives 1.72 million metric tons. Currently 10 percent of this material, or 172,800 metric tons is disposed in boilers. If between a 10 and 40 percent increase in the incidence of this practice is probable, then an increase of 17,180 to 68,730 metric tons would be projected.

3.3.3 Pulping Processes and Petroleum Refining

Quantities of petroleum refining waste firing and pulp manufacturing sulfite liquor firing are not anticipated to change between 1980 and 1985. Increases of petroleum refining production (<1 percent/yr) and the practice of refinery waste burning are considerably less than the inaccuracy of the original practice estimate of 7.6 million metric tons/yr. Production of pulping sulfite liquor is anticipated to decrease, although recovery and burning of other hazardous byproducts should increase slightly.

3.3.4 Miscellaneous Industries

To complete the quantification of the projected increase in the practice of boiler disposal of hazardous wastes, several SIC code industries were evaluated for significant possible increases not covered by oils or solvents. Two restrictions eliminated most industry segments. By limiting waste streams considered to those with high energy content, wastewater sludges and low organic content mixed wastes were excluded. The second limitation was to exclude all segments whose combustible waste was primarily waste oil or solvents, covered above. Excluded were the major combustible waste generating segments of SIC codes 26, 29, 30, 32, 34, 35, 37 and tentatively 22. Only segments covered by SIC codes 28 (organic chemicals) and 31 (leather products) appeared to have significant other combustible wastes. Food and kindred products (SIC code 20) was excluded since various background documents are contradictory, identifying the subject waste as either hazardous (Ref. 27) (290,000 metric tons/yr) or completely nonhazardous (Ref. 2). Some food processing waste is hazardous but is mainly composed of items not easily burned in boilers. These include pesticide residues in sludges, contaminated containers, and dyes. Hazardous organic materials used to process foods (i.e., coffee extracting) are disposed of in small quantities.

A detailed review of SIC codes 28 and 31 industries to determine the quantity of boiler disposable waste at facilities that have boilers would be quite time consuming. However, a short approach was followed for this report that was primarily based on previously compiled data (Ref. 5,14). For the chemical industry, the initial data included 14 million metric tons, wet weight, of hazardous wastes in SIC code 28 industries. The dry weight content was estimated to be 4.7 million metric tons. Of this material, 2.0 million metric tons was rated good or excellent for boiler disposal.[1] Based on usage, rather than waste generation, the solvent content of this waste (previously accounted for) amounts to 169,000 metric tons. The remaining 1.83 million metric tons is currently disposed of by other methods.

A factor was assigned to each current disposal technique which represented the probability that generators would discontinue disposal by the current process and burn the material in boilers instead. This factor is based on the premise that one or more of the following near-term pressures on the hazardous waste producer were likely to encourage such a change: savings in fuel, a change in existing equipment and systems, or a change in operating personnel and procedures. A current disposal method would be assigned a probability factor of 10 percent if it did not meet any of these criteria. 33 percent if it met one criterion, and 50 percent if it met two criteria.

Processes with minimal environmental impact to generators, hauling offsite to landfills or incineration, and controlled incineration were considered the processes least likely to change to boiler disposal, since no near-term pressures were likely to encourage this change. A 10 percent probability was assigned to these. Onsite deep-well and landfill disposal are more capacity-limited and expensive to operate. A factor of 33 percent was assigned here since only one of the three factors favor it. Uncontrolled incineration is expected to face some plant changes because of the phase II requirements; thus, a full 50 percent would be expected to change disposal to boiler feed due to economy.

Process	Percent waste treated (Ref. 14)	Probability factor for disposal change
Landfill	14.8	0.33
Deep well	2.0	0.33
Uncontrolled incineration	48	0.5
Controlled incineration	21.6	0.1
Landfill offsite	3.6	0.1
Incineration offsite	1.5	0.1
Weighted average	--	0.32

By applying the composite conversion (from other disposal to boilers) probability factor for organic chemical hazardous wastes (0.32) to the quantity of nonsolvent hazardous wastes identified (Ref. 27) (3.31 million metric tons after adjustment to include all the current nonsolvent total SIC code 28 hazardous wastes), a total of 1.06 million metric tons is obtained. New production of waste is projected to be 4.3 percent per year (same as industry growth) and amounts to 704,500 metric tons by 1985.

Using the increased boiler disposal fractions of 10 and 40 percent, the annual amounts are:

Source	Increased use as fuel, 1980 to 1985 (metric tons per yr)	
Organic chemicals (nonsolvent SIC code 28)	10%	40%
Conversion from other disposal	96,360	385,440
New production	70,450	281,800
Total	166,810	667,240

The leather finishing and tanning industry (SIC code 31) has been assessed for hazardous wastes in several studies. Currently, EPA is changing the listing of all chromium as a major toxicant and substituting only hexavalent chromium. Also, some tanning wastes were removed from the list (November 1980). Only an estimated 5 percent of the existing tanneries have coal or other solid fuel boilers that could fire their hazardous waste solids. Furthermore, the quantity of hazardous wastes by definition for this industry has decreased by over half by the chromium component clarification -- down to 218,000 metric tons (Ref. 5). Future growth is not expected for this industry. Using 10 and 40 percent waste disposal fractions and the prior factors, the annual amounts are:

Source	Increased use as fuel, 1980 to 1985 (metric tons per yr)	
	10%	40%
Leather tanning and finishing (SIC code 31)	1,100	4,400

3.3.5 Evaluation of Quantification Approach

The approach presented above appears more reasonable than using a purely SIC code summary approach, as the latter requires better composite data than are available. An example has been selected to illustrate this point.

Supporting data for EPA's Environmental Impact Statement (April 1980) contained an estimated value of 2.87 million metric tons for SIC code 26 industries without identification of its composition. EPA's background document on Hazardous Waste Generation (Ref. 5) identified SIC code 26 as having 1.22 million metric tons/yr of total hazardous waste. No wastes were included on the "EPA Waste List" or as "characteristic wastes." It was further estimated that the range of values was 0 to 1.76 million metric tons and was generally composed of wastewater treatment sludges with onsite disposal (97 percent) in lagoons. However, among the itemized hazardous wastes inventoried by states, the following materials were listed:

- Halogenated solvents
- Nonhalogenated solvents
- Unknown solvents
- Mixture of solvents
- Other organic liquids
- Ink solids and sludges
- Metal finishing sludge
- Paper sludge
- Nonemulsified waste oil
- Lab chemicals

- Nonhalogenated tank residue
- Textile coating solutions
- Rinse water and wastewaters
- Other mixed waste

This list indicates that solvents and waste oil do compose a part of SIC code 26 waste, about 60 percent by weight, though these categories are not specifically called out in EPA's waste generation inventory.

For one state, the inventoried waste was two-third's organic and the remainder inert or metallic. On the other hand, wastewater sludges, estimated at approximately 1.6 million metric tons/yr (based on 1980 projected production of 62 million tons) are rarely hazardous and not included in most inventories. (These wastes are usually landfilled onsite after several years storage or sent to publicly owned treatment works. A small quantity of material is reclaimed and some burned both for fuel and simply incinerated for consolidation.) Thus, if the EPA waste generation data base and categories were used, no hazardous wastes would be listed as possible boiler disposed. However, by alternatively selecting waste oils and solvents, specifically called out in some state inventories, considerable material would be listed as candidates for such a practice.

Similar difficulties would be encountered if EPA generator values were selected for SIC code 32. Here the EPA data base (Ref. 5) specifically tabulates solvents and acid waste. However, the total quantity of hazardous waste estimated here is lower than the quantity of solvents listed in state inventories (Ref. 9,10,24).

SIC code 28 state inventories likewise exceed the EPA generator values due principally to the contribution of spent solvents. Thus, the selection of generic materials (oil and solvent) provided a better base for projecting waste burning practices than would have been accomplished using types of industry data.

3.3.6 Summary of Waste Quantities

The quantity of hazardous wastes that is projected to be disposed in boilers is summarized as follows:

Source	Increased use as fuel, 1980-1985 (metric tons per yr)	
	10%	40%
Waste oils	289,640	443,180
Solvents	44,180	176,730
Organic chemicals	166,810	667,240
Tanning	1,100	4,400
Total	501,730	1,291,550

If only 10 percent of the wastes identified as likely candidates for boiler disposed materials are so disposed, it will amount to an increase in half a million metric tons/yr additional boiler disposal in 1985 over current levels of about 18.8 million metric tons. If 40 percent of these wastes become boiler disposed, a projected additional 1.3 million metric tons will be boiler disposed.

4. Industrial Boiler Population

This section establishes the industrial boiler population inventory required to support the technical overview of hazardous waste disposal in industrial boilers. The objectives were twofold. The first was to assemble an inventory of the industrial boiler population, specifically broken down by capacity, design type, fuel fired, and regional location. This inventory could then be combined with the regional inventory of hazardous waste generation described in Section 3 to provide what might be termed the "matrix of opportunity" for hazardous waste combustion in industrial boilers.

The second objective was to select and describe the characteristics of a set of standard boilers representative of those most widely used by industry. These descriptions include specifying the capacity, design type, and fuel fired so that the standard boilers would be representative of installed capacity.

4.1 INDUSTRIAL BOILER INVENTORY

In assembling this inventory, attention was focused solely on industrial boilers with heat input capacities greater than 2.9 MW (10 million Btu/hr). Commercial and institutional boilers were not included since these establishments generate negligible quantities of hazardous waste. The capacity limitation is applied since smaller units are generally natural-gas-or oil-fired firetube, or cast iron units used for space heating. Such units do not easily lend themselves to either waste firing or to the fuel blending required in waste cofiring.

Establishing an industrial boiler population inventory is complicated by several factors:

- Virtually all industries in the U.S. use boilers for space heat and process steam
- Industrial boilers are sold by a large number of manufacturers, almost all of which consider their sales data proprietary
- Unit retirement rates are unknown; many units have been in operation well over 50 years
- Identifying the fuel fired is difficult because (1) many units were sold to fire several fuels, and (2) many other units that were sold to fire a single fuel have since been modified to fire another

Perhaps the most comprehensive and well-documented attempt to establish an industrial boiler inventory was performed by Battelle (Ref. 28,19). This 1971 population inventory has long been regarded as the best available. A more recent characterization of the industrial boiler population has been assembled by PEDCo (Ref. 30). The PEDCo effort essentially used Battelle's work as a baseline, and updated the 1971 inventory to 1977 using sales data from the American Boiler Manufacturers Association (ABMA) and the Hydronics Institute. Most of the data presented in this section were derived from PEDCo's effort.

As noted in the introduction to this section, the desired boiler inventory characterizes the installed population by capacity, design type, fuel, and regional location. The PEDCo 1977 inventory provides information with respect to the first three, but has no regional breakdown. To provide the desired regional data, results of another PEDCo study were used (Ref. 31). This second study compiled National Emission Data System (NEDS) information on industrial boilers and gives a state-by-state breakdown of units reported in the NEDS base.

Unfortunately, the NEDS-based information accounts for only about 16,000 industrial boilers with capacity greater than 2.9 MW (10 million Btu/hr). The aggregate capacity of these units comprises 530 GW (1.8 x 10^{12} Btu/hr). In contrast, the national inventory contains about 43,000 units and 730 GW (2.5 x 10^{12} Btu/hr) of capacity (Ref. 30). In addition, the NEDS-based data do not provide a design type breakdown. The fuel type categories also differ. For example, the NEDS-based data includes process gas; the national inventory does not. Finally, the size range categories differ; the NEDS-based data has only two size ranges, 2.9 to 29 MW (10 to 100 million Btu/hr) and greater than 29 MW (100 million Btu/hr), whereas the national inventory has seven size ranges.

To prepare the boiler inventory by region and capacity and adjust the NEDS information to approximate the national inventory, the two data systems were combined using the following guidelines:

- The national inventory was reorganized by design type, capacity, and fuel categories as used in the NEDS-based inventory
- The reorganized inventory was distributed among the states (both number of units and capacity) according to the fractional distribution (by capacity and fuel) as defined by the NEDS-based inventory
- Each state's inventory was distributed between design type according to the national fractional distribution
- This state-by-state inventory was combined into an EPA regional inventory (both number of units and capacity)

It should be emphasized that this approach gives only an approximation to an actual regional inventory because of its underlying assumptions. The most critical assumptions implicit in this approach are:

- The NEDS-based inventory gives an accurate representation of the statewide fractional distribution of units and capacity, by capacity range and fuel. This implies that the NEDS undercounting (compared to the national inventory) of units and capacity does not vary state to state.
- The design type distribution of boilers does not vary state by state and can be represented by the national inventory

Neither of these assumptions is absolute. Therefore, the regional characterization obtained in this manner must be considered to be an approximation.

Tables 13 and 14 present the results of this regional industrial boiler population characterization. Table 13 categorizes units in the 2.9 to 29 MW (10 to 100 million Btu/hr) heat input capacity range. Table 14 gives a corresponding breakdown for units with capacities greater than 29 MW (100 million Btu/hr). Table 15 presents a summary of regional population by capacity only.

4.2 TYPICAL BOILERS

To establish if industrial boilers are appropriate for the disposal of hazardous wastes, the combustion characteristics of industrial boilers in terms of their temperature-time profiles must be defined. The temperature-time profile of a boiler depends on several variables, most notably design type, capacity, operating load, and fuel fired. Therefore, to define industrial boiler combustion characteristics, a set of "typical" units was defined to represent the key features of the nation's boiler population.

Selecting the set of typical boilers to be characterized in this study was guided by work done by PEDCo and documented in the national boiler population inventory study (Ref. 30). The set of typical boilers selected as representative of the installed population by fuel and design type is given in Table 16.

The 4.4-MW firetube boiler firing either natural gas or distillate oil represents a larger than average firetube unit but is typical of the capacity range of interest in this study. The 44-MW packaged watertube units firing residual oil, distillate oil, or natural gas represent typical large industrial boilers firing liquid or gas fuels. The 44-MW spreader-stoker-fired watertube unit is typical of larger stokers currently being sold. The 59-MW pulverized-coal unit is representative of industrial pulverized-coal units, though it is on the lower capacity range of a typical pulverized-coal unit, most of which are used for utility electricity generation. This set of typical boilers spans the range of equipment found in the industrial steam-generating sector.

Utilizing the data from Section 3 on the number of waste generators and Section 4 on the boiler inventory, Table 17 can be generated. If the compounds from class B (Section 3) are used as supplemental fuel, an analysis of the availability of boiler capacity to consume the waste can be conducted.

TABLE 13. REGIONAL DISTRIBUTION OF U.S. INDUSTRIAL BOILERS IN THE 2.9 TO 29 MW CAPACITY RANGE (MW (MILLION BTU/HR))

		Watertube boilers							Firetube boilers					
EPA region		Spreader stoker	Underfeed stoker	Overfeed stoker	Residual oil	Distillate oil	Natural gas	Process gas	Coal	Residual oil	Distillate oil	Natural gas	Process gas	Total
I	Number of units Capacity	5 51 (174)	14 148 (505)	2 22 (75)	1,253 11,262 (38,430)	22 168 (573)	106 1,513 (5,163)	-- -- (--)	7 38 (130)	747 3,474 (11,850)	45 211 (720)	102 631 (2,150)	-- -- (--)	2,303 17,518 (59,770)
II	Number of units Capacity	2 7 (24)	8 24 (82)	2 6 (20)	1,344 17,021 (58,081)	98 981 (3,350)	314 4,945 (16,870)	-- -- (--)	4 12 (41)	803 5,250 (17,910)	206 1,240 (4,231)	299 2,062 (7,036)	-- -- (--)	3,080 31,548 (107,650)
III	Number of units Capacity	141 2,264 (7,725)	432 6,548 (22,340)	104 1,659 (5,661)	1,131 15,215 (51,918)	145 1,736 (5,924)	572 8,399 (28,660)	20 298 (1017)	232 1,628 (5,555)	674 4,692 (16,010)	306 2,185 (7,456)	546 3,503 (11,950)	20 125 (427)	4,323 48,252 (164,650)
IV	Number of units Capacity	169 1,907 (6,507)	517 5,526 (18,860)	126 1,396 (4,764)	1,455 14,897 (50,833)	413 3,931 (13,410)	2,614 30,257 (103,250)	50 989 (3,375)	276 1,370 (4,675)	868 4,595 (15,680)	870 4,910 (16,750)	2,499 15,287 (52,164)	48 421 (1,437)	9,905 85,486 (291,710)
V	Number of units Capacity	407 6,254 (21,340)	1,241 18,137 (61,889)	302 4,621 (15,770)	940 13,387 (45,681)	210 2,998 (10,230)	1,642 24,029 (81,995)	153 2,429 (8,289)	666 4,480 (15,290)	561 4,130 (14,090)	452 3,473 (11,850)	1,564 11,540 (39,378)	145 1,013 (3,537)	8,283 96,491 (329,260)
VI	Number of units Capacity	1 4 (14)	2 11 (38)	-- -- (--)	37 461 (1,573)	26 335 (1,143)	1,241 15,816 (53,969)	99 1,593 (5,436)	1 4 (14)	23 143 (488)	51 423 (1,443)	1,182 6,619 (22,590)	96 672 (2,293)	2,759 26,081 (89,000)
VII	Number of units Capacity	16 263 (897)	50 763 (2,604)	12 183 (624)	288 2,676 (9,131)	155 1,158 (3,951)	1,228 10,664 (36,389)	10 145 (495)	26 191 (652)	137 825 (2,815)	327 1,453 (4,958)	1,170 5,156 (17,590)	10 60 (205)	3,369 23,537 (80,310)
VIII	Number of units Capacity	4 38 (130)	10 109 (372)	1 12 (41)	109 1,687 (5,757)	17 146 498	222 3,013 (10,280)	44 734 (2,505)	5 29 (99)	66 521 (1,778)	34 184 (628)	212 1,257 (4,289)	42 306 (1,044)	766 8,036 (27,420)
IX	Number of units Capacity	-- -- (--)	-- -- (--)	-- -- (--)	673 5,296 (18,070)	131 467 (1,594)	573 3,616 (12,340)	12 144 (491)	-- -- (--)	402 1,635 (5,579)	278 869 (2,965)	522 2,217 (7,565)	12 60 (205)	2,603 14,304 (48,810)
X	Number of units Capacity	4 37 (126)	14 104 (355)	3 26 (89)	121 1,928 (6,579)	41 655 (2,235)	146 2,850 (9,725)	2 51 (174)	7 28 (96)	72 595 (2,030)	84 822 (2,805)	140 1,190 (4,061)	2 21 (72)	636 8,307 (28,350)
Total	Number of units Capacity	749 10,825 (36,940)	2,288 31,370 (107,050)	552 7,925 (27,040)	7,291 83,830 (286,060)	1,258 12,575 (42,910)	8,658 105,102 (358,330)	390 6,383 (21,780)	1,224 7,780 (26,550)	4,353 25,860 (88,240)	2,653 15,770 (53,810)	8,236 49,462 (168,780)	375 2,678 (9,140)	38,027 359,560 (1,226,930)

TABLE 14. REGIONAL DISTRIBUTION OF U.S. INDUSTRIAL BOILERS WITH CAPACITY GREATER THAN 29 MW (MW (MILLION BTU/HR))

EPA region		Watertube boilers[a] Pulverized coal	Spreader stoker	Underfeed stoker	Overfeed stoker	Residual oil	Distillate oil	Natural gas	Process gas	Total
I	Number of units	1	2	--	--	172	--	18	--	193
	Capacity	62	184	--	--	8,749	--	1,217	--	10,212
		(212)	(628)	(--)	(--)	(29,850)	(--)	(4,153)	(--)	(34,850)
II	Number of units	52	15	6	3	183	5	35	3	302
	Capacity	3,727	2,751	1,082	621	9,661	309	17,286	131	35,568
		(12,720)	(9,387)	(3,692)	(2,119)	(32,970)	(1,054)	(58,990)	(447)	(121,370)
III	Number of units	167	60	23	12	171	18	86	50	587
	Capacity	12,763	4,030	1,584	911	10,383	877	3,936	4,647	39,126
		(43,550)	(13,750)	(5,405)	(3,109)	(35,430)	(2,976)	(13,430)	(15,860)	(133,510)
IV	Number of units	160	106	38	22	332	37	331	27	1,053
	Capacity	12,643	6,887	2,708	1,552	22,768	1,746	20,271	1,282	69,857
		(43,140)	(23,500)	(9,241)	(5,296)	(77,690)	(5,958)	(69,170)	(4,375)	(238,380)
V	Number of units	278	321	124	64	242	48	430	20	1,527
	Capacity	23,909	13,605	5,617	3,224	17,132	2,316	23,040	3,337	92,180
		(81,590)	(46,420)	(19,170)	(11,000)	(58,460)	(7,903)	(78,620)	(11,390)	(314,550)
VI	Number of units	--	--	--	--	57	43	481	90	671
	Capacity	--	--	--	--	7,349	3,055	56,889	10,997	78,290
		(--)	(--)	(--)	(--)	(25,080)	(10,420)	(154,120)	(37,530)	(267,150)
VII	Number of units	32	31	11	7	38	18	104	11	252
	Capacity	1,824	1,730	680	389	2,060	705	4,974	662	13,024
		(6,224)	(5,903)	(2,320)	(1,327)	(7,029)	(2,406)	(16,970)	(2,260)	(44,440)
VIII	Number of units	15	11	3	2	22	5	97	12	117
	Capacity	4,072	834	134	77	1,077	318	2,453	770	9,735
		(13,890)	(2,846)	(457)	(263)	(3,675)	(1,085)	(8,370)	(2,627)	(33,220)
IX	Number of units	--	3	1	1	79	17	60	22	183
	Capacity	--	225	89	51	5,084	733	3,596	996	10,774
		(--)	(768)	(304)	(174)	(17,350)	(2,500)	(12,270)	(3,399)	(36,760)
X	Number of units	14	13	4	2	68	9	97	3	160
	Capacity	1,075	1,059	416	240	3,897	401	2,717	254	10,059
		(3,668)	(3,614)	(1,420)	(819)	(13,300)	(1,368)	(9,271)	(867)	(34,320)
Total	Number of units	719	562	210	113	1,364	200	1,639	238	5,045
	Capacity	60,075	31,305	12,310	7,065	88,160	10,455	136,379	23,076	368,825
		(205,000)	(106,820)	(42,000)	(24,110)	(300,830)	(35,680)	(465,370)	(78,740)	(1,258,550)

[a]No firetube boilers have capacities greater than 29 MW

TABLE 15. REGIONAL DISTRIBUTION OF U.S. INDUSTRIAL BOILERS WITH CAPACITY GREATER THAN 2.9 MW (MW (MILLIION BTU/HR)

Capacity range[a]		EPA Region										
		I	II	III	IV	V	VI	VII	VIII	IX	X	Total
2.9-29 (10-100)	Number of units	2,303	3,080	4,323	9,905	8,283	2,759	3,369	766	2,603	636	38,027
	Capacity	17,518 (59,770)	31,548 (107,650)	48,252 (164,650)	85,486 (291,710)	96,491 (329,260)	.26,081 (89,000)	23,537 (80,310)	8,036 (27,420)	14,304 (48,810)	8,307 (28,350)	359,560 (1,226,930)
>29 (>100)	Number of units	193	302	587	1,053	1,527	671	252	117	183	160	5,045
	Capacity	10,212 (34,850)	35,568 (121,370)	39,126 (133,510)	69,857 (238,380)	92,180 (314,550)	78,290 (267,150)	13,024 (44,440)	9,735 (33,220)	10,774 (36,760)	10,059 (34,320)	368,825 (1,258,550)
Total	Number of units	2,496	3,382	4,910	10,958	9,810	3,430	3,621	833	2,786	796	43,072
	Capacity	27,730 (94,620)	67,116 (229,020)	87,378 (298,160)	155,343 (530,090)	188,671 (643,810)	104,371 (356,150)	36,561 (124,750)	17,771 (60,640)	25,078 (85,570)	18,366 (62,670)	728,385 (2,485,480)

[a]Thermal input

TABLE 16. TYPICAL INDUSTRIAL BOILERS

Heat input capacity (MW (million Btu/hr))	Fuel	Design type
4.4 (15)	Distillate oil	Packaged -- scotch firetube
4.4 (15)	Natural gas	Packaged -- scotch firetube
44 (150)	Residual oil	Packaged -- watertube
44 (150)	Distillate oil	Packaged -- watertube
44 (150)	Natural gas	Packaged -- watertube
44 (150)	Stoker coal	Field-erected -- spreader stoker -- watertube
59 (200)	Pulverized coal	Field-erected -- watertube

TABLE 17. GEOGRAPHIC DISTRIBUTION OF BOILERS AND WASTE GENERATORS

	EPA Region										
	I	II	III	IV	V	VI	VII	VIII	IX	X	Total
Number of boilers 2.9-29 MW (10-100 x million Btu/hr)	2,303	3,080	4,323	9,905	8,283	2,759	3,369	766	2,603	636	38,027
Number of boilers >29 MW (>100 x million Btu/hr)	193	302	587	1,053	1,527	671	252	117	183	160	5,045
Total boilers	2,496	3,382	4,910	10,958	9,810	3,430	3,621	833	2,786	796	43,072
Number of Waste Generators	3,370	5,260	5,080	6,920	9,800	8,270	2,270	1,570	4,315	1,500	48,355

Total industrial boiler population (including size <10 x million Btu/hr) 238,000

By making certain assumptions, the boiler capacity available to destroy class B hazardous wastes can be approximated. There are 43,000 boilers with aggregate capacity of 728 GW (2.5 x 10^{12} Btu/hr), for an average boiler size of 17 MW (58 x million Btu/hr). The assumptions necessary for this evaluation are

- 10 percent of the total available boiler capacity will be selected for utilization to destroy hazardous waste
- The hazardous wastes will represent 5 percent of the total Btu input to the boiler
- Combustion of these hazardous wastes yield approximately 33 MJ/l (120,000 Btu/gal) and the hazardous wastes have an average density of 0.95 kg/l
- Boilers will operate 300 days/year and 24 hours/day

Using the class B hazardous waste quantities generated, the calculated maximum waste quantity destructible (as limited by these assumptions) is shown in Table 18. Though each region may not have sufficient boiler capacity to consume all class B waste, it is projected that sufficient boiler capacity exists to destroy significant quantities.

TABLE 18. GEOGRAPHIC DISTRIBUTION OF CLASS B HAZARDOUS WASTE DESTRUCTION CAPACITY (1,000 METRIC TONS/HR)

	EPA Region										
	I	II	III	IV	V	VI	VII	VIII	IX	X	Total
Combustible constituents (class B) generation	19	109	173	535	176	570	45	7	112	19	1,765
Quantity of hazardous waste destructible	87	118	172	384	343	120	127	29	97	28	1,508

5. Characterization of Industrial Boilers

The efficient destruction of combustible hazardous waste achievable in industrial boilers is primarily dependent on the temperature and residence time profiles of the combustion products as they pass through the furnace. The destruction efficiency of any species is maximized when the gas temperature and residence time are maximized. Since industrial boilers are nonadiabatic reactors, the temperature of the combustion products peaks in the flame zone and is reduced as heat is absorbed by the furnace walls. Thus, a representative temperature-time profile through the combustion zone is necessary to estimate the destruction potential for hazardous waste constituents.

This section presents design and operating characteristics of typical industrial boiler types and estimates bulk gas temperatures and residence time profiles in the combustion chamber of these boilers. The temperature and residence times necessary to reach the required destruction efficiencies are coupled with selected hazardous waste characteristics in Section 6. As a prelude to the determination that destruction of a hazardous waste in a boiler represents a viable disposal option, the physical compatability between the boiler and proposed waste must be carefully evaluated. This evaluation is primarily concerned with the mechanical changes required to allow firing of the waste. This includes facility design, retrofit considerations, fuel requirements, burner considerations, and process feasibility. Section 5.3 discusses the key elements for this evaluation.

5.1 DESCRIPTION OF BOILERS, FUELS, AND OPERATING CHARACTERISTICS

The temperature-time profile of combustion gases in boiler furnaces varies with boiler design and operating characteristics such as furnace geometry, burner configuration, fuel type, heat input, and excess air level. As presented in Section 4, seven individual boiler design/fuel combinations were selected for consideration in this study. The following discussion summarizes the design and operating factors for each boiler/fuel combination.

Table 19 lists the boiler design types, the heat input capacities, and the ultimate analyses of fuel assumed. Table 20 summarizes furnace dimensions and fuel and combustion air flowrates and temperatures for each unit. Furnace dimensions were obtained in part from published design data (Ref. 32,33) and from telephone contacts with major boiler manufacturers

TABLE 19. MODEL INDUSTRIAL BOILERS AND FUELS USED IN THE ANALYSIS

Assigned identi- fication	Boiler	Capacity MW (million Btu/hr)	Fuel							Fuel higher heating value (MJ/kg (Btu/lb))	Typical operating load factor (% capacity)
			Type	% C	% H_2	% S	% O_2	% Ash	% H_2O		
A	Packaged scotch firetube	4.4 (15)	Distillate oil	87.2	12.5	0.3	Nil	--	--	42.6 (18,300)	45
B	Packaged scotch firetube	4.4 (15)	Natural gas	74.7	12.5	--	--	--	--	53.3 (22,900)	45
C	Packaged watertube	44 (150)	Residual oil	86	11.0	3.0	0.3	0.05	--	42.3 (18,200)	55
D	Packaged watertube	44 (150)	Distillate oil	87.2	12.5	0.3	Nil	--	--	42.6 (18,300)	55
E	Packaged watertube	44 (150)	Natural gas	74.7	23.3	--	--	--	--	53.3 (22,900)	55
F	Field erected watertube spreader stoker	44 (150)	Bituminous coal[a]	72.8	4.8	2.2	6.2	9.0	3.5	30.4 (13,080)	60
G	Field erected watertube pulverized coal	58.6 (200)	Bituminous coal[a]	72.8	4.8	2.2	6.2	9.0	3.5	30.4 (13,080)	60

[a]ASTM Rank: Class II Group 3.
Ash fushion temperature -- 1,511K (2,260°F) at initial deformation.

TABLE 20. DESIGN AND OPERATING CHARACTERISTICS OF MODEL BOILERS

Assigned Identification	Boiler	No. of Burners or Feeders	Approximate Furnace Dimensions [m (ft)] Length	Width	Height	Furnace Heating Surface [m^2 (ft^2)]	Furnace Volume [m^3 (ft^3)]	Fuel Flow [kg/s (lb/hr)]	[kg/s (10^3 lb/hr)]	Temperature [°K (°F)]	Temperature [°K (°F)]	Excess O_2 in Furnace (%)	Flue Gas Mass Flow [kg/s (10^3 lb/hr)]	Flue Gas Volumetric Flow [m^3/s (scfm)]	Other Boiler Descriptions
A	Packaged scotch firetube	1	4.6 (15)	0.76 D* (2.5 D)		11 (118)	2.1 (74)	0.10 (820)	1.71 (13.6)	300 (80)	300 (80)	3.0	1.83 (14.5)	1.35 (2,850)	Dryback, three pass
B	Packaged scotch firetube	1	4.6 (15)	0.76 D (2.5 D)		11 (118)	2.1 (74)	0.08 (655)	1.59 (12.6)	300 (80)	300 (80)	3.0	1.66 (13.2)	1.18 (2,490)	Dryback, three pass
C	Packaged watertube	1	11.9 (39)	2.0 (6.5)	2.7 (9)	117 (1,260)	68 (2,400)	0.96 (1,580)	15.3 (121)	355 (180)	450 (350)	3.0	16.3 (129)	13.2 (28,000)	D-type, refractory base
D	Packaged watertube	1	11.9 (39)	2.0 (6.5)	2.7 (9)	117 (1,260)	68 (2,400)	0.95 (7,540)	15.8 (125)	300 (80)	450 (350)	3.0	16.6 (132)	13.2 (28,000)	D-type, refractory base
E	Packaged watertube	1	11.9 (39)	2.0 (6.5)	2.7 (9)	117 (1,260)	68 (2,400)	0.77 (6,120)	15.0 (119)	300 (80)	450 (350)	3.0	15.8 (125)	11.8 (24,900)	D-type, refractory base
F	Field erected watertube spreader stoker	4	5.2 (17)	4.3 (14)	9.8 (32)	155 (1,670)	212 (7,500)	1.35 (10,700)	18.5 (147)	300 (80)	450 (350)	6.0	19.9 (158)	14.7 (31,100)	Effective grate area = 20 m^2 (215 ft^2)
G	Field erected watertube pulverized coal	4	5.5 (18)	5.2 (17)	15.8** (52)	286 (3,080)	317 (11,200)	1.78 (14,140)	23.3 (185)	339 (150)	477 (400)	5.0	25.2 (200)	18.2 (38,500)	Single wall, dry bottom

*D = diameter.

**Total furnace height from bottom hopper to convective section. Height above main burner zone is 7.9 m (26 ft) for a combustion volume of 107 m^3 (3,800 ft^3) assuming horizontal firing.

(Ref. 34-37). Although furnace dimensions reported here are typical of these boiler designs, they are not to be considered specific to any industrial boiler manufacturer.

The furnace gas exit temperature is a vital design consideration. A furnace exit temperature of 1,300K (1,880°F) indicates that about one half of the heat released by the fuel is absorbed by the furnace walls prior to the combustion gas entering the convective region of the boiler. In the case of coal combustion, the design flue gas temperature at the exit of the furnace is heavily dependent on coal rank and its ash properties. For example, the design exit temperature at full load is normally 110K (200°F) less than the fusion temperature of the ash in the coal. This design criterion is imposed to minimize fouling of the convective passes by deposition of molten coal ash on the heat transfer tubes. This design exit temperature, together with coal combustion characteristics, determines the overall size of the furnace and, thus, the bulk residence time of the combustion products. Gas exit temperature was also verified through contacts with boiler manufacturers (Ref. 34-37).

Table 21 summarizes the estimated peak temperatures in the primary combustion zone and design flue gas exit temperatures at full load operation. Based on the high and low gas temperatures, the bulk gas residence time and heat absorbed in the furnace were calculated. These values are refined in Section 5.3 with the application of a mathematical model to calculate temperature profiles.

5.2 TEMPERATURE-TIME PROFILE CALCULATIONS

At the onset of this task, a review of published boiler literature was conducted to compile empirical temperature and heat transfer data throughout industrial boiler furnaces. This literature consisted mainly of technical brochures from industrial boiler manufacturers and field test reports performed under numerous domestic and foreign government-sponsored environmental and boiler performance studies. This review showed the near absence of pertinent data. Further telephone contacts with major industrial boiler manufacturers indicated that although empirical data and predictive mathematical schemes to estimate the temperature profile were available from selected manufacturers, this information was considered proprietary or not easily available within the timeframe of this study. Technical information released by boiler manufacturers consisted of typical furnace dimensions and approximate exit temperatures as indicated in Section 5.1.

Therefore, to estimate temperature-time profiles, it was necessary to devise a computational scheme based on proven techniques and thermodynamic relationships. The following discussion describes one such scheme used to calculate temperature-residence time profiles of the combustion products throughout the radiant section of each industrial boiler described in Section 5.1. Limitations of the model and assumptions used are also discussed.

TABLE 21. CALCULATED RESIDENCE TIMES AND FURNACE TEMPERATURES

Assigned boiler identification	Fuel	Percent of operating capacity	Total heat input (MW (million Btu/hr))	Furnace heat release rate (kW/m^2 (million Btu/hr-ft^2))	Furnace heat liberation rate (kW/m^3 (10^3 Btu/hr-ft^3))	Average flame or bed temperature in active combustion zone (°K (°F))	Average furnace residence time (s)	Furnace exit temperature (°K (°F))	Heat adsorbed in the furnace (MW (million Btu/hr))
A	Distillate oil	100	4.4 (15.0)	400 (127)	2,100 (203)	2,144 (3,400)	0.22	1,589 (2,400)	1.52 (5.2)
B	Natural gas	100	4.4 (15.0)	400 (127)	2,100 (203)	2,255 (3,600)	0.23	1,589 (2,400)	1.17 (5.7)
C	Residual oil	100	44.0 (150)	375 (119)	652 (63)	2,255 (3,600)	0.79	1,283 (1,850)	25.1 (85.8)
D	Distillate oil	100	44.0 (150)	375 (119)	652 (63)	2,200 (3,500)	0.77	1,366 (2,000)	24.4 (83.2)
E	Natural gas	100	44.0 (150)	375 (119)	652 (63)	2,283 (3,650)	0.82	1,477 (2,200)	23.7 (81.0)
F	Bituminous coal	100	44.0 (150)	287 (91)	207 (20)	1,722 (3,000)	1.90	1,400 (2,060)	16.9 (57.6)
G	Bituminous coal	100	59.2 (202)	208 (66)	186 (18)	2,060 (3,250)	1.82	1,400 (2,060)	25.3 (86.3)

5.2.1 Mathematical Model

One of the more complete and widely used mathematical models of radiation heat transfer in turbulent diffusion flames is the zone method of analysis (Ref. 38). In this model, the furnace of a given boiler is divided into a number of zones for which steady-state energy balances are developed. This procedure assumes that the zones are small enough to be isothermal and that the flue gas properties within the zone are constant. Thus, the larger the number of zones, the more precise the profile. The simultaneous solution of the energy balance equations for all zones leads to the temperature distribution in the furnace and, consequently, the evaluation of the heat flux in the furnace.

An underlying assumption in this approach is that the gas in the furnace is perfectly mixed in radial or lateral directions. This assumption, though not conservative, generally introduces small uncertainties in model predictions.

Since the model assumes that a given zone is isothermal, each zone is represented by a single effective radiating temperature. In reality, the gas temperature varies at any one cross section in the furnace, as illustrated in Figure 1. The gas temperature near the furnace walls can be significantly less than the mean radiating temperature, by about 100°K (180°F). This aspect of temperature distribution across a given zone was considered when estimating the destruction efficiency of hazardous wastes injected into the furnace.

The thermal zone model (Ref. 38) has been shown to be suitable for both nonluminous flames (those from natural gas combustion) having low emissivities as well as luminous flames (those from oil and coal combustion) having high emissivities. The presence of suspended soot and carbon particles in a hot gas gives the flame its luminous characteristics and contributes to its high radiating property.

5.2.1.1 Energy Balance and Temperature Calculations

A computational routine based on a steady-state energy balance within each zone was devised in order to calculate the gas temperature profile throughout the furnace and evaluate the heat flux characteristics. With the aid of the computer, the energy balance calculations were performed simultaneously for each isothermal zone in the furnace until a gas temperature profile consistent with known radiative and convective heat fluxes was obtained. This section briefly discusses the energy balance computational scheme used for the temperature profile calculations and presents the primary assumptions and limitations of the model as it was used in this study.

Figure 2 shows a plan view schematic of a packaged, single burner, watertube boiler (44 MW, 150 million Btu/hr), representing boiler type D in Tables 19 through 21. As illustrated, the furnace has been divided into six equal cubical zones. The shape of the zone is dictated by the geometry of

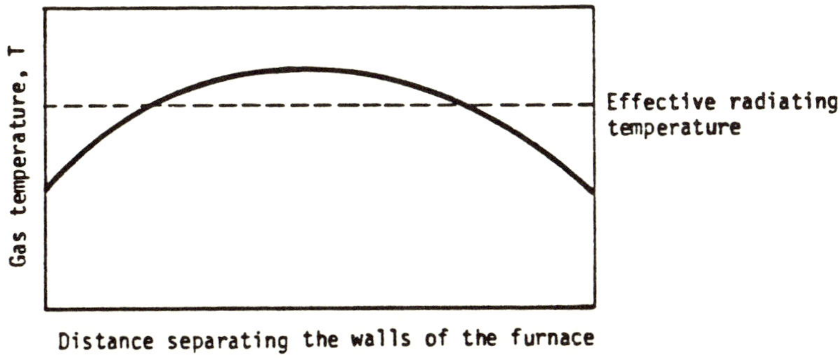

Figure 1. Temperature distribution across a heat generating and absorbing medium contained between furnace walls (Ref. 39).

the furnace, and its size is dictated by the effective volume through which most of the combustion products pass. Figure 3 illustrates the various heat fluxes entering and leaving a zone. The heat fluxes are identified as follows.

<u>In-Fluxes</u>

$\dot{Q}_{ch}$ = Chemical heat released by combustion of the fuel. Total chemical heat in the fuel may be released in more than one zone.

$\dot{Q}_{rec+1}$ = Heat entering the zone because of recirculation of hot gases from the next zone

$\dot{q}_r$ = Radiative heat transfer from adjacent zones

$\dot{q}_{wr}$ = Radiative heat transfer from the hot gray furnace walls

$\dot{Q}_{sens}$ = Sensible heat entering the zone from preceding zone

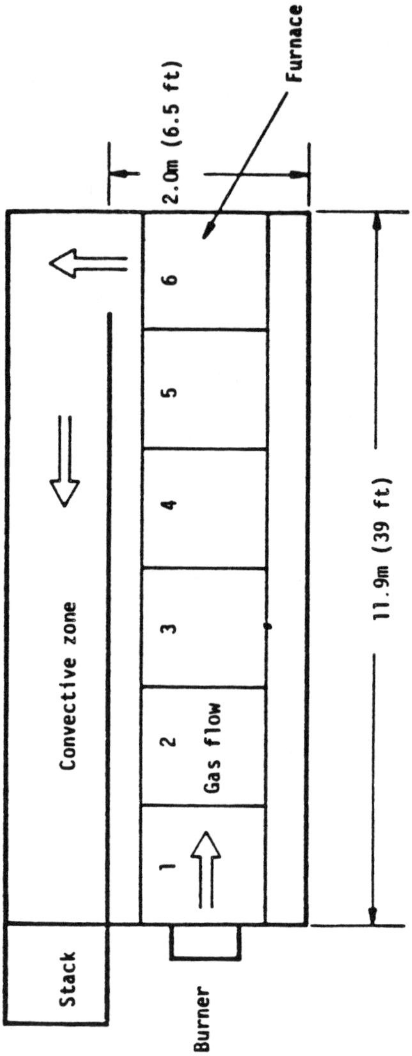

Figure 2. Plan view schematic of package watertube boiler 44 MW (150 million Btu/hr) capacity; furnace height = 2.7m (9 ft).

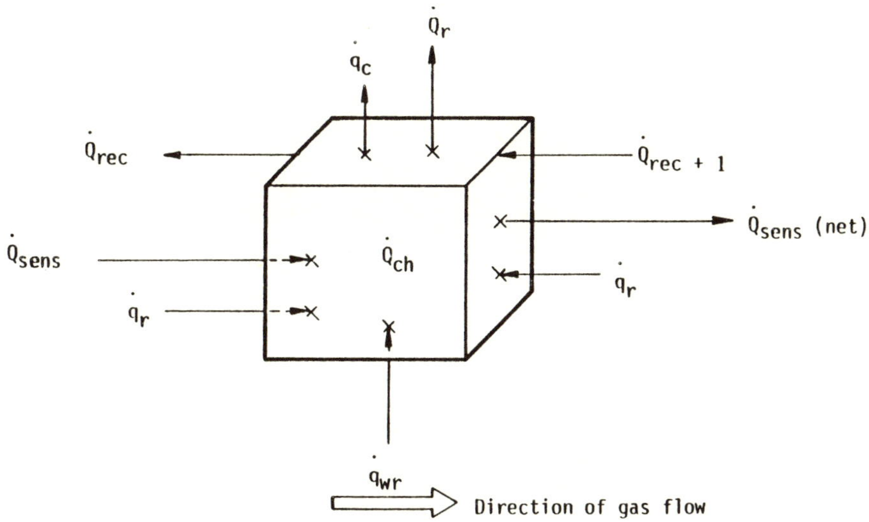

Figure 3. Schematic of a zone and heat fluxes.

Out-Fluxes

$\dot{Q}_r$ = Radiative heat transfer to the furnace walls enclosing the zone

$\dot{q}_c$ = Convective heat transfer to the furnace walls enclosing the zone

$\dot{Q}_{rec}$ = Heat leaving the zone because of recirculation of hot gases into preceding zone

$\dot{Q}_{sens\ (net)}$ = Net sensible heat leaving the zone and entering next

Thus, the energy balance equation takes the following form:

$$\dot{Q}_{sens\ (net)} = \dot{Q}_{sens} + \dot{Q}_{rec\ +1} \dot{q}_r + \dot{q}_{wr} + \dot{Q}_{ch} - \dot{Q}_{rec} - \dot{Q}_r - \dot{q}_c \qquad (5\text{-}1)$$

This expression for net sensible heat allows for a solution for the bulk gas temperature leaving one zone and entering the next. Given an initial set of conditions, an iterative routine was used to solve for the gas temperature as a function of the axial distance along the furnace. The criteria for convergence was that the temperature change at any position on successive iterations was less than 14K (25°F). The calculated exit flue gas temperature was then compared with the boiler manufacturer suggested exit temperature. When these temperatures differed significantly (e.g., ±65K (±100°F)), the input assumptions were reevaluated.

Three sets of initial conditions have a major effect on the temperature profile of combustion gases in the furnace. These are the emissivity of the gas in each zone, the convective heat transfer in each zone, and the percent of gas recirculating from one zone into the preceding one. The gas cloud emissivity and its variation through the furnace is the most significant parameter for determining the temperature profile because of the significant role radiation plays in the transfer of heat to the furnace walls and, thus, the cooling rate of the gas.

Gas Cloud Emissivities

Calculations of flame emissivities for a mixture of gas and soot was developed as follows (Ref. 38):

$$\varepsilon_m = \sum_n a_{gn} \ (T)(1 - \exp(-K_n L) \qquad (5\text{-}2)$$

Where a_{gn} is the fractional amount of energy in the n spectral regions where a gas of absorption coefficient K_n exists, and L is the path length representing the linear dimension that defines the cloud size. The absorption coefficient, K_n, is given for both gas and soot in each spectral region as:

$$K_n = k_{sn} C + k_{gn}(p_c + p_w) \quad (5\text{-}3)$$

where k's represent specific absorption coefficient with subscript s for soot and g for gas, and p_c and p_w are the partial pressures for CO_2 and H_2O, respectively. For gas clouds from combustion of natural gas $p_w/p_c \approx 2.0$, while for oil and coal combustion $p_w/p_c \approx 1.0$.[40] C is the concentration of soot particles in mass per unit volume.

A number of investigators have found considerable variation in the absorption coefficient depending on the type of flame, combustion history, flow, and mixing characteristics (Ref. 41). During this study, absorption coefficients used to develop gas cloud emissivities were estimated based on published data from experimental measurements and known combustion properties of different fuels. This greatly simplified the application of the model without significant impact on the temperature profile solutions. The zone emissivity was then calculated based on the relationship:

$$\varepsilon = 1 - \exp(-fkL) \quad (5\text{-}4)$$

where L is the path length defined in the zone method as the linear dimension used to define the cloud size, and f is the zone shape factor listed in Table 22 for each of the zone shapes used in this study. Resulting emissivities compared well with those published in a number of studies specific to both pilot-and full-scale industrial furnaces (Ref. 39, 42-44).

Convective Heat Transfer

Natural and forced convection heat transfer plays a less significant role than radiation in a boiler furnace. In the burner zone, where flame radiation is highest due to the presence of soot particles and high temperature radiating gases, convection accounts for less than 10 percent of the heat flux (Ref. 39). As the temperature is reduced and radiative properties of the gases decrease with burnout of soot and reduced temperature along the axial path of the furnace, convective heat transfer plays an increasing role. However, it still generally remains less significant than radiation.

Natural and forced convection between a fluid and an adjacent surface having a temperature gradient ΔT is expressed as:

TABLE 22. APPROXIMATE EMISSIVITY RELATIONS FOR A UNIFORM ISOTHERMAL CLOUD (REF 39)

Cloud or zone shape	Linear dimension used to define cloud size	Shape factor f		
		kL = 0	kL = 1	kL = 2
Convex cloud of arbitrary shape (used for stoker)	Ratio of volume to surface	4	3.4	3.4
Cube (used for all watertube units except the stoker)	Edge length	0.67	0.59	0.58
Cylinder (used for firetube boiler)	Diameter	0.67	0.64	0.61

$$\dot{q}_c = Nu\ k\ \Delta T/L \quad (5\text{-}5)$$

where Nu is the dimensionless Nusselt number given by the equations (Ref. 39):

$$Nu = 0.13\ (Gr)(Pr)^{1/2} \quad (5\text{-}6)$$

for watertube boilers and

$$Nu = 0.023\ (Re)^{0.8}\ (Pr)^{0.4}\ (1 + Re^{0.7}) \quad (5\text{-}7)$$

for firetube boilers, and k is the thermal conductivity of the gas. Gr, Pr, and Re are the Grashof, Prandtl, and Reynolds numbers, respectively, evaluated at the temperature near the combustor waterwall surface. These Nusselt numbers were used in the convective heat flux model.

Chemical Heat Release

The heat released by the combustion of the fuel occurs in more than one zone. Thus, it is relatively important to estimate how much of the fuel is burned in each zone. The heat of reaction can be represented by an exponential curve which has a value of zero at the burner throat and

99-percent completion by the end of the flame. Most of the heat is liberated near the burner to achieve a stable flame pattern.

As a guide, the mathematical model assumed a flame envelope of 0.2 m^3 (7 ft^3) for each MW of thermal heat input (Ref. 44). The length of the flame is a function of its diameter, which in turn varies according to the degree ofswirl imparted by the burner register settings. For the firetube units, the flame diameter was assumed to be 0.61m (2 ft). For the larger capacity watertube boilers, the flame diameter was assumed to be 1.37m (4.5 ft). These dimensions resulted in flame lengths of about 3.05m (10 ft) for the firetubes and about 5.9m (20 ft) for the package watertube boilers, corresponding to over half the furnace length. These results are in good agreement with visual observation of burner flames in industrial boiler furnaces.

Gas Recirculation

Recirculation of hot combustion gases is of particular importance because it serves to carry heat to the incoming reactants, thus raising the temperature to ignition levels. Recirculation in small industrial boiler furnaces such as firetube boilers can be as high as 50 percent of the total mass throughput (Ref. 38). The magnitude is a function of the furnace geometry and degree of swirl imparted to the combustion air.

In the mathematical model, no attempt was made to calculate the magnitude of the recirculating combustion products from one zone into the preceding one. The magnitude was crudely estimated based on published data for applications of the zone method for small industrial boilers (Ref. 38). For the boilers studied in this program, the magnitude of the recirculating gas never exceeded 30 percent and only affected the near burner region where turbulence is normally highest. This simplification in estimating the recirculation of the flue gas did not affect the usefulness of the solutions.

5.2.1.2 Bulk Residence Time Calculations

The time an elemental gas volume or particle in the flue gas remains in the furnace depends on the path it follows in passing through the furnace. The plug flow model modified by a degree of mixing is a good approximation to the gas flow in most boiler furnaces. In the plug flow model, the basic assumptions are that the combustion products flow through the furnace without any longitudinal mixing and that the temperature, composition, and velocity are uniform at any one cross section of the furnace.

The residence time distribution in a plug flow reactor modified by a degree of mixing is illustrated in Figure 4. Here, F(t) is the fraction of gas residing in the furnace for less than the time t, and $\bar{t}$ is the bulk mean residence time (defined as the furnace volume divided by the volume flowrate). Experimental measurements indicate that some gas will exit the furnace by a time of $t = 0.5\bar{t}$. Virtually all the flue gas would have exited by a time $t = 2\bar{t}$ (Ref. 35). The minimum residence time of $0.5\bar{t}$ is important

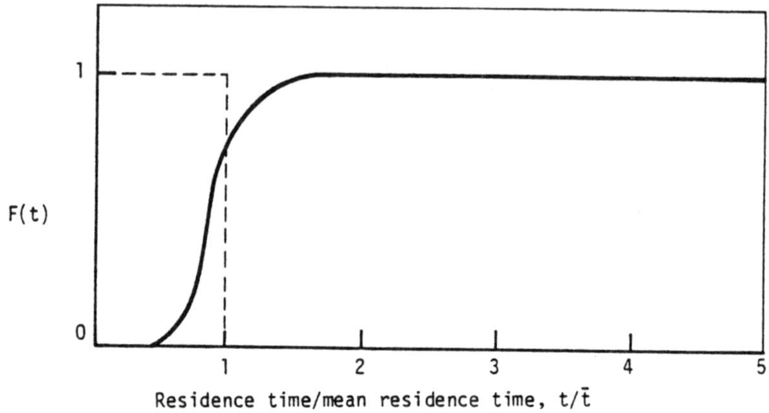

Figure 4. Residence time distribution in plug flow modified by a degree of mixing (Ref. 39)

in estimating the degree of combustion efficiency of the fuel as well as the degree of destruction of the hazardous wates considered in this study.

Residence times reported in the following sections are mean residence times ($\bar{t}$) calculated based on a total furnace volume from the burner region to the exit plane of the furnace. In reality, the total volume of the furnace is not utilized by the combustion products, which results in a portion of the gas exiting at a time much less than $\bar{t}$. Longitudinal mixing and the presence of recirculating zones are the primary cause for portions of the gas remaining for times much greater than or less than $\bar{t}$. In fact, a small fraction of the flue gas may reside in the furnace almost indefinitely due to stagnant recirculating zones or "dead space."

5.3 INDUSTRIAL BOILER TEMPERATURE-TIME PROFILES

This section presents the temperature-time profiles obtained using the isothermal zone method described previously. Appendices A and B, which illustrate the division of the furnaces into zones and the computer listings of calculated results, respectively, are provided as general support material. Figures 5 through 18 illustrate the temperature-time profiles and the heat fluxes in the furnaces of the seven boilers surveyed in this study. Both full and half boiler heat input operating conditions are presented. The excess air (EA) levels were increased for the half load cases to account for the generally higher air/fuel ratios used during reduced boiler load operation.

The trends of both temperature and heat absorption results are similar for all the boilers. The temperature of the combustion products showed a

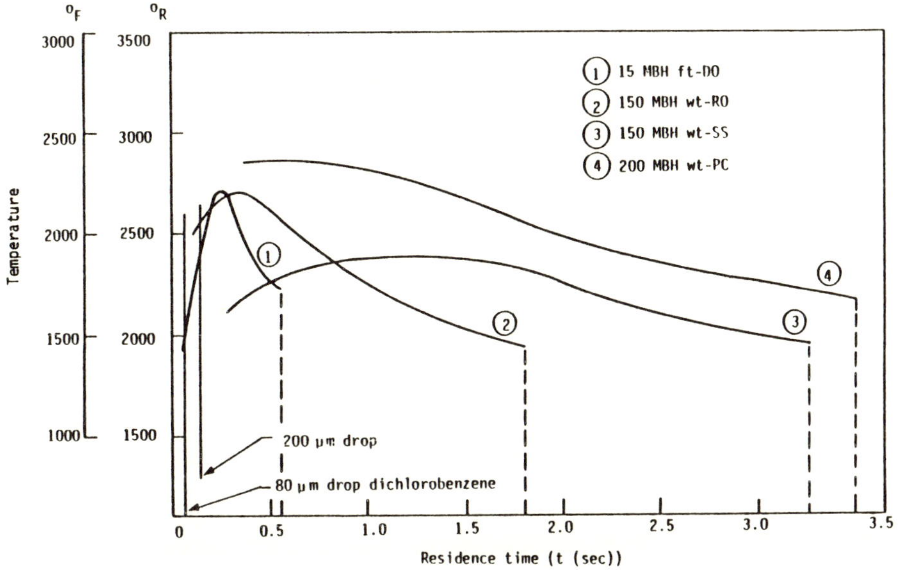

Figure 5. Temperature-time profile for the 58.6-MW (200 million Btu/hr) pulverized-coal-fired boiler (boiler G).

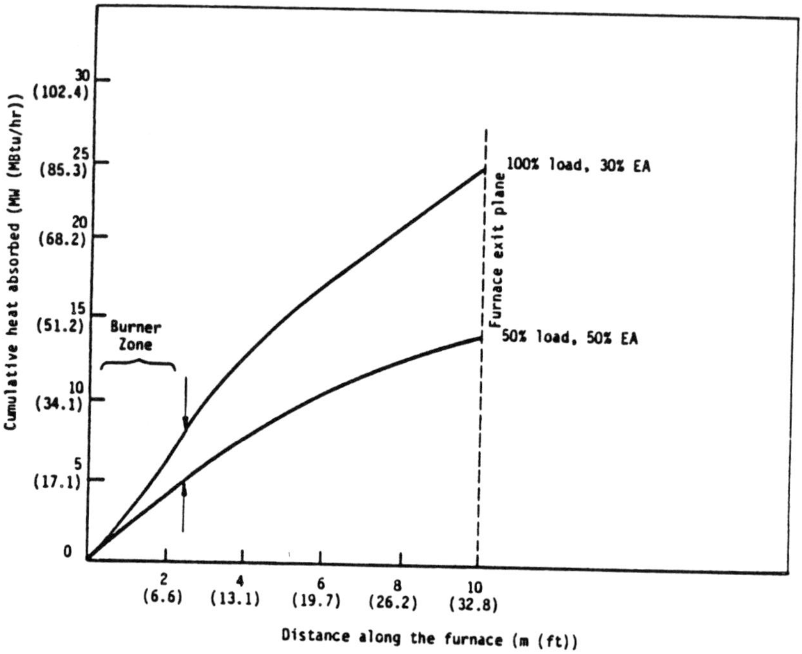

Figure 6. Cumulative heat absorption for the 58.6-MW (200 million Btu/hr) pulverized-coal-fired boiler (boiler G).

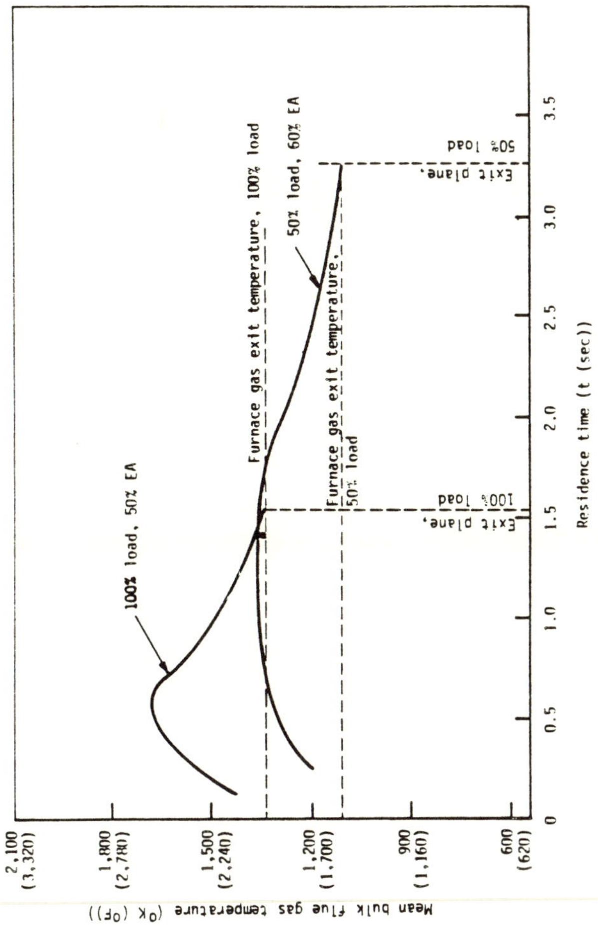

Figure 7. Temperature-time profile for the 44-MW (150 million Btu/hr) spreader stoker boiler (boiler F).

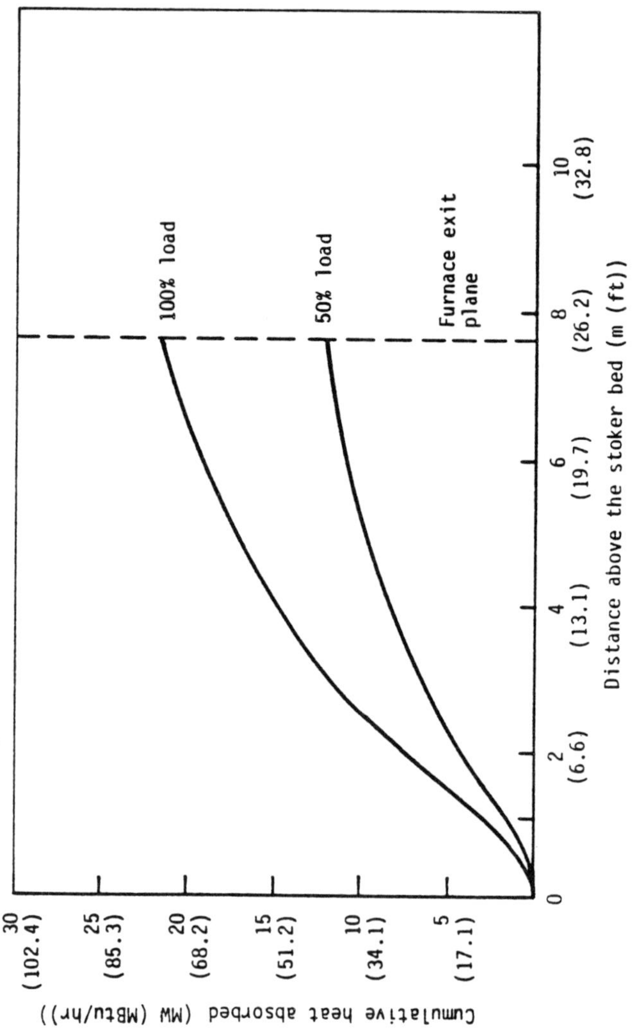

Figure 8. Cumulative heat absorption for the 44-MW (150 million Btu/hr) spreader-stoker boiler (boiler F).

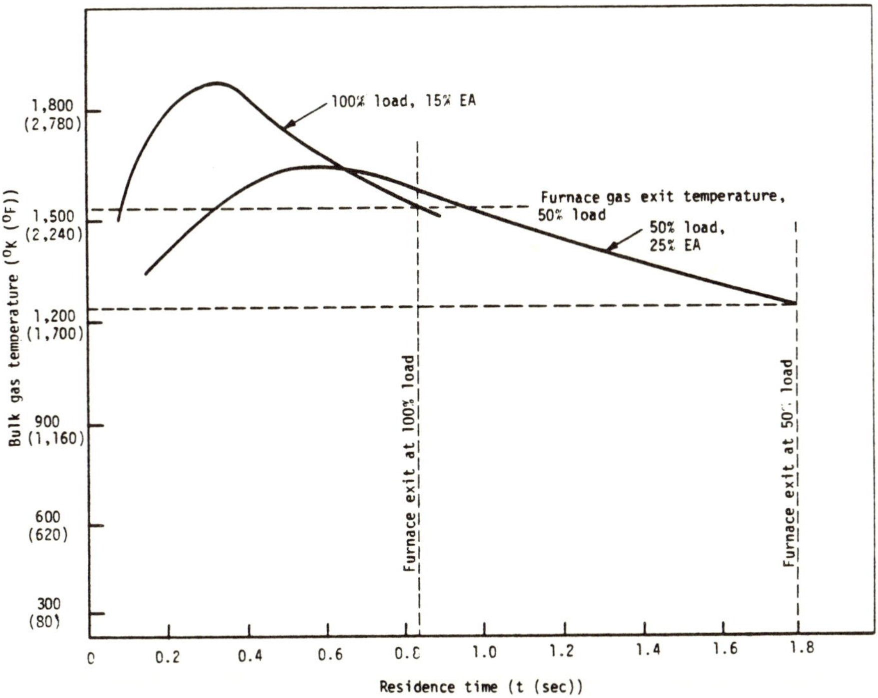

Figure 9. Temperature-time profile for the 44-MW (150 million Btu/hr) residual-oil-fired packaged boiler (boiler C).

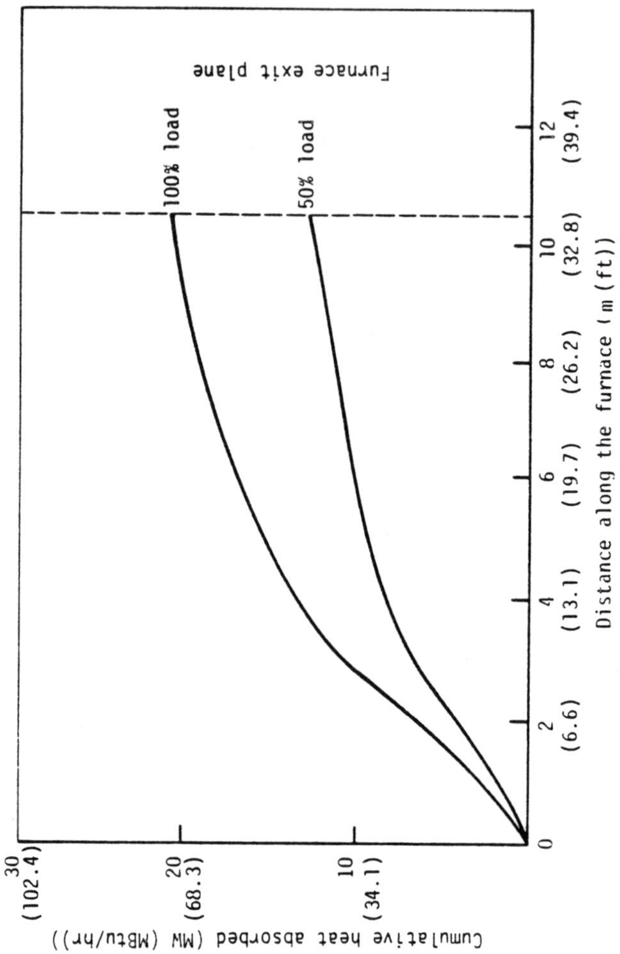

Figure 10. Cumulative heat absorption for the 44-MW (150 million Btu/hr) residual-oil-fired packaged boiler (boiler C).

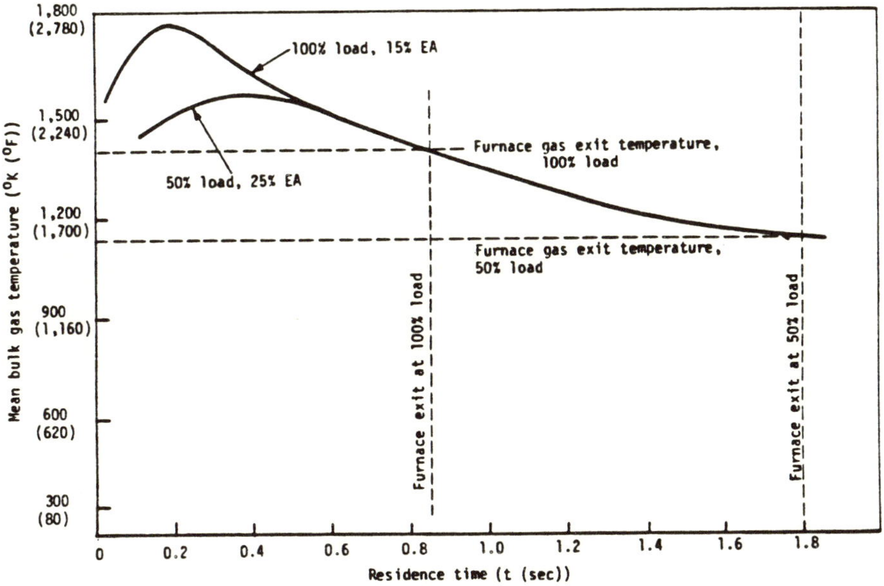

Figure 11. Temperature-time profile for the 44-MW (150 million Btu/hr) distillate-oil-fired packaged boiler (boiler D).

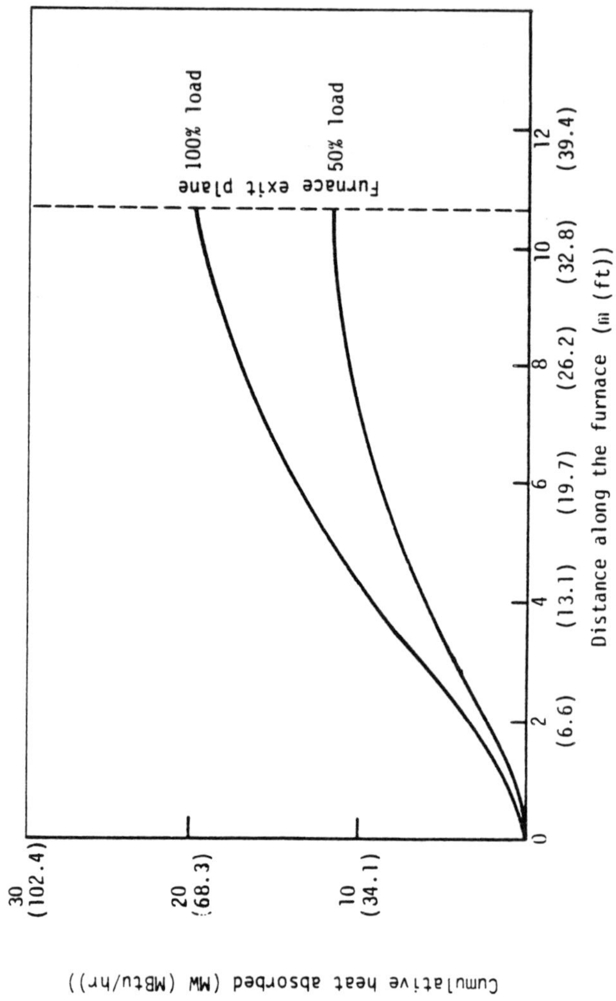

Figure 12. Cumulative heat absorption for the 44-MW (150 million Btu/hr) distillate-oil-fired packaged boiler (boiler D).

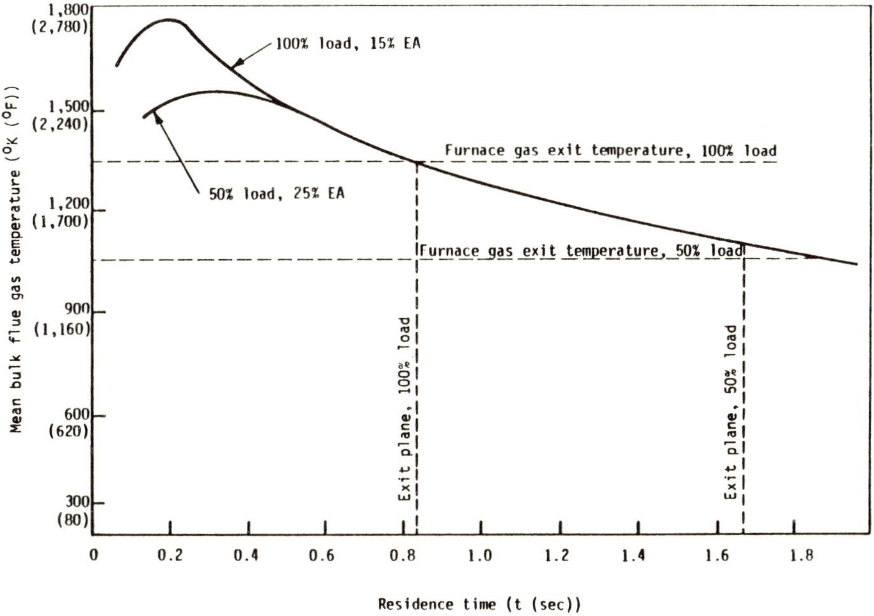

Figure 13. Temperature-time profile for the 44-MW (150 million Btu/hr) natural-gas-fired packaged boiler (boiler E).

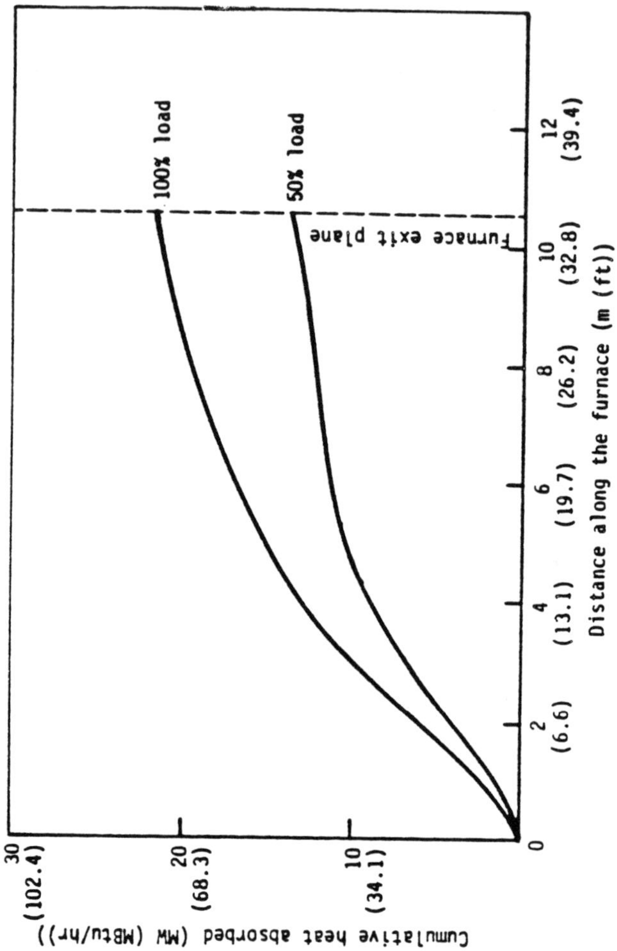

Figure 14. Cumulative heat absorption for the 44-MW (150 million Btu/hr) natural-gas-fired packaged boiler (boiler E).

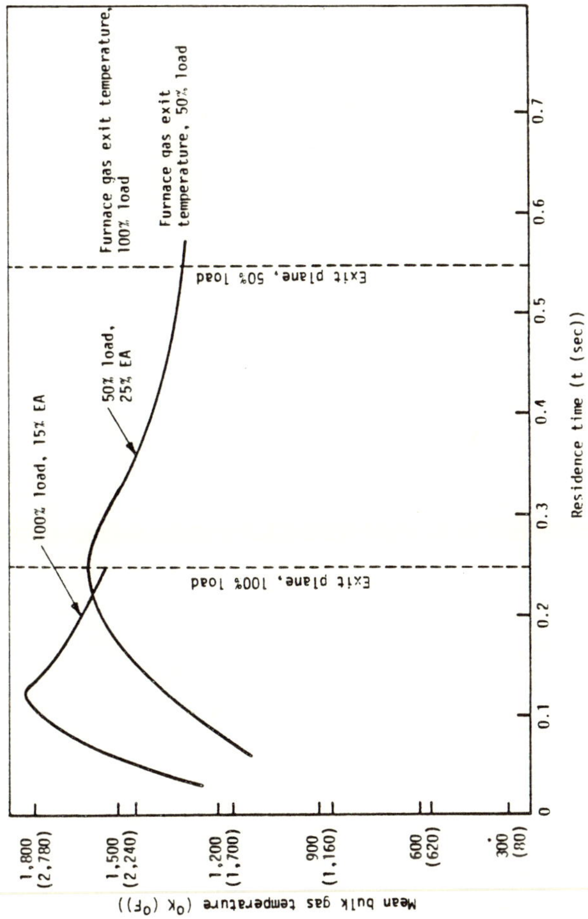

Figure 15. Temperature-time profile for the 4.4-MW (15 million Btu/hr) distillate-oil-fired firetube boiler (boiler A).

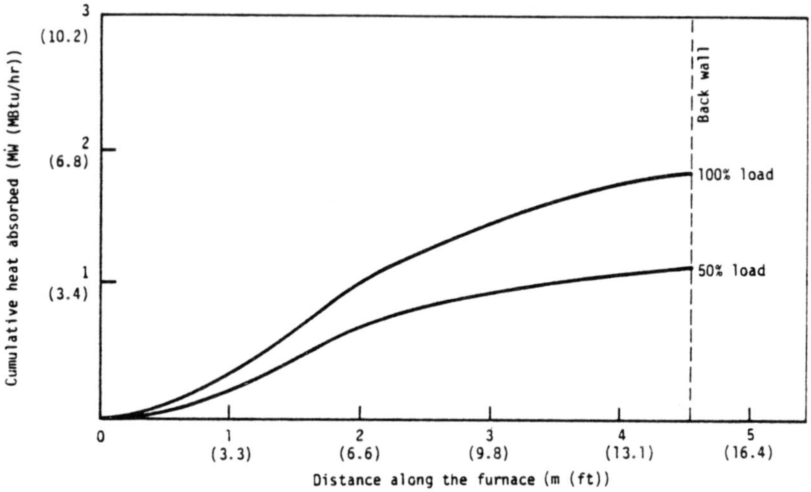

Figure 16. Cumulative heat absorption for the 4.4-MW (15 million Btu/hr) distillate-oil-fired firetube boiler (boiler A).

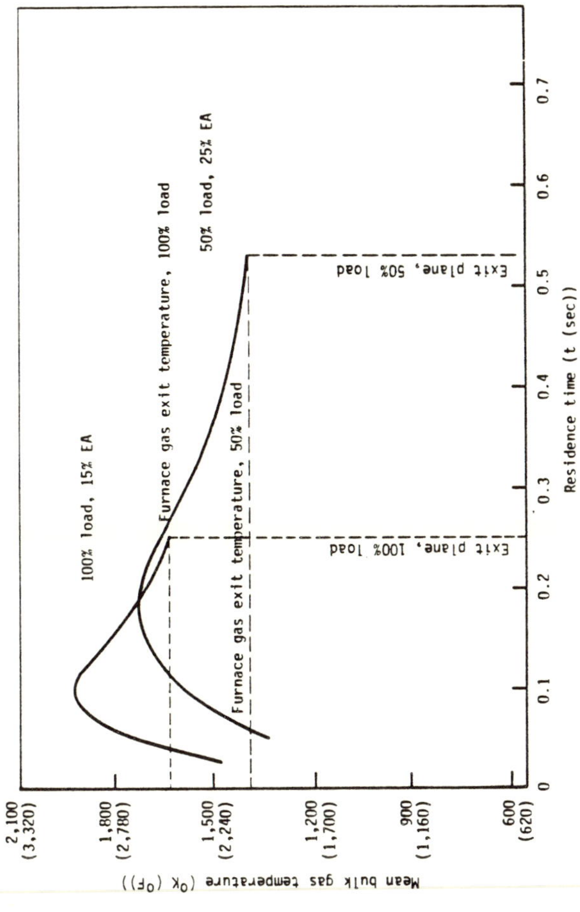

Figure 17. Temperature-time profile for the 4.4-MW (15 million Btu/hr) natural-gas-fired firetube boiler (boiler B).

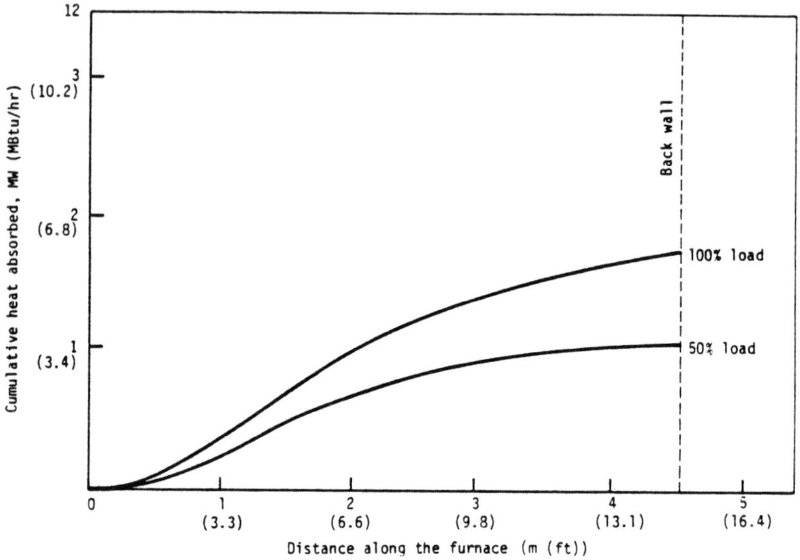

Figure 18. Cumulative heat absorption for the 4.4-MW (15 million Btu/hr) natural-gas-fired firetube boiler (boiler B).

peak in the near burner zone after which the temperature decreased nearly linearly to the furnace exit level. The heat flux to the furnace walls was highest in the burner zone where the heat release is high and turbulent mixing occurs. Downstream of this primary combustion region, the rate of heat absorption showed a decrease as the temperature and radiant heat transfer decreased. These results compare quite well with published data (Ref. 39,43,44)

The location of the temperature peak is primarily a function of the amount of heat released and the extent of combustion gas recirculation. In the case of the pulverized-coal-fired boiler (boiler G), the first computational zone encompassed the entire burner region (see Figure A-1, Appendix A). About 90 percent of the heat release was assumed to take place in the burner region. Gas recirculation from the second zone into the first was considered minimal since most of the turbulent mixing takes place within the burner region. Consequently, the gas temperature and the rate of heat absorption peaks in the first zone as illustrated in Figures 5 and 6.

For the spreader stoker unit (boiler F), Figures 7 and 8, the total heat release was distributed over the first three computational zones. Over 50 percent of the heat was assumed to be liberated in the first zone, close to the coal bed, while the remainder of the heat release was assumed to take place within the throwing region of the spreader.

Gas recirculation from the second to the first zone caused by the overfire combustion air jets, occurred at and above the stoker grate. The peak temperatures occurred at approximately 2.7m (8.8 ft) above the fuel bed which had a temperature of about 1,300K (2,030°F). Reducing boiler load to half of the heat input capacity resulted in the peak temperature occurring at approximately 3.8m (12.4 ft) above the stoker grate.

For both the package watertube and firetube units (boilers A through E), (Figures 9 to 18), the heat release was distributed over the calculated length of the flames. Gas recirculation and turbulent mixing were generally assumed to take place over half of the furnace length. Both the temperature-time profiles and heat flux distributions for these units are very similar, with the peak temperature occurring at a distance of about 2.7 to 4.6m (9 to 15 ft) and 2.3m (7.5 ft) downstream of the burner for the watertube and firetube units, respectively.

Tables 23 and 24 summarize the results of the heat transfer calculations outlining both peak and exit gas temperatures, bulk residence times, and heat transfer data for the full load and half load cases. A minimum residence time (0.5 times the mean time) is also listed based on arguments presented in Section 5.2. The highest temperatures occur as expected in the distillate-oil-and natural-gas-fired boilers because of the lower radiative properties of the combustion products from these fuels. Shortest residence times occur in the firetube boilers, which show the lowest percent of heat absorbed in the furnace.

TABLE 23. SUMMARY OF BOILER TEMPERATURE-TIME PROFILE DATA -- 100-PERCENT LOAD

Boiler	Gas temperature (°K (°F))		Residence time at furnace exit (sec)		Heat flux (MW (million Btu/hr))	
	Peak	Furnace exit	Mean, $\bar{t}$	Minimum $0.5\bar{t}$	Total heat absorbed in furnace	Percent of input
A: 4.4 MW (15 million Btu/hr) Natural-gas-fired firetube	1,815 (2,810)	1,600 (2,425)	0.25	0.13	1.56 (5.3)	36
B: 4.4 MW (15 million Btu/hr) Distillate-oil-fired firetube	1,710 (2,615)	1,490 (2,215)	0.25	0.13	1.62 (5.5)	37
C: 44 MW (150 million Btu/hr) Natural-gas-fired packaged watertube	1,815 (2,810)	1,475 (2,200)	0.83	0.42	19.3 (65.7)	44
D: 44 MW (150 million Btu/hr) Distillate-oil-fired packaged watertube	1,715 (2,630)	1,340 (1,955)	0.84	0.42	21.1 (72.0)	48
E: 44 MW (150 million Btu/hr) Residual-oil-fired packaged watertube	1,700 (2,600)	1,295 (1,875)	0.83	0.42	21.4 (73.1)	49
F: 44 MW (150 million Btu/hr) Spreader-stoker watertube	1,620 (2,450)	1,325 (1,925)	1.57	0.79	20.2 (69.0)	46
G: 58.6 MW (200 million Btu/hr) Pulverized-coal-fired watertube	1,790 (2,770)	1,435 (2,120)	1.75	0.88	22.3 (76.0)	38

TABLE 24. SUMMARY OF BOILER TEMPERATURE-TIME PROFILE DATA -- 50-PERCENT LOAD

Boiler	Gas temperature (°K (°F))		Residence time at furnace exit (sec)		Heat flux (MW (million Btu/hr))	
	Peak	Furnace exit	Mean, $\overline{t}$	Minimum $0.5\overline{t}$	Total heat absorbed in furnace	Percent of input
A: 4.4 MW (15 million Btu/hr) Natural-gas-fired firetube	1,630 (2,475)	1,335 (1,940)	0.53	0.27	1.0 (3.43)	46
B: 4.4 MW (15 million Btu/hr) Distillate-oil-fired firetube	1,515 (2,270)	1,235 (1,764)	0.55	0.28	1.03 (3.51)	47
C: 44 MW (150 million Btu/hr) Natural-gas-fired packaged watertube	1,575 (2,370)	1,210 (1,715)	1.78	0.89	11.7 (39.8)	53
D: 44 MW (150 million Btu/hr) Distillate-oil-fired packaged watertube	1,505 (2,250)	1,085 (1,495)	1.81	0.91	12.6 (43.0)	57
E: 44 MW (150 million Btu/hr) Residual-oil-fired packaged watertube	1,490 (2,220)	1,045 (1,425)	1.85	0.93	13.0 (44.5)	59
F: 44 MW (150 million Btu/hr) Spreader-stoker watertube	1,315 (1,910)	1,080 (1,490)	3.2	1.6	11.3 (38.6)	51
G: 58.6 MW (200 million Btu/hr)Pulverized-coal-fired watertube	1,585 (2,390)	1,195 (1,690)	3.5	1.8	12.9 (44.0)	44

5.4 EVALUATION OF CALCULATED TEMPERATURE-TIME PROFILES

The temperature-time profile solutions presented in this study should be used with care. The generic approach and assumptions used have resulted in the development of a model that, when properly applied, will lead to the generation of information. In evaluating the destruction efficiency of hazardous wastes in typical industrial boiler furnaces such as the ones considered in this study, it should be emphasized that a conservative use of these data be adopted. This conservative use consists essentially of taking the lowest temperature profile, such as the temperature gas profile close to the furnace waterwalls, and the least gas residence time. This combination, although inconsistent (i.e., the lowest temperature gas next to the furnace wall will have the longest residence time), represents conservatism. These conservative values can then be compared to the temperature and residence times required for thermal destruction of hazardous wastes as will be presented in Sections 6 and 7. If these data predict that a certain hazardous waste is destroyed in a given boiler type, a more extensive evaluation should be undertaken. This should include an assessment of of the boiler compatibility with the waste, secondary and fugitive emissions projections retrofit considerations, and a detailed economic analysis of the benefit of disposing of hazardous waste in industrial boilers over other disposal technologies. An overview of these considerations is presented in Section 8.

6. Theoretical Approach to Thermal Destruction of Hazardous Wastes

The purpose of this section is to develop a mathematical expression predicting the time for the 99.99-percent destruction of a given hazardous waste as a function of a given boiler time and temperature profile. The approach will be to:

- Establish a physical/chemical model of the processes through which the hazardous waste must pass
- Generate a mathematical expression for each process
- Predict conservative time functions for each process
- Sum the time functions to establish the required residence time at boiler conditions
- Compare predicted residence times against experimental values

The word "conservative" implies that all assumptions should tend to maximize the time required for a particular process. Thus, the predicted total time for the conservative model should always exceed any experimentally determined times. This will ensure that if the model shows a particular boiler to be applicable, it can be assumed to achieve the required destruction.

6.1 THE QUALITATIVE MODEL

The model used consists of a series of steps. The steps are:

- Heating the hazardous waste to its boiling temperature
- Vaporizing the waste
- Heating the hazardous waste vapor to the boiler ambient temperature
- Destroying the waste constituents by one of two processes
  - -- Thermal cracking of the strongest bond
  - -- First-stage oxidation

Five simplistic destruction models incorporating these processes are represented schematically in Figure 19. The conservative nature of this qualitative picture is readily apparent. In each of the first four cases, it is assumed that:

- The material is brought to its boiling point <u>before</u> any other process takes place (i.e., volatilization from the surface)
- The material is completely vaporized <u>before</u> any succeeding process starts

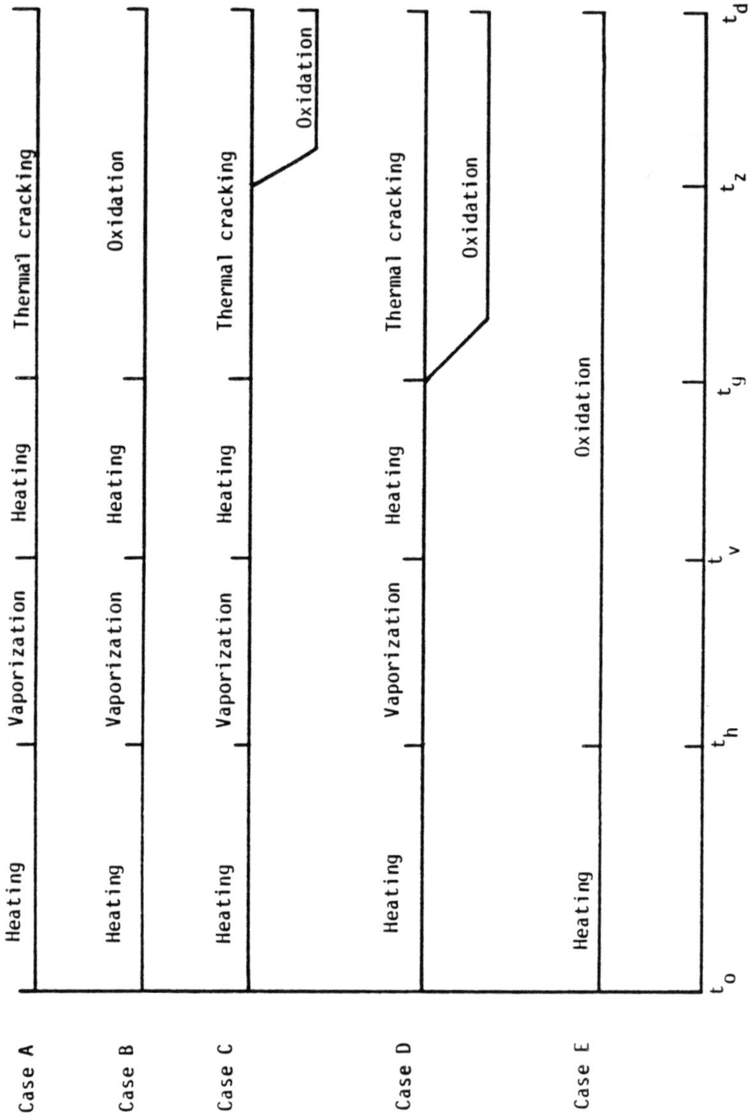

Figure 19. Processes for thermal destruction.

- All gas phase materials are heated to boiler ambient temperature before any destruction takes place

It is known that material ignites and burns (i.e., oxidizes) on its surface before complete volatilization occurs, and that diffusion limited burning of an expanding sphere of gas is also a normal boiler process. This process is depicted in case E of Figure 19 and is the least conservative.

6.2 THE QUANTITATIVE MODEL

A most conservative semiquantitative model for the destruction of hazardous waste materials can be easily formulated providing certain critical data are available. Without these data, conservative assumptions can be made that will also allow conservative prediction of degree of hazardous waste destruction.

The physical parameters determining destruction efficiency are time, temperature, and turbulence. The values of these parameters required to meet a certain efficiency are dependent on the destruction process (thermal cracking or oxidation) and the kinetic constants of the individual compound. If the time, temperature, and turbulence (which is related to oxygen availability) requirements to meet the required destruction efficiency of a given compound can be expressed mathematically, and if these can be compared to the time, temperature, and turbulence profile of a given boiler, then the ability of the boiler to achieve the desired destruction efficiency can be predicted.

The construction of the semiquantitative predictive model consists of:

- A series of assumptions
- A qualitative model chosen from those laid out in the previous section
- A theoretical mathematical expression generated that predicts the time and temperature requirements for 99.99-percent destruction
- A boiler time and temperature expression generated that can be compared to compound's time and temperature requirements

6.2.1 Choice of Model

The destructive processes of significance in thermal destruction are thermolytic cracking (pyrolysis) and oxidation. Another process, certain to occur in a boiler environment, is flame oxidation. Flame process rates are much faster than either thermal pyrolysis or oxidation reactions. However, so little rate data exist for these processes, that they are not considered herein.

Without reference to products of incomplete combustion, the parameter of concern is the disappearance of a particular hazardous waste component. The first step of destruction by pyrolysis is the homolytic thermolysis of a bond. In oxidation, it is either the radical abstraction of a hydrogen atom by oxygen, or the insertion of an oxygen radical into the molecule.

Pyrolysis is first-order kinetics (Ref. 45), while first-stage oxidation can be approximated by pseudo-first-order kinetics. (Ref. 46) Pyrolysis is the slower of the two processes. Thus, case A of the qualitative models is chosen as the most conservative approach.

6.2.2 Mathematical Formulation for Rate of Destruction

Assuming that the hazardous waste is in the gas phase, the rate of disappearance of a hazardous waste constituent by pyrolysis can be expressed as:

$$\frac{dC_A}{dt} = -kC_A \tag{6-1}$$

where C_A is the concentration of the material at time t and k is the temperature dependent rate constant. Solution of this equation leads to:

$$\ln \frac{C_A}{C_{A_o}} = -k\Delta t \tag{6-2}$$

where C_{A_o} is the initial concentration of material.

The rate constant k is given by the Arrhenius rate expression.

$$k = A \exp(-E/RT) \tag{6-3}$$

where

A = Arrhenius pre-exponential frequency factor (sec^{-1})
E = Energy of activation for bond rupture (cal/mole)
R = Gas constant (1.987 cal/mole-K)
T = Temperature (K)

For 99.99-percent destruction, the concentration term is:

$$\ln \frac{C_A}{C_{A_o}} = -9.21 \tag{6-4}$$

Substitution into Equation (6-2) and solving for Δt gives:

$$\ln \Delta t = \ln \frac{9.21}{A} + \frac{E}{RT} \tag{6-5}$$

which is just the time and temperature expression necessary for 99.99-percent destruction.

6.2.3 Boiler Time-Temperature Relationship to Destruction Efficiency

A typical boiler time and temperature profile is shown in Figure 20. If isotherms (lines of equal temperature) are drawn across the profile, and their intersection dropped to the time coordinate, increments of time for which the gas is above a given temperature can be determined, as shown in Figure 21.

If these time increments are plotted against their corresponding isotherms, a curve relating time at or above a given temperature for the boiler can be obtained. During the temperature rise portion in the boiler, the time-temperature relationship is approximated by:

$$T = a + b \ln t_1 \qquad (6\text{-}6)$$

while during the fall it is:

$$T = c \exp Dt_2 \qquad (6\text{-}7)$$

The incremental time above a given temperature, Δt_{Tm} is $t_2 - t_1$. Algebraic manipulation gives:

$$\Delta t_T = \frac{1}{D} \ln \frac{T}{c} - \exp \frac{T-a}{b} \qquad (6\text{-}8)$$

Plotting T versus ΔtT gives the solid curve in Figure 22. Using Equation (6-5) and plotting T versus Δt for a given compound gives the dashed lines in Figure 22.

If any point on the compound line falls below the boiler curve, the boiler has sufficient time at high enough temperature to achieve 99.99-percent destruction.

In Figure 22 three hypothetical compound time-temperature requirements have been plotted. Compound A demonstrates a case where 99.99-percent destruction most probably will not be achieved. Compound B shows a case where destruction will be achieved if the material follows the mean time-temperature profile of the boiler. Compound C presents an example where destruction will always be achieved.

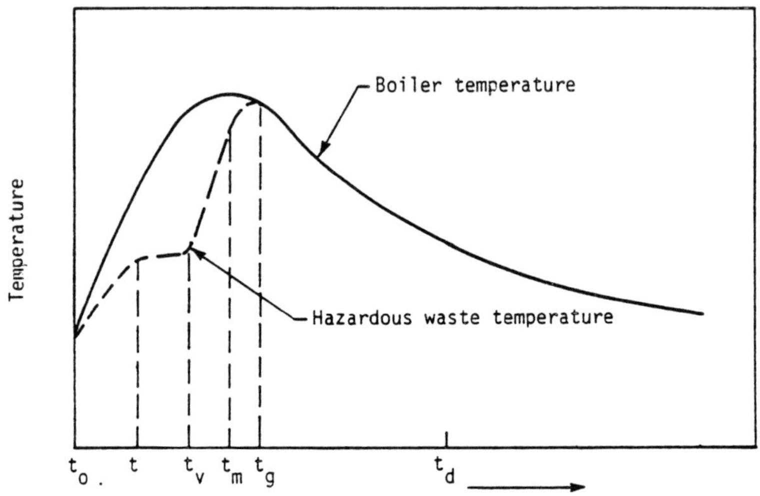

Figure 20. Waste material time-temperature path superimposed on boiler temperature profile.

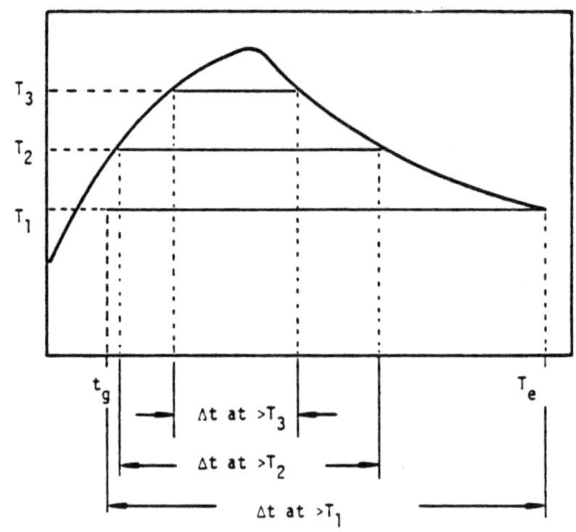

Figure 21. Determination of time-above-temperature points.

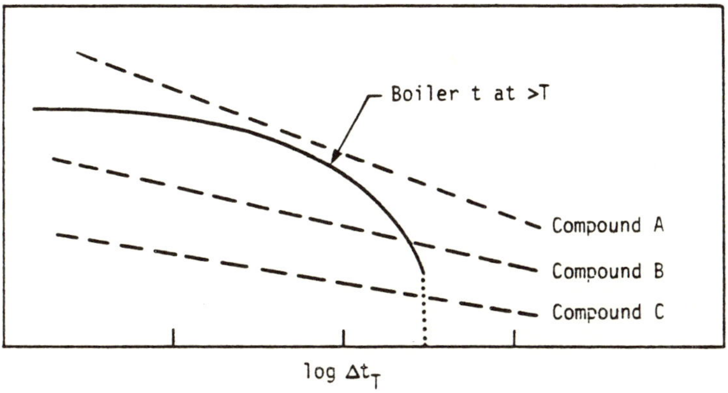

Figure 22. Hypothetical boiler time-above-temperature curve (solid curve) and compound destruction time-temperature curves (dashed curve).

6.3 MAKING PREDICTIONS

The following two subsections give two conservative approaches to destructibility predictions using the above model. The first is the most conservative case, and immediately classifies the compounds (or wastes) into a yes-no-maybe category. The second, extended conservative case refines the maybe category. Both cases assume the material is in the gas phase by the time of maximum temperature under the given boiler conditions. The determination of the validity of this assumption is discussed in Section 7.4.

One can assume that all volatile material is vaporized by the time (t_m) the maximum mean temperature in the boiler (T_m) is reached, since the maximum rate of heat release (combustion) has occurred. Also, as discussed in Section 5.2 the shortest residence time in a boiler is approximately one-half the mean residence time (t_e)(Ref. 39), and the lowest temperature path has a temperature maximum (T'_m) about 100K below T_m (Ref. 39).

6.3.1 Most Conservative Case

If a component is in the gas phase at time t_m in the boiler, and if the fastest path through the boiler is one-half the mean residence time, it is axiomatic that the component will be at or above the exit temperature for at least the time ($0.5\ t_e - t_m$). Given, then, the following information:

- Boiler mean residence time
- Boiler exit temperature
- Boiler mean time to maximum temperature
- Component kinetic A and E values

the material can be 99.99 percent destroyed if:

$$\ln\,(0.5\ t_e - t_m) > \ln \frac{9.21}{A} + \frac{E}{RT_e} \qquad (6\text{-}9)$$

A solution to the relationship between boiler time-above temperature and 99.99 percent destruction can be found that predicts components for which 99.99 percent destruction will certainly be achieved. This is accomplished by recognizing that for some particular set of E and A values the will be tangent to (or equal to the derivatives of) the boiler component curve time-above-temperature curve. The solution is:

$$\left[\left(\frac{E}{2R}\right)\left(-1 + \sqrt{1 + \frac{4RT_m}{E}}\right) - T_m\right]\left(\frac{t_e - t_m}{T_e - T_m}\right) > \frac{9.21}{A}\ \exp\left[2\left(-1 + \sqrt{1 + \frac{4RT_m}{E}}\right)^{-1}\right] \qquad (6\text{-}10)$$

Thus, for any set of component and boiler parameters meeting this relationship, the material will meet 99.99 percent destruction. All other cases are of the "maybe" category. Their relationships are further developed in Section 7.5, and demonstrated graphically for various hazardous waste stream and boiler parameters.

6.3.2 Extended Conservative Approach

In reality, the material will always be at temperatures above the exit in temperature. Because the rate of destruction is exponentially related to temperature, this should be taken into account when estimating destruction and removal efficiencies. One approach is to plot the curve of Equation (6-8) and the thermodynamic destruction curve and compare. However, with the assumption all waste constituents are in the gas phase at t_m, the equation can be simplified to use readily available parameters.

The temperature decay after T_m is approximately linear with time to the exit temperature (T_e). Thus, the increment of time, Δt, that a component will be above temperature, T, is given by:

$$\Delta t = (T - T_m)\frac{t_e - t_m}{T_e - T_m} \qquad (6\text{-}11)$$

The absolute minimum time increment, Δt, that a given material is above a given temperature, T, in the boiler is thus found by substituting 0.5 t_e for t_e, and $T_m' = T_m - 100K$ for T_m, with the temperature bounds $T_m' > T \geq T_e$. This is the "fast path" temperature profile.

Since the increment of time required for 99.99-percent destruction is given by Equation (6-5), if:

$$\ln \Delta t > \ln \frac{9.21}{A} + \frac{E}{RT} \qquad (6\text{-}12)$$

(where Δt is calculated by Equation (6-11) for fast path values), the boiler will achieve 99.99-percent destruction. It is easiest to simply plot Equations (6-11) and (6-5) to see if Equation (6-5) falls below Equation (6-11) at any point. A plot similar to Figure 22 will be obtained.

The extended conservative estimate of a boiler's capacity to destroy a given hazardous waste is achieved in the following manner:

- For given boiler conditions, the time increment is plotted against the temperature (Equation (6-11)
- For a given hazardous waste compound, the curve of Equation (6-5) is plotted on the same graph
- If any point of the boiler curve falls above the compound curve, the boiler will give 99.99-percent destruction of the compound

6.4 STREAM SELECTION

Stream selection for evaluation with respect to boiler destructibility was based on three principal criteria:

- Those streams intuitively thought to be able to be destroyed (i.e., those streams that were primarily organic in nature)
- Streams that were reasonably well characterized (with respect to composition and physical properties) in the available literature
- Streams covering a wide variety of classes of organic compounds

Table 25 lists the streams identified for evaluation and the classes of compounds contained. Table 26 gives composition information on each stream. (Note: Stream K012, acrylonitrile still bottoms, is no longer classified as a hazardous waste stream. It is used here because of the compound class.)

In actuality, there are several perturbations affecting the position of the boiler curve. This occurs because the curve, as presented, is determined from the mean temperature profile, the mean residence time, and (for nongaseous hazardous wastes) the time and temperature requirements to vaporize the waste.

6.5 COMPOSITE COMPOUND CURVES FOR SELECTED WASTE STREAMS -- RELATIVE INCINERABILITY

Tables 27 and 28 present physical and kinetic data for compounds associated with the chosen hazardous waste streams. The data were obtained from a limited literature search. As can be seen, there are large gaps in the available information and, furthermore, the data are often reported in a manner that is not directly applicable. For instance:

- Densities -- some values are for solids
- Heat capacities -- very few reported temperature dependencies
- Kinetic rate expressions
  - -- Pyrolysis data are usually for a specific bond cleavage as opposed to compound disappearance
  - -- Oxidation data rarely give oxygen concentration dependence. Most work has been conducted in air as opposed to 2 to 5 percent excess oxygen.

With the preceding deficiencies in mind, the predictions of waste stream destruction in Section 7 must be recognized as examples of the approach and not firm predictions of destructibility.

Our approach is to use conservative worst case characteristics in establishing a waste stream composite compound curve and comparing it to boiler curves. For the kinetic rate expression, the smallest Arrenhius pre-exponent (A) and largest energy of activation (E) are chosen from the available compound data on a given stream. This represents a very conservative approach in that very small frequency factors are rarely associated with very large activation energies. Where kinetic data are not

TABLE 25. CLASSES OF ORGANIC COMPOUNDS ASSOCIATED WITH PARTICULAR HAZARDOUS WASTE STREAMS

EPA hazardous waste number	Aromatics	Oxygenated aromatics	Chlorinated aromatics	Chlorinated aliphatics	Olefins	Unsaturated carbonyls	Polynuclear aromatics	Oxygenated aliphatics
F003	X							X
F004		X						
F005	X	X						X
K012						X		
K015			X					
K018			X	X	X			
K022	X	X					X	
K030			X	X	X			

TABLE 26. COMPOUNDS APPEARING IN SELECTED HAZARDOUS WASTE STREAMS

Stream	Components	
F003	Spent nonhalogenated solvents: xylene, ethyl benzene, cyclohexanone, acetone, ignitable solids (10 to 50 percent of original waste)	
F004	Spent nonhalogenated solvents: cresol, nitrobenzene, ignitable solids (10 to 50 percent of original waste)	
F005	Spent nonhalogenated solvents: toluene, methyl ethyl ketone, carbon disulfide, pyridine, "still bottom residues" (10 to 50 percent of original waste)	
K012	Still bottoms from final purification of acrylonitrile	
	Acrolein	6 g/kg waste
	Acrylamide	6 g/kg waste
	Acrylonitrite	0.1 g/kg waste
K015	Still bottoms from the distillation of benzyl chloride	
	• By photochlorination	
	Benzal chloride	20 g/kg waste
	Trichlorobenzenes	5 g/kg waste
	Tetrachlorobenzene	5 g/kg waste
	Pentachlorobenzene	5 g/kg waste
	Hexachlorobenzene	5 g/kg waste
	• By catalytic chlorination	
	Benzal chloride	100 g/kg waste
	Chlorobenzene	10 g/kg waste
	Trichlorobenzenes	10 g/kg waste
K018	Heavy ends from ethyl chloride production	
	Ethyl chloride	3 g/kg waste
	1,2-dichloroethane	20 g/kg waste
	Trichloroethane	30 g/kg waste
	Hexachlorobutadiene	40 g/kg waste
	Hexachlorobenzene	40 g/kg waste

(continued)

TABLE 26. (continued)

Stream	Components	
K022	Distillation tars from phenol production by cumene process	
	Acetophenone	1.9 g/kg waste
	Phenol	0.75 g/kg waste
	Cumyl phenol	0.85 g/kg waste
	Other tars	50-100 g/kg waste
K030	Column bottoms or heavy ends from combined production of trichloroethylene and perchloroethylene	
	Hexachlorobutadiene	230 g/kg waste
	Hexachlorobenzene	20 g/kg waste
	Hexachloroethane	10 g/kg waste
	Tetrachloroethane	10 g/kg waste
	Ethylene chloride	10 g/kg waste
	Chlorobutadiene	10 g/kg waste

TABLE 27. PHYSICAL DATA FOR COMPOUNDS IN SELECTED HAZARDOUS WASTE STREAMS (REF. 45)

Hazardous waste number	Chemical compound	Melting point (°C)	Boiling point (°C)	Specific gravity (20°C)	Heat capacity (cal/g-°C)	ΔH°_{vap} (cal/g)	Auto-ignition temperature (°K)	Remarks
F003	o-xylene	-25.0	144.0	0.880	0.411 (0.451)*	94.33		*30°C (39°C)
	m-xylene	-48.0	139.0	0.860	0.393+; 0.40++; (0.387)	93.44		+0°C ++9°-40°C (16°-35°C)
	p-xylene	13.0	138.0	0.861	0.383* 0.397+ (0.428)	92.54		*0°C +30°C (40.8°)
	Ethylbenzene	-95.0	136.0	0.867	0.392 (0.407)	87.75		0°C (30°C)
	Cyclohexanone	-45.0	155.0	0.947	0.431*			*15°-18°C
	Acetone	-95.35	56.2	0.7899	0.514-0.538	134.74	738	See K022
F004	o-cresol	31.0	191.0	1.0273	0.497*	115.62	832	*0°-20°C
	m-cresol	11.0	201.0	1.0336	0.551**	124.85	832	**21°-197°C
	p-cresol	35.0	202.0	1.0178		126.03	832	
	Nitrobenzene	5.7	210.9	1.205	0.358; 0.339; 0.330 (0.343); 0.394	79.08	756	10°C, 30°C, 50°-70°C (90°C), 120°C
F005	Toluene	-95.0	110.8	0.866	0.364; 0.534 (0.44)	101.83	809	10°C, 85°C, (12-99)
	Pyridine	-42	115.0	0.982	0.405; 0.431 (0.395)	107.36	816	20°C, 21°-108°C (0-20)
	Methylethyl ketone	-85.9	79.6	0.805	0.549	105.93	747	20°-78°C
	Carbon disulfide	-108.6	46.3	1.263		89.29	373	
K012	Acrolein	-87.7	52.5	0.841			507	
	Acrylamide		84.85				797	
	Acrylonitrile	-82	78-79	0.811			754	
K015A	Benzylchloride	-39	179	1.10	0.323			
	Trichlorobenzene	17-63.5	208.5- 219	1.574			844	
	Tetrachlorobenzene	47-138	246- 254	1.734				
	Pentachlorobenzene	86	277	1.834				
	Hexachlorobenzene	228	326	2.044		30.55		
K015B	Benzylchloride	-39	179	1.10	0.323			
	Chlorobenzene	-45.2	131	1.107		77.6	911	
	Trichlorobenzene	63.5	208.5				844	
K018	Ethylchloride	-139.0	13.0	0.917	0.426 (0.367)	92.9	792	-28° to +4°C (0°C)
	1,2-dichloroethane	-35.3	83.7	1.256			686	
	Hexachlorobenzene	228.31	309.0	2.044		30.55		
	Trichloroethane	--	74.1	1.325	0.266		856, 955	20°C
	Hexachlorobutadiene						883	

(continued)

TABLE 27. (continued)

Hazardous waste number	Chemical compound	Melting point (°C)	Boiling point (°C)	Specific gravity (20°C)	Heat capacity (cal/g-°C)	$\Delta H°_{vap}$ (cal/g)	Auto-ignition temperature (°K)	Remarks
K022	Acetone	-94.6	56.5	0.792	0.514*; 0.538**	105-130	738	*3°-22.4°C; **24.2°-49.4°C
	Phenol	42.0	181.4	1.07	0.561 (s)	126.41	988	
	Cumene	-96.9	152.5	0.862		89.77		
K030	Trichloroethylene	-73.0	87.2	1.466	0.223	57.24	683	20°C
	Hexachlorobenzene	228.0	309.0	2.044		30.55		
	Hexachloroethane	186.0	186.0	2.09	0.174 (s)	26.06		
	Ethyl chloride	-139.0	13.0	0.917	0.426 (0.367)	92.93	792	-28° to +4°C (0°C)
	Chloro-buta-1,3-diene	--	59.4	0.958				
	Chloro-buta-1,2-diene	--	88.0	0.991				

TABLE 28. KINETIC DATA FOR COMPOUNDS OCCURRING IN SELECTED HAZARDOUS WASTE STREAMS (REF. 45, 46, 54)

	Oxidation (in air)		Pyrolysis	
	A (sec^{-1})	E (kcal/g-mole)	A (sec^{-1})	E (kcal/g-mole)
Aromatics				
Benzene	7.42×10^{21}	95.9		
Chlorobenzene				
Dichlorobenzene				
Trichlorobenzene				
Tetrachlorobenzene				
Pentachlorobenzene				
Hexachlorobenzene	1.9×10^{16}	72.6		
Ethylbenzene			10^{13}	63
m-xylene			4.2×10^{12}	77.5
p-xylene			4.8×10^{12}	76
o xylene			4.8×10^{12}	75
Toluene	2.28×10^{13}	56.5	2.1×10^{12}	77.5
Methyl naphthalene			1.5×10^{13}	73.5
Propylbenzene			3.2×10^{13}	57.5
Bromobenzene			2×10^{13}	70.9
Benzylbromide			10^{13}	50.5
Biphenyl	3.55×10^{16}	70.7		
Tetrachloro biphenyl	6.8×10^{16}	77		
Pentachloro biphenyl	1.1×10^{16}	70.0	7.44×10^{9}	53.6
Decachloro biphenyl	3.46×10^{12}	55		
Hexachloro biphenyl	2.7×10^{16}	71		
Aliphatics				
Methane	1.68×10^{11}	52.1		
Ethane	5.65×10^{14}	63.6	2×10^{17}	84
Propane	5.25×10^{19}	85.2		
Ethyl chloride			4×10^{14}	60.8
1,1 dichloroethane			1.3×10^{12}	49.5
1,2 dichloroethane	4.82×10^{11}	45.6	10^{13}	70

(continued)

TABLE 28. (continued)

	Oxidation (in air)		Pyrolysis	
	A (sec^{-1})	E (kcal/g-mole)	A (sec^{-1})	E (kcal/g-mole)
Aliphatics (continued)				
1,1,1 Trichloroethane			3.2×10^{12}	47.9
Propyl chloride			2.8×10^{13}	55.0
Isopropyl chloride			2.5×10^{13}	50.5
1,2 Dichloropropane			6.3×10^{13}	54.9
Propyl bromide			4×10^{13}	47.7
Isopropyl bromide			10^{13}	50.7
Olefinics				
Ethylene	1.37×10^{12}	50.8		
Propylene	4.63×10^{8}	34.2		
Butene	3.74×10^{14}	58.2	5×10^{12}	63
Vinyl chloride	3.57×10^{14}	63.3		
Vinyl bromide			5×10^{12}	47.5
Miscellaneous				
Ethanol	5.37×10^{11}	48.1		
t-Butanol			4.8×10^{14}	66
t-Amyl Alcohol			3.2×10^{13}	60
Ethyl acetate			3.2×10^{12}	47
Isopropyl acetate			10^{13}	45.0
t-Butyl acetate			2.2×10^{13}	40.5
Ethyl acrylate	2.19×10^{12}	46		
Acrylonitrile	2.18×10^{12}	52.1		
Acrolein	3.3×10^{10}	35.9		
Acetic anhyride	10^{12}	34.5		

TABLE 29. COMPOSITE COMPOUND PARAMETERS[a] BY WASTE STREAM

	Melting point (°C)	Boiling point (°C)	Specific gravity[b]	Heat capacity (cal/g-°C)	$\Delta H°_{vap}$ (cal/g)	A (sec-1)	E (kcal/g-mole)	Mode
F003	13	155	0.947	0.538	134.7	5×10^{12}	78	Pyrolysis
F004	35	211	1.205	0.551	126	(10^{10})[c]	(70)[c]	(Pyrolysis)
F005	-42	115	1.263	0.549	106	2.1×10^{12}	77.5	Pyrolysis
K012	-82	85	0.841	(0.6)	(140)	3.3×10^{10}	52.1	Oxidation
K015	228	326	1.834[b]	0.323	77.6	1.9×10^{16}	72.6	Oxidation
K018	228	326	1.834	0.426	91.3	3.2×10^{12}	70	Pyrolysis
K022	42	181	1.07[b]	0.561	126	(10^{10})[c]	(70)[c]	(Pyrolysis)
K030	228	326	2.09[b]	0.426	92.9	10^{13}	70	Pyrolysis

[a]Values in parentheses are arbitrary.
[b]Solid density.
[c]Guess.

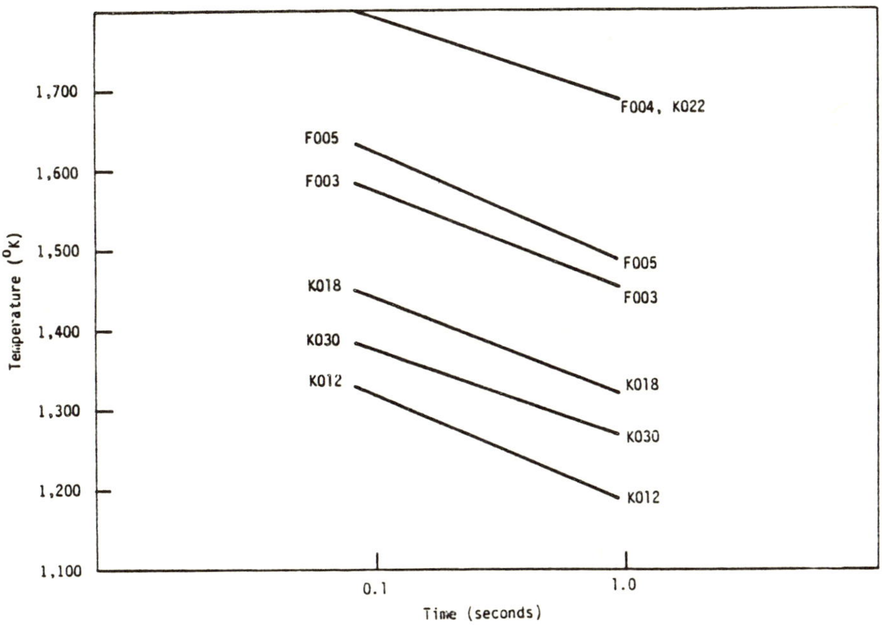

Figure 23. Plots of time-temperature relationships required for 99.99-percent destruction of selected hazardous waste streams (assuming destruction is by the conservative pyrolysis mode and using the lowest A and highest E values for compounds in the stream).

available, an arbitrary A of 10^{10} and an E of 70 kcal/mole has been used. (As noted above, high E values (>70) and low A values (>10^{10}) do not occur with the same compound.)

The composite compound parameters for selected waste streams is given in Table 29. Figure 23 plots the gas phase time-temperature requirement for each stream where pyrolytic data are available. Pyrolytic data are used because they are the most conservative, and the only oxidation data are for air (20-percent oxygen), as opposed to 2 to 5 percent oxygen.

6.6 AUTOIGNITION TEMPERATURES AND DESTRUCTIBILITY

Autoignition temperatures are more widely available than kinetic destruction data. Autoignition temperature has been shown to be a good predictor of relative combustibility for 15 compounds (Ref. 46). There is currently insufficient data on oxidative destruction of compounds to verify this predictor for other compounds.

There is a much wider data base available on pyrolytic destruction. However, pyrolytic destruction proceeds by mechanisms different from oxidative destruction. Even though oxidative destruction can be orders of magnitude faster than pyrolytic destruction, the relative rates between a series of compounds can be reversed. For example, the rate constants for oxidation of ethane and 1,2-dichloroethane are approximately 7.1 and 52 sec^{-1} at 1,000K. For pyrolysis, the relative order is reversed, being about 0.09 and 0.005 sec^{-1}.

With the previous caveats in mind, a table of the relative destructibility of compounds, based on the K.C. Lee autoignition temperature predictor (Ref. 46) has been prepared (Table 30). A second table has been prepared (Table 31), grouping compounds into structural classes and indicating the limiting compound (most refractory) for destruction.

TABLE 30. ORDER OF INCREASING DESTRUCTIBILITY BY TEMPERATURE REQUIRED FOR 99.99-PERCENT DESTRUCTION AT 1-SECOND RESIDENCE TIME

Order	Compound	T (°F) needed at 1 sec	Destruction rate data
1	Methyl chloride	1,550	
2	Methane	1,535	X
3	Phenol	1,484	
4	Methylene chloride	1,466	
5	Pyridine	1,455	
6	Chlorobenzene	1,413	
7	Dichlorobenzene	1,408	
8	Hexachlorobenzene	1,405	X
9	Ethane	1,401	X
10	Vinyl chloride	1,374	X
11	Ethyl chloride	1,358	
12	Benzene	1,358	X
13	Cresol	1,331	
14	Ethylene	1,329	X
15	Toluene	1,327	X
16	Nitrobenzene	1,327	
17	Hexa chlorobutadiene	1,325	
18	Trichlorobenzene	1,320	
19	Vinylidene chloride	1,313	
20	Acetophenol	1,310	
21	Propane	1,305	X
22	1,2,2-trichloro-1,1,2-trifluoroethane	1,277	
23	Dichloropropane	1,266	
24	Trichloroethane	1,260	
25	Phthalic anhydride	1,240	
26	Isobutanol	1,229	
27	Trichloroethylene	1,222	
28	Naphthalene	1,221	
29	Methyl ethyl ketone	1,213	
30	Dichloroethane	1,212	X
31	Epichlorohydrin	1,143	
32	Maleic anhydride	1,123	
33	Bis(2-chloroethyl) ether	1,115	
34	Methyl isobutyl ketone	1,109	
35	Trichloropropane	1,060	
36	Benzotrichloride	982	
37	Carbon disulfide	751	

TABLE 31. WORST CASE COMPOUND FOR DESTRUCTION BY STRUCTURAL CLASS

Compound	Limiting (more refractory) compound with data	Reasoning
Methane	Methane	Data
Methyl chloride	Methane	UCT
Methylene chloride	Methane	UCT
Chloroform	Methane	Analogy
Carbon tetrachloride	Methane	Analogy
Carbon disulfide	Dichloropropane	UCT
Ethane	Ethane	Data
Ethyl chloride	Vinyl chloride	UCT
Dichloroethanes	Dichloroethanes	Data
Trichloroethanes	Propane	UCT
Tetrachloroethanes	Propane	Analogy
Pentachloroethanes	Propane	Analogy
Hexachloroethanes	Propane	Analogy
Trichlorotrifluoroethane	Propane	Analogy
Ethylene	Ethylene	Data
Vinyl chloride	Vinyl chloride	Data
Vinylidine chloride	Vinyl chloride	UCT
Trichloroethylene	Propane	UCT
Perchloroethylene	Propane	Analogy
Bis(chloromethyl)ether	Dichloroethanes	Analogy with Bis(2-chloromethyl)ether
Propane	Propane	Data
Chloropropane	Propane	TP
Dichloropropane	Propane	UCT
Trichloropropane	Dichloroethanes	UCT
Dichloropropanols	Propane	Analogy
Epichlorohydrin	Dichloroethanes	UCT
Isobutanol	Propane	UCT
Chlorobutadiene	Toluene	Analogy
Hexachlorobutadiene	Toluene	UCT
Maleic Anhydride	Dichloroethanes	UCT
Bis(2-chloroethyl)ether	Dichloroethanes	UCT

(continued)

UCT -- Union Carbide predicted temperature
TP -- Thermal pyrolysis data
Analogy -- Structural correlation and speculation

TABLE 31. (continued)

Compound	Limiting (more refractory) compound with data	Reasoning
Methyl ethyl ketone	Dichloroethanes	UCT
Methyl isobutyl ketone	Dichloroethanes	UCT
Benzene	Benzene	Data
Chlorobenzene	Hexachlorobenzene	UCT
Dichlorobenzene	Hexachlorobenzene	UCT
Trichlorobenzene	Hexachlorobenzene	UCT
Tetrachlorobenzene	Hexachlorobenzene	Analogy
Pentachlorobenzene	Hexachlorobenzene	Analogy
Hexachlorobenzene	Hexachlorobenzene	Data
Pyridine	Methane	UCT
Nitrobenzene	Toluene	UCT
Phenol	Methane	UCT
Chlorophenol	Methane	Analogy
Dichlorophenol	Methane	Analogy
Trichlorophenol	Methane	Analogy
Tetrachlorophenol	Methane	Analogy
Pentachlorophenol	Methane	Analogy
Nitrophenols	Methane	Analogy
Toluene	Toluene	Data
Benzalchloride	Toluene	Analogy
Benzotrichloride	Dichlorethane	UCT
Cresols	Benzene	UCT
Cresylic acid	Benzene	Analogy
Nitrocresols	Benzene	Analogy
Dinitrocresols	Benzene	Analogy
Acetophenone	Toluene	UCT
Phthatic anhydride	Propane	UCT
Dimethyl phenols	Methane	Analogy
Cumyl phenol	Methane	Analogy
Naphthalene	Propane	UCT
Naphthoquinone	Propane	Analogy

UCT -- Union Carbide predicted temperature
TP -- Thermal pyrolysis data
Analogy -- Structural correlation and speculation

7. Use of Industrial Boilers to Burn Hazardous Waste

Boiler and waste characteristics must be compared to determine the suitability of a boiler to burn hazardous wastes. The boiler characteristics developed in Section 5 give the bulk temperature and mean residence time profiles for selected industrial boilers. The destruction curves developed in Section 6 give the time-temperature requirements for the destruction of a given waste. By overlaying the results of Sections 5 and 6, an evaluation can be made of which boilers are suitable for cofiring hazardous wastes and which wastes are amenable to cofiring.

The purpose of this section is to apply the predictive model of Section 6 to selected hazardous waste streams and boiler conditions. The approach is to:

- Present boiler parameters
- Determine effects of hazardous waste characteristics on boiler time-above-temperature curves
- Verify the assumption that hazardous waste constituents are in the gas phase hazardous waste at t_m
- Calculate boiler time above temperature curves
- Compare compound and boiler curves
- Make destructibility prediction

7.1 BOILER PARAMETERS

The boiler parameters for seven boiler classes and operational paths are given in Table 32. This table is derived from the boiler time-temperature profiles of Section 5, and curve fitting using the approximation formulas of Section 6.2.3. To reiterate, the curve fitting formula for temperature rise over the time interval t_0 to t_m (the temperature rise portion) is:

$$T = a + b \ln t \qquad (7\text{-}1)$$

where t_0 = 0.001 sec, T_0 = 400K, t ranges from t_0 to t_m and T ranges from T_0 to T_m. For the temperature fall portion, the formula is:

$$T = c \exp(Dt) \qquad (7\text{-}2)$$

where t ranges from t_m to t_e, and T ranges from T_m to T_e. Intermediate points for the curve fit were obtained by manual reading of the

TABLE 32. BOILER TEMPERATURE PARAMETERS

Boiler	a	b	c	D	t_m (sec)	T_m (°K)	t_e (sec)	T_e (°K)
G: 58.6 MW (200 million Btu/hr) Pulverized-coal-fired watertube								
Full load -- mean path	2,296	274	1,993	-0.15	0.20	1,790	1.73	1,435
-- fast path	2,335	280	1,726	-0.21	0.10	1,690	0.88	1,435
Half load -- mean path	1,822	205	1,759	-0.09	0.39	1,585	3.5	1,195
-- fast path	1,815	205	1,526	-0.14	0.20	1,485	1.8	1,195
F: 44 MW (150 million Btu/hr) Spreader-stoker watertube								
Full load -- mean path	1,817	203	1,865	-0.22	0.59	1,620	1.57	1,325
-- fast path	1,756	196	1,653	-0.28	0.30	1,520	0.79	1,325
Half load -- mean path	1,393	143	1,627	-0.12	0.75	1,315	3.2	1,080
-- fast path	1,348	137	1,260	-0.10	0.38	1,215	1.6	1,080
E: 44 MW (150 million Btu/hr) Residual-oil-fired packaged watertube								
Full load -- mean path	2,252	266	1,876	-0.40	0.19	1,700	0.83	1,295
-- fast path	2,200	261	1,709	-0.66	0.10	1,600	0.42	1,295
Half load -- mean path	1,842	208	1,689	-0.27	0.41	1,490	1.85	1,045
-- fast path	1,679	185	1,511	-0.40	0.21	1,390	0.93	1,045
D: 4.4 MW (150 million Btu/hr) Distillate-oil-fired packaged watertube								
Full load -- mean path	2,269	269	1,921	-0.38	0.20	1,715	0.84	1,340
-- fast path	2,223	264	1,712	-0.58	0.10	1,615	0.42	1,340
Half load -- mean path	1,824	205	1,725	-0.25	0.41	1,505	1.81	1,085
-- fast path	1,698	188	1,518	-0.37	0.21	1,405	0.91	1,085
C: 4.4 MW (150 million Btu/hr) Natural-gas-fired packaged watertube								
Full load -- mean path	2,122	250	2,145	-0.41	0.35	1,815	0.83	1,475
-- fast path	2,149	253	1,920	-0.63	0.18	1,715	0.42	1,475
Half load -- mean path	1,781	194	1,910	-0.26	0.72	1,575	1.78	1,210
-- fast path	1,662	183	1,687	-0.37	0.36	1,475	0.89	1,210

(continued)

TABLE 32. (continued)

Boiler	a	b	c	D	t_m (sec)	T_m (°K)	t_e (sec)	T_e (°K)
B: 4.4 MW (15 million Btu/hr) Distillate-oil-fired packaged firetube								
Full load -- mean path	2,357	285	2,011	-1.14	0.12	1,710	0.25	1,490
-- fast path	2,441	296	1,721	-1.11	0.06	1,610	0.13	1,490
Half load -- mean path	1,850	214	1,862	-0.68	0.26	1,515	0.55	1,235
-- fast path	1,840	209	1,592	-0.91	0.13	1,415	0.18	1,235
A: 4.4 MW (15 million Btu/hr) Natural-gas-fired packaged firetube								
Full load -- mean path	2,756	341	2,175	-1.19	0.11	1,815	0.25	1,600
-- fast path	2,619	321	1,820	-0.99	0.06	1,715	0.13	1,600
Half load -- mean path	2,193	261	1,960	-0.69	0.15	1,630	0.52	1,335
-- fast path	2,181	258	1,620	-0.72	0.08	1,530	0.27	1,335

time-temperature profile plots in Section 5. The curve constants (a, b, c, D) were calculated using a least squares curve fit program on a hand-held programmable calculator. The regression fit (r^2) ranged from 0.94 to 1.0. Reported time and temperature values (T_m and T_e in Table 32) are taken directly from the zonal heat balance calculations and are not those calculated from the curve fit parameters.

7.2 EFFECTS OF STREAM PHYSICAL PROPERTIES ON BOILER TIME-ABOVE-TEMPERATURE CURVES

In the development of the model in Section 6, it was assumed that all hazardous waste constituents were in the vapor phase at time t_m. The validity of this assumption is dependent on the physical properties of the waste stream and the droplet size of the atomized waste. This assumption will now be verified for various streams and boiler conditions.

The method of verification is to calculate the time necessary to reach full volatilization under given boiler conditions. The approach is again conservative. The material will first be raised from some ambient starting temperature to its boiling point, and then vaporized. (In reality, there will be extensive surface boiling before the bulk of the material reaches its boiling point). The two following subsections develop the equations for these calculations.

7.2.1 Initial Heating Time

The initial heating time, t_h, is the time necessary to heat the material to its boiling point (assuming it is a liquid). The amount of heat required is:

$$Q = m \int_{T_{am}}^{T_{bp}} C_p(T)\, dT \tag{7-3}$$

where

Q = heat required (cal)

m = mass of hazardous waste (g)

$C_p(T)$ = heat capacity of the hazardous waste as a function of its temperature (cal/g-K)

T_{am} = injection temperature of material

T_{bp} = boiling point of hazardous waste

The rate of heating is:

$$\frac{dQ}{dt} = hA \left[T_b (t) - T_h (t)\right] \tag{7-4}$$

where

h = convective heat transfer coefficent of the hazardous waste ($\mathrm{cal/cm^2 \cdot S \cdot {}^\circ K}$)

A = surface area of hazardous waste droplet ($\mathrm{cm^2}$)

$T_b(t)$ = boiler ambient temperature at time t (°K)

$T_h(t)$ = bulk hazardous waste droplet temperature at time t (°K)

The free convective heat transfer coefficient is chosen as the most conservative rate of heat transfer within the liquid hazardous waste.

Combining and rearranging Equations (7-3) and (7-4):

$$\int_{t_o}^{t_h} \Delta T \, dt = \frac{m}{hA} \int_{T_{am}}^{T_{bp}} C_p(T) \, dT \sim \frac{m\overline{C}_p}{hA} (T_{bp} - T_{am}) \tag{7-5}$$

where $\Delta T = T_b(t) - T_h(t)$, and $\overline{C}_p$ is the average waste heat capacity over the temperature range T_{am} to T_{bp}.

From Section 5, for the temperature rise portion of the boiler, the time-temperature relationship in the boiler is approximated by:

$$T_b(t) = a + b \ln t \tag{7-6}$$

The difference, ΔT, between the boiler temperature and the waste temperature is:

$$\Delta T = T_b(t) - T_h(t) \tag{7-7}$$

If at t_o, the quantity $\Delta T = 0$ (conservative, since waste temperature will be be below boiler temperature at its injection point), then it can be shown that:

$$\Delta T = \frac{t}{t_h} (a + b \ln t_h - T_{bp}) \tag{7-8}$$

Integrating the right side of this equation with respect to time, (dt), over the interval t_o to t_h results in:

$$\int_{t_o}^{t_h} \Delta T \, dt = \frac{t^2}{t_h} (a+b \ln t_h - t_{bp}) \Big|_{t_o}^{t_h} \qquad (7\text{-}9)$$

Inserting the right side of Equation (7-9) into Equation (7-5) and solving for Δt_h gives:

$$t_h (a + b \ln t_h - T_{bp}) = \frac{m\bar{C}_p}{hA} (T_{bp} - T_{am}) \qquad (7\text{-}10)$$

where t_h = time to heat the material (sec) to its boiling point.

Assuming spherical droplets, the mass (m) and surface area (A) are:

$$m = \frac{\rho 4\pi \, r_o^3}{3}$$

and

$$A = 4\pi \, r_o^2$$

respectively, where

ρ = density of material (g/cm^3)

r_o = initial droplet radius (cm)

Replacing in Equation (7-10):

$$t_h (a + b \ln t_h - T_{bp}) = \frac{\rho r_o}{3h} \bar{C}_p (T_{bp} - T_{am}) \qquad (7\text{-}11)$$

7.2.2 Volatilization Time

The total amount of heat Q required to vaporize a droplet of hazardous waste (when it is at its boiling point) is just the heat of vaporization multiplied by the mass. That is:

$$Q = \frac{\rho 4\pi\ r_o}{3}\ \Delta H_{vap} \tag{7-12}$$

The rate of heat transfer is:

$$\frac{dQ}{dt} = hA\ \Delta T(t) \tag{7-13}$$

Taking the derivative of Equation (7-12) with respect to r and equating:

$$\frac{dQ}{dt} = h\ (4\pi r^2)\ \ \Delta T(t) = \frac{4\pi\rho}{3}\ \Delta H_{vap}\left(3\ r^2\ \frac{dr}{dt}\right) \tag{7-14}$$

Simplifying, canceling terms, and integrating:

$$\int_{t_h}^{t_v} \Delta T(t)\ dt = \frac{\rho\Delta H_{vap}}{h}\int_{o}^{r_o} dr \tag{7-15}$$

Remembering that $\Delta T(t) = T_B(t) - T_h(t)$ and at this point $T_h(t) = T_{bp}$, then $\Delta T(t) = a + b \ln t - T_{bp}$ and the integral is evaluated as:

$$(a - b - T_{bp})\ \Delta t_v + b\ (t_v \ln t_v - t_h \ln t_h) = \frac{\rho r_o\ \Delta H_{vap}}{h} \tag{7-16}$$

where $\Delta t_v = t_v - t_h$, and it is assumed that complete volatization occurs before the droplet passes the peak temperature of the boiler profile.

To determine t_v, Δt_v must be added to t_h and:

$$t_v\left[a + b \ln t_v - (b + T_{bp})\right] = \frac{\rho r_o}{h}\left[\Delta H_{vap} + \frac{\bar{C}_p}{3}(T_{bp} - T_{am})\right] - bt_h \tag{7-17}$$

If $t_v > t_m$ (time to maximum temperature), a slightly different form of the equation is used. This form is:

$$t_v = \frac{1}{D} \ln \left[\frac{M + T_m}{c}\right] \qquad (7\text{-}18)$$

where

$$M = D \left\{ \frac{\rho r o}{h} \left[\Delta H_{vap} + \frac{\bar{C}p}{3} (T_{bp} - T_{am})\right] - t_m (T_m - T_{bp}) - b(t_h - t_m) \right\}$$

and c and D are boiler temperature profile constants for the temperature fall portion of the boiler profile.

7.3 CALCULATION OF TIME TO VAPORIZATION AND COMPARISON TO BOILER MAXIMUM TEMPERATURE

Hazardous waste "composite compound" data were presented in Table 29. Values needed for calculation of time to vaporization are density (ρ), boiling point, heat capacity (C_p), and heat of vaporization (ΔH_{vap}). Times to vaporization are increasing functions of these parameters. Therefore, the largest of each parameter is taken from the stream data set to form the properties of the "composite compound."

Vaporization times for two droplet diameters, 80 µm and 200 µm are calculated. The 80-µm droplet represents the mass mean diameter of most fuel (assumed to be comparable to waste) droplets, while less than 1 percent are above 200 µm. The time to vaporization for the 200 µm droplet was calculated for all streams under one boiler condition (pulverized coal, full load, mean path). For that stream with the longest t_v (K030), calculations were performed for all boilers.

The time of heating and time of vaporization values were calculated by iterative closure of the equations relating time and total heat input (Equations (7-11) and (7-16)). The two values were found independently and then summed. The results are tabulated in Table 33.

As was assumed, the time to vaporization was less than t_m in all but one case (200-µm droplet under full-load, fast-path conditions in a pulverized-coal boiler). Firetube boilers will probably not meet the assumption for fast path conditions, nor will the distillate-oil-fired watertube boilers under full-load, fast-path conditions. Conversely, it is not probable that the largest droplets will take the fastest path.

TABLE 33. TIME TO VAPORIZATION (SEC) FOR SELECTED HAZARDOUS WASTE STREAMS UNDER VARIOUS BOILER CONDITIONS (80 μm DIAMETER DROPLETS)

Boiler	F003	F005	K012	K018	K030 80 μm	K030 200 μm	Boiler t_m
G: 58.6 MW (200 million Btu/hr) Pulverized-coal-fired watertube							
Full load -- mean path	0.03/0.06[a]	0.03/0.07[a]	0.03/0.06[a]	0.05/0.10[a]	0.05	0.11	0.20
Full load -- fast path	0.03/0.07	0.03/0.07	0.03/0.06	0.05/0.09	0.05	0.11	0.10
Half load -- mean path	0.04	0.04	0.04	0.06	0.07	0.14	0.39
F: 44 MW (150 million Btu/hr) Spreader-stoker watertube							
Full load -- mean path	0.04	0.04	0.04	0.06	0.07	0.14	0.59
Half load -- mean path	0.06	0.05	0.05	0.09	0.10	0.20	0.75
E: 44 MW (150 million Btu/hr) Residual-oil-fired packaged watertube							
Full load -- mean path	0.03	0.03	0.03	0.05	0.05	0.11	0.19
Half load -- mean path	0.04	0.04	0.04	0.06	0.07	0.14	0.41
D: 44 MW (150 million Btu/hr) Distillate-oil-fired packaged watertube							
Full load -- mean path	0.03	0.03	0.03	0.05	0.05	0.11	0.20
Half load -- mean path	0.04	0.04	0.04	0.06	0.07	0.11	0.41
C: 44 MW (150 million Btu/hr) Natural-gas-fired packaged watertube							
Full load -- mean path	0.04	0.04	0.03	0.05	0.06	0.12	0.35
Half load -- mean path	0.04	0.04	0.04	0.06	0.07	0.14	0.72
B: 4.4 MW (15 million Btu/hr) Distillate-oil-fired packaged firetube							
Full load -- mean path	0.03	0.03	0.03	0.05	0.05	0.10	0.12
Half load -- mean path	0.04	0.04	0.04	0.06	0.07	0.14	0.26
A: 4.4 MW (15 million Btu/hr) Natural-gas-fired packaged firetube							
Full load -- mean path	0.03	0.03	0.03	0.04	0.04	0.09	0.11
Half load -- mean path	0.04	0.04	0.03	0.05	0.06	0.11	0.15

[a]For 200 μm diameter droplets

7.4 BOILER PARAMETERS AND DESTRUCTIBILITY (MOST CONSERVATIVE)

The boiler parameters needed for application of the model are given in Table 32. The most conservative approach to predicting boiler applicability will use the time difference $t_e - t_m$ and the exit temperature T_e (see Section 6.2.2). For a given waste stream, the pyrolysis data are used because they are more conservative than oxidation data, and the available oxidation data were obtained using air containing 20 percent oxygen as opposed to 2 to 5 percent oxygen, which more closely approximates actual conditions in boilers.

A model boiler example with five hypothetical compounds is shown in Figure 24. The boiler zone is generated from the following four data points (the assumption is made that the material is in gas phase at t_m; this has been verified in Section 7.3):

- The mean elapsed time the material is above exit temperature under full-load conditions
- The shortest elapsed time the material is above exit temperature at full-load
- The mean elapsed time the material is above exit temperature under half-load conditions
- The shortest elapsed time at half-load conditions

The area thus inscribed is called the boiler kinetic operational zone (BKOZ) above exit temperature. The points within the BKOZ represent the amount of time the gas is above the exit temperature, and the exit temperatures that can be obtained for a boiler operating between full and half load. These represent the range of thermodynamic points for calculating destruction efficiencies. If the time-temperature curve for 99.99-percent destruction of a given compound is overlayed on the BKOZ plot, the following two conclusions can be reached:

- Any compound whose kinetic curve lies above the BKOZ will never reach 99.99-percent destruction by most conservative estimates (e.g., compound A; Figure 24)
- Any compound whose plot lies below the BKOZ will always achieve 99.99-percent destruction under the given boiler conditions (e.g., compound E; Figure 24)

A third conclusion can be made that any compound whose plot lies completely beneath the upper bound line (T_e at full-load and 0.5 t_e to t_e at full-load) will always achieve 99.99-percent destruction under full-load conditions (e.g., compounds C, D, and E; Figure 24).

A corollary is that any compound whose plot lies completely above the lower bound line (T_e at half load and 0.5 t_e to t_e at half load) will not reach 99.99-percent destruction at half load conditions (e.g., compounds A, B, and C; Figure 24), if one were to rely on exit temperature alone for the destruction.

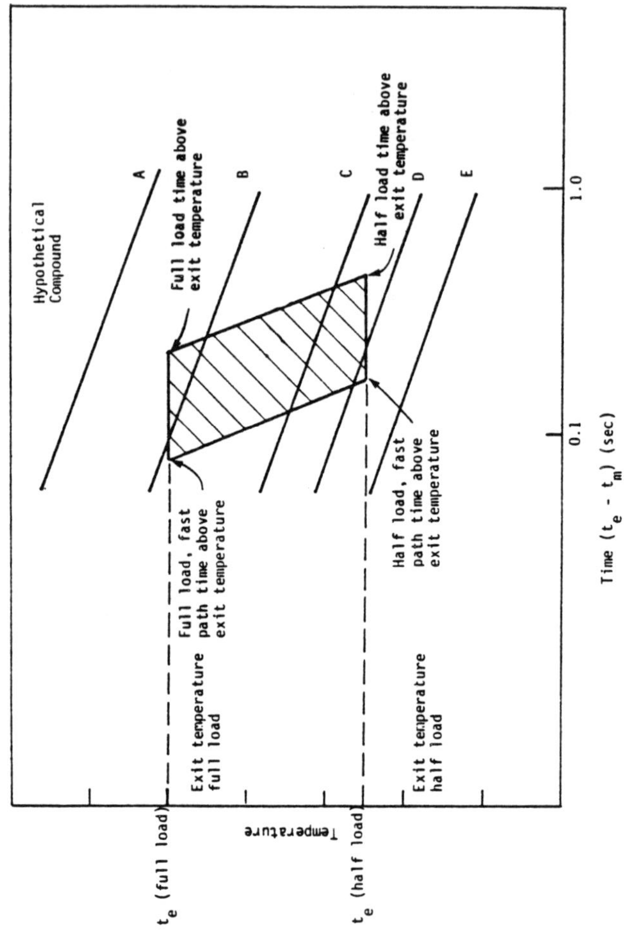

Figure 24. Example of overlaying compound curve and boiler curve -- most conservative case.

Figures 25 through 27 illustrate this approach for seven hazardous waste streams and seven types of boilers. The results are tabulated in Table 34. To reiterate these predictions are based on:

- The assumption that all material is in the gaseous state at t_m in the boiler
- Incomplete thermodynamic data
- No attempt to predict products of incomplete combustion (PIC's)
- No attempt to predict synergistic effects

Conversely, the kinetic data are for pyrolysis (which is much slower than oxidation) and the model does not account for destruction at temperatures above boiler exit temperatures.

In summary, this section predicts that:

- Stream K012 can be destroyed in all boilers studied
- Streams K018 and K030 can be destroyed in most of the boilers studied

7.5 EXTENDED APPROXIMATION TO DESTRUCTIBILITY

In this section, an extended BKOZ is developed to account for the higher destruction rates achieved at temperatures above the boiler exit temperature. The extension of the BKOZ is straightforward, but requires some care in interpretation. The problem is that the kinetic curves are not truly linear over more than one to two orders of magnitude in time, but tend to break upward at shorter times. Furthermore, essentially NO kinetic data exist for times shorter than 0.1 second.

Recall that in Section 6.2.2 a boiler time above temperature curve was developed. By constructing such curves from two points of the most conservative BKOZ, the extended BKOZ is generated. The two points of interest are the upper right (full load, mean path) and lower left (half load, fast path). An example is shown in Figure 28.

It is seen immediately that whereas under most conservative estimates it was questionable whether compound B would reach 99.99-percent destruction efficiency under full load, the kinetic plot falls below the full load, fast path curve and the waste is, indeed, predicted to be destructible. The same argument for half load conditions holds for compound D. Furthermore, it is apparent that compound C, predicted not to be destructible to 99.99-percent efficiency under half load conditions, can be destroyed.

The extended BKOZ for each boiler type under study is plotted in Figures 29 through 35. The appropriate hazardous waste stream kinetic curves are overlayed and a summary of the results (for the extended case) is given in Table 35.

TABLE 34. APPLICABILITY OF VARIOUS BOILERS TO THE GAS PHASE PYROLYSIS OF GIVEN STREAMS -- MOST CONSERVATIVE ESTIMATE (USING TIME ABOVE EXIT TEMPERATURE)

Boiler	Hazardous waste number						
	F003	F004	F005	K012	K018	K022	K030
G: 58.6 MW (200 million Btu/hr) Pulverized-coal-fired watertube							
Full load	X	N	N	A	A	N	A
Half load	N	N	N	A	N	N	P
F: 44 MW (150 million Btu/hr) Spreader-stoker-fired watertube							
Full load	N	N	N	A	N	N	A
Half load	N	N	N	P	N	N	P
E: 44 MW (150 million Btu/hr) Residential-oil-fired packaged watertube							
Full load	N	N	N	A	N	N	N
Half load	N	N	N	P	N	N	N
D: 44 MW (150 million Btu/hr) Distillate-oil-fired packaged watertube							
Full load	N	N	N	A	N	N	A
Half load	N	N	N	P	P	N	P
C: 44 MW (150 million Btu/hr) natural-gas-fired packaged watertube							
Full load	N	N	N	A	A	N	A
Half load	N	N	N	P	P	N	P

(continued)

TABLE 34. (continued)

Boiler	Hazardous waste number						
	F003	F004	F005	K012	K018	K022	K030
B: 4.4 MW (15 million Btu/hr) Distillate-oil-fired packaged firetube							
Full load	N	N	N	A	A	N	A
Half load	N	N	N	X	P	N	P
A: 4.4 MW (15 million Btu/hr) Natural gas-fired packaged firetube							
Full load	X	N	N	A	A	N	A
Half load	N	N	N	A	P	N	X

A -- Applicable
P -- Probably applicable under conditions between full load and half load
X -- Marginally applicable (composite compound line passes below mean path, but above fast path)
N -- Not applicable (composite compound line is above mean path)

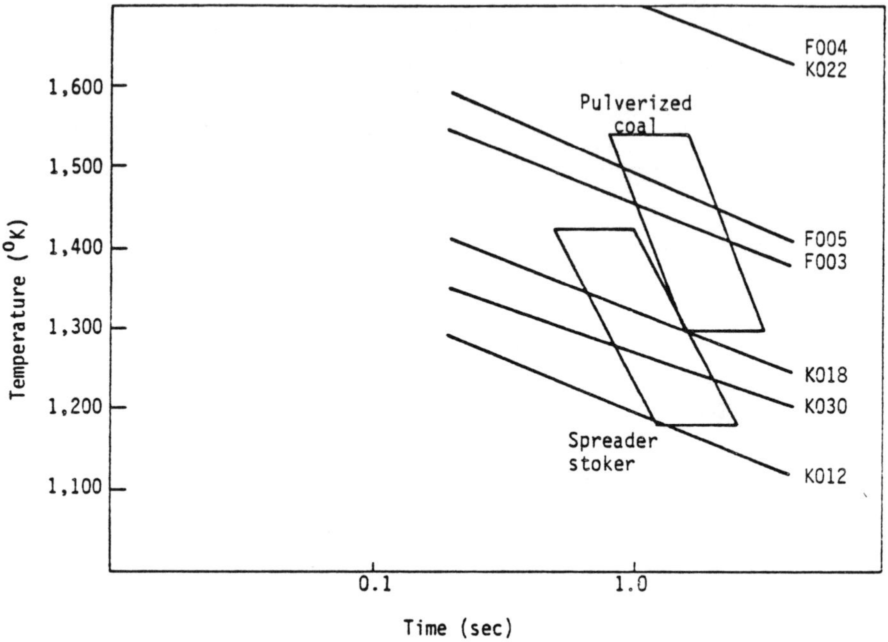

Figure 25. Stream destructibility predictions for pulverized-coal and spreader-stoker boilers.

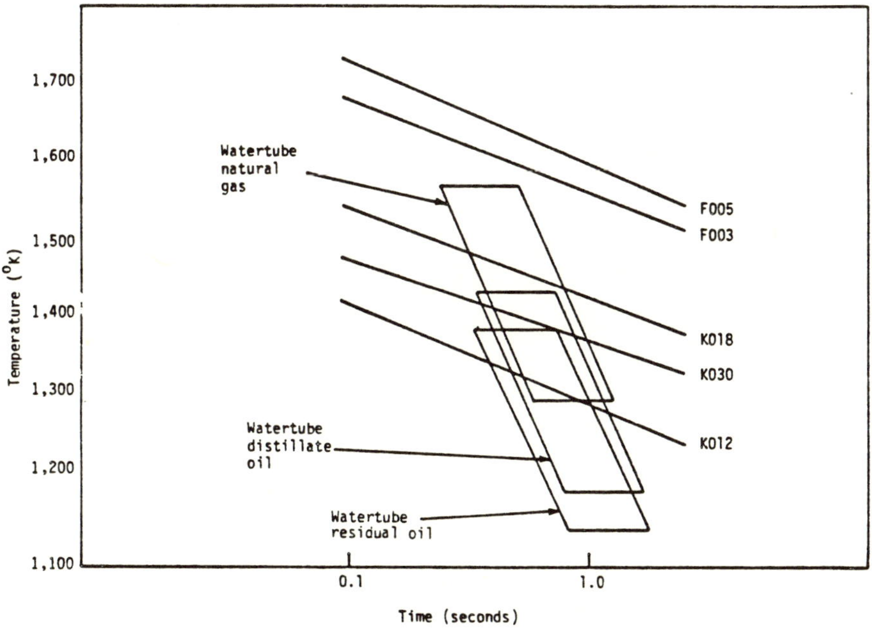

Figure 26. Stream destructibility predictions for watertube boilers.

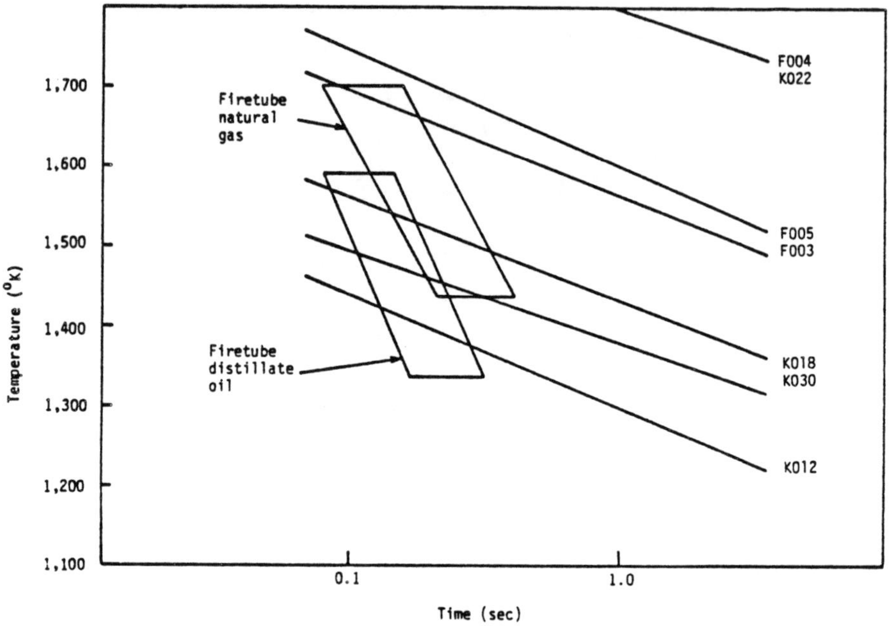

Figure 27. Stream destructibility predictions for firetube boilers.

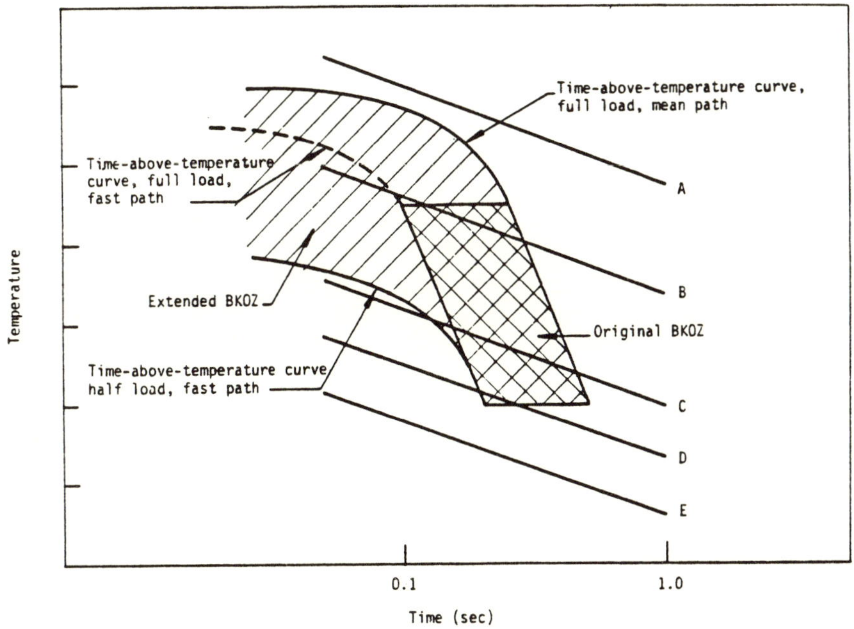

Figure 28. Example of an extended BKOZ.

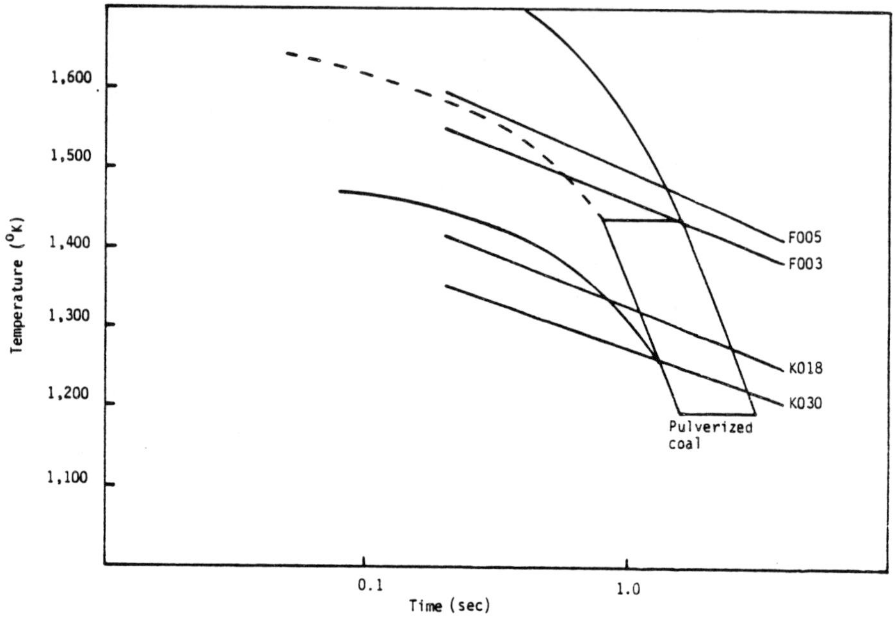

Figure 29. Extended BKOZ for the pulverized-coal boiler.

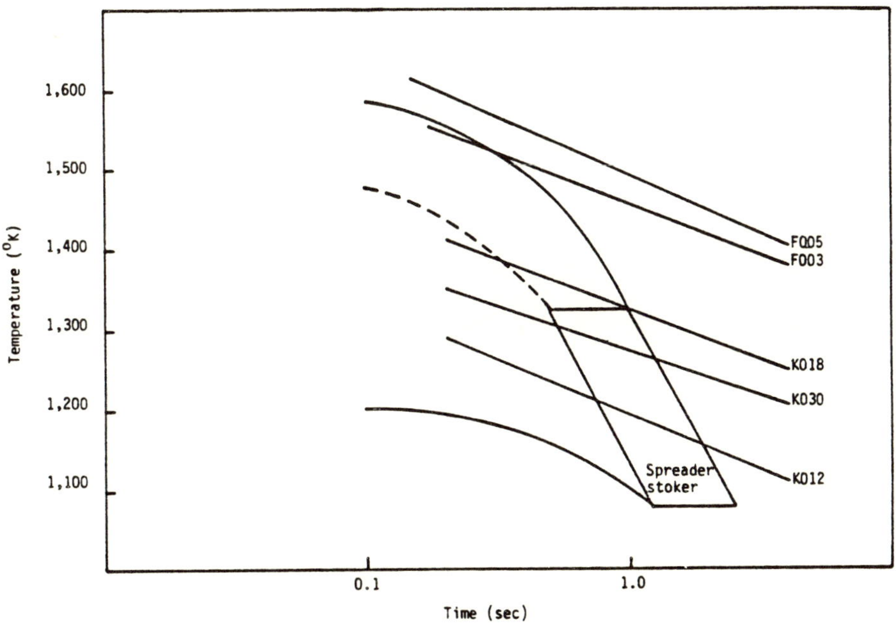

Figure 30. Extended BKOZ for the spreader-stoker boiler.

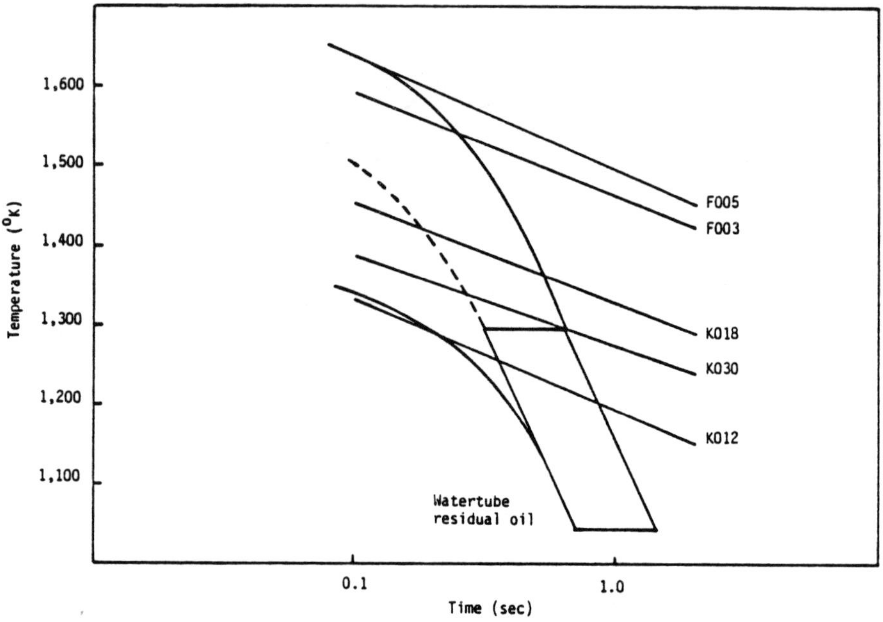

Figure 31. Extended BKOZ for the residual oil-fired watertube boiler.

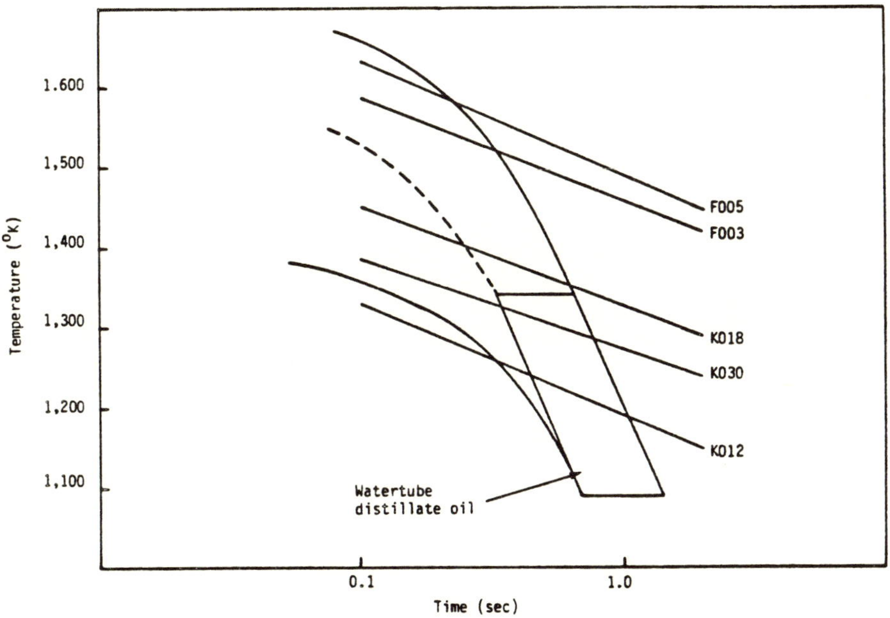

Figure 32. Extended BKOZ for the distillate-oil-fired watertube boiler.

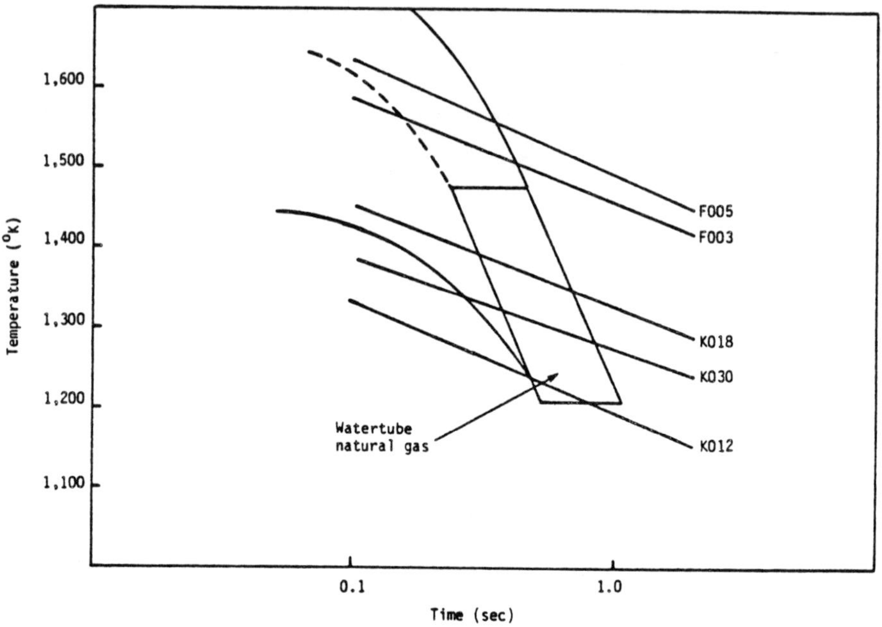

Figure 33. Extended BKOZ for the natural-gas-fired watertube boiler.

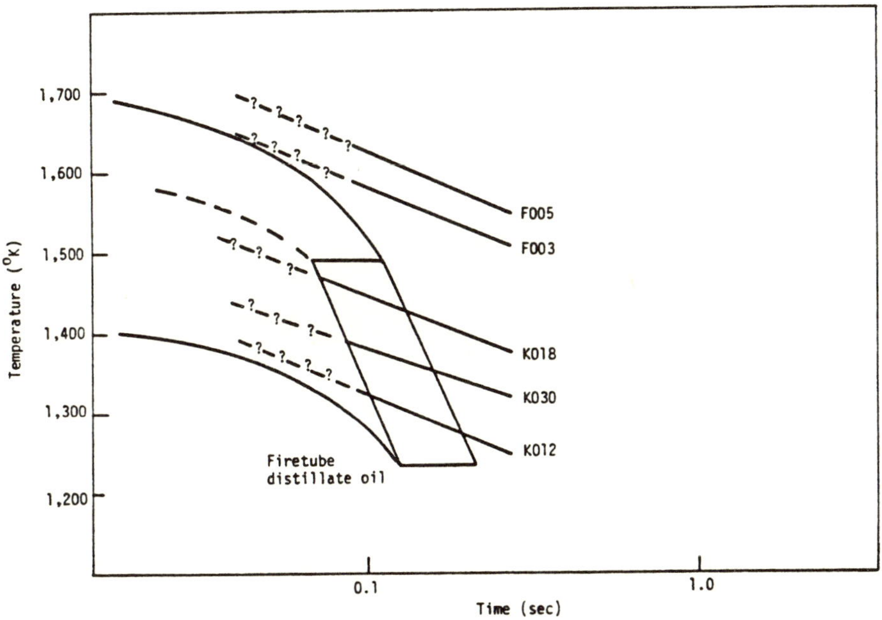

Figure 34. Extended BKOZ for the distillate-oil-fired firetube boiler.

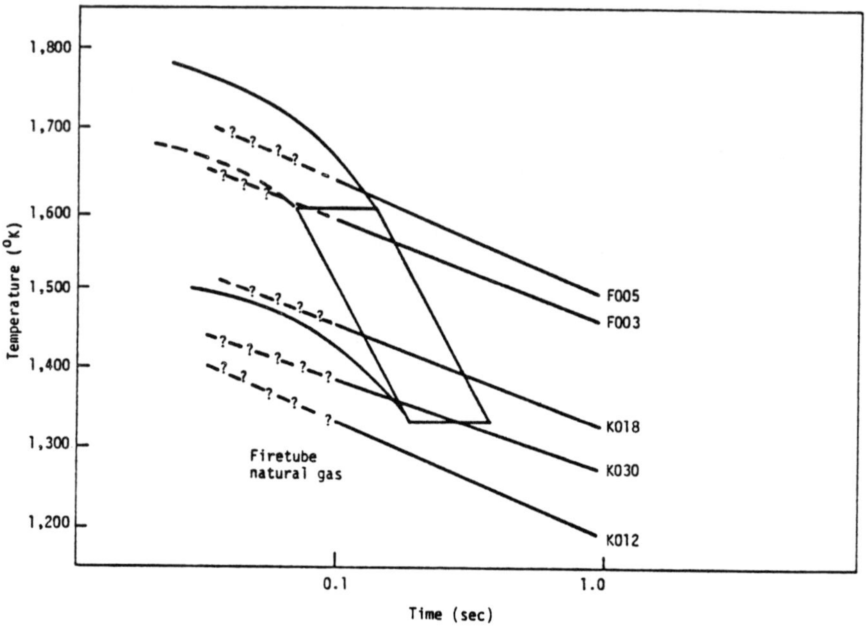

Figure 35. Extended BKOZ for the natural-gas-fired firetube boiler.

TABLE 35. APPLICABILITY OF VARIOUS BOILERS TO THE GAS PHASE PYROLYSIS OF GIVEN STREAMS -- EXTENDED CONSERVATIVE ESTIMATE (USING TIME ABOVE TEMPERATURE)

Boiler	Hazardous waste number						
	F003	F004	F005	K012	K018	K022	K030
G: 58.6 MW (200 million Btu/hr) Pulverized coal-fired watertube							
Full load	A	N	X	A	A	N	A
Half load	N	N	N	A	A	N	A
F: 44 MW (150 million Btu/hr) Spreader stoker-fired watertube							
Full load	X	N	N	A	A	N	A
Half load	N	N	N	X	N	N	N
E: 44 MW (150 million Btu/hr) Residential oil-fired packaged watertube							
Full load	X	N	N	A	A	N	A
Half load	N	N	N	A	N	N	X
D: 44 MW (150 million Btu/hr) Distillate oil-fired packaged watertube							
Full load	X	N	X	A	A	N	A
Half load	N	N	N	A	X	N	X
C: 44 MW (150 million Btu/hr) Natural gas-fired packaged watertube							
Full load	A	N	X	A	A	N	A
Half load	N	N	N	A	X	N	A

(continued)

TABLE 35. (continued)

Boiler	Hazardous waste number						
	F003	F004	F005	K012	K018	K022	K030
B: 4.4 MW (15 million Btu/hr)							
Distillate oil-	A	N	X	A	A	N	A
fired packaged firetube	N	N	N	A	A	N	A
Full load	N	N	N	A(?)	A(?)	N	A(?)
Half load	N	N	N	X(?)	N	N	X(?)
A: 4.4 MW (15 million Btu/hr) Natural gas-fired packaged firetube							
Full load	A(?)	N	X(?)	A	A	N	A
Half load	N	N	N	A	X	N	A

A -- Applicable
X -- Marginally applicable (composite compound line passes below mean path, but above fast path)
N -- Not applicable (composite compound line is above mean path)

8. Secondary Impacts of Hazardous Waste Cofiring

Incremental emissions arising from the cofiring of hazardous wastes may occur in all three emission categories: solids, liquids, and gases. Solids will arise primarily from contaminated containers, bottom and flyash, slag, and treatment sludges. Liquids (contaminated water) will be associated with plant runoff and scrubber effluents. Gas emissions will occur in the flue gas (along with uncontrolled flyash) and as fugitive emissions around handling areas.

Modifying and operating an industrial boiler for hazardous waste cofiring may thus cause numerous impacts on facility design and operation. This section reviews the two major impacts, source emissions and source operation.

8.1 FLUE GAS AND FUGITIVE EMISSIONS

The bulk of emissions will occur in the flue gas. These will fall into the classification of criteria pollutants (CO, SO_x, NO_x, particulates) and products of incomplete hazardous waste combustion (PIC's). The PIC's, in turn, may be gaseous, condensed liquids, or condensed solids, depending on exit gas temperatures.

Fugitive emissions will occur during the handling of wastes with a high content of volatile components. Under good operating procedures, these concentrations should be negligible.

8.1.1 Flue Gas Emissions

The majority of the organic hazardous wastes suitable for thermal destruction may be classified on the basis of their elemental composition or potential products of complete combustion. Criteria pollutant emissions (CO, NO_x, and SO_x) as well as other emissions such as HCl, HCN, and P_2O_5, will be dependent on this elemental composition. Table 36 lists hazardous wastes elemental compositions, examples of origin of each waste class and the combustion products from the assumed complete oxidation-decomposition of the waste.

Actual wastes will contain particulate material such as dirt and scale, as well as trace elements from corrosion, catalysts, and feedstock sources. The inorganic content of the hazardous wastes will be oxidized, and depending

TABLE 36. ELEMENTAL COMPOSITION AND PRODUCTS OF COMPLETE COMBUSTION

Waste class	Chemical elements	Examples (waste origin)	Combustion products after total decomposition
1	Carbon and hydrogen	Tars from production of styrene	CO_2, H_2O, N_2, NO_x
	Carbon, hydrogen and oxygen	Off-specification phenol	CO_2, H_2O, N_2, NO_x
2	Carbon, hydrogen and nitrogen	Solid residue from manufacture of aromatic amines	CO_2, H_2O, N_2, NO_x
	Carbon, hydrogen, nitrogen and oxygen	TDI manufacture reactor tar bottoms	CO_2, H_2O, N_2, NO_x
3	Carbon, hydrogen and chlorine	Vinyl chloride monomer manufacturing wastes	CO_2, H_2O, N_2, HCl(g), NO_x
	Carbon, hydrogen, chlorine and oxygen	Phenolic tar from 2,4-D manufacture	CO_2, H_2O, N_2, HCl(g), NO_x
4	Carbon, hydrogen, chlorine and nitrogen	Nitrochlorobenzene manufacturing wastes	CO_2, H_2O, N_2, HCl(g), NO_x
	Carbon, hydrogen and sulfur	Petroleum refining sour waste	CO_2, H_2O, N_2, NO_x, SO_x
	Carbon, hydrogen and fluorine	Fluorinated herbicide wastes	CO_2, H_2O, N_2, HF(g), NO_x
	Carbon, hydrogen and bromine	Ethylene bromide manufacturing waste	CO_2, H_2O, N_2, NO_x, HBr
	Carbon, hydrogen and phosphorus	Malathion	CO_2, H_2O, N_2, NO_x, P_2O_5
	Carbon, hydrogen and silicon	Tetraethyl orthosilicate wastes	CO_2, H_2O, N_2, NO_x, Particulate (silica)
	Carbon, hydrogen and sodium	Refinery spent caustic	CO_2, H_2O, N_2, NO_x, Particulate (Na_2O)

upon the volatility of the wastes, the combustion products will contain gaseous or particulate inorganic oxides.

The thermal decomposition properties of all wastes within one class will not be the same. Within each class, solid hazardous wastes may degrade differently from liquids, and liquids containing a high percentage of combustible hydrocarbons will decompose differently from slurries containing a high percentage of water but which are low in combustible hydrocarbons. The incentive for examining the method of disposal of hazardous wastes by thermal destruction is to minimize the objectionable environmental impacts with alternate disposal methods. Foremost among the concerns in the thermal destruction method is the possibility of PIC formation in the boiler.

Studies of PIC formation are almost exclusively limited to the formation of polycyclic organic matter POM's and the combustion products of chlorinated aromatics. The emphasis on chlorinated aromatics arises from the deep concern for chlorinated dibenzofurans and chlorinated dibenzo-p-dioxins. Although it has been conclusively shown that such PIC's do form under thermal destruction conditions for several classes of compounds, there is still much debate over the survival of such compounds as they pass through a combustion zone. From the limited test data available (almost exclusively on incinerators designed for complete combustion) these compounds survive in the flue gas or on flyash in only the smallest measureable concentrations, if at all. Conversely, there are positive indications of PIC's associated with polychlorinated aromatics occurring in bottom ash. For other classes of organic compounds, there is essentially no available quantitative data on PIC formation.

Assuming that the chemical components of combustible hazardous wastes are carbon, hydrogen, oxygen, nitrogen, chlorine, sulfur, etc., oxidation of these elements will produce incremental emissions of CO_2, H_2O, N_2, NO_x, SO_x, HCl and other gases. The relative contribution of incremental criteria pollutant emissions from thermal destruction of hazardous wastes in the boiler is dependent upon the ratio of hazardous wastes to primary fuel that is being utilized in the industrial boiler. The impacts of these emissions to the environment will again depend upon the efficiency of the pollution control device that will be used for the boiler.

Warnings of accidental release of hazardous wastes to the atmosphere can be provided by online continuous analyzer and monitoring instrumentation. Real-time determination of PIC's and hazardous waste components is not feasible. Typical analyzers that may be employed are:

CO_2 and CO -- Nondispersive infrared
O_2 -- Paramagnetic
HC -- Flame ionization detector
NO/NO_x -- Chemiluminescence

Set point limits of these analyzer/monitors should be linked to the boiler instrumentation panel for audio and visual signals. Carbon dioxide (CO_2) and oxygen (O_2) readings provide information on excess air in the combustion products. Carbon monoxide (CO) and hydrocarbon (HC) data indicate to what extent the combustibles have been oxidized. Hydrocarbon and carbon monoxide concentration data normally are used to determine if destruction of the wastes has been completed.

The major emissions from fossil fuel burning boilers are particulates, sulfur oxides (SO_x), nitrogen oxides (NO_x), hydrocarbons, and carbon monoxide (CO). Polycyclic organic matter (POM's) are among the hydrocarbon species that may be discharged into the atmosphere from the boiler. Many hazardous wastes contain high nitrogen levels. Such nitrogenous wastes when oxidized in the boiler's combustion chamber may produce increased quantities of nitrogen oxides, though 100 percent conversion of waste nitrogen to NO_x would not be expected as this is not the case for fuel bound nitrogen conversion in coal and oil firing. In addition to gaseous pollutant emissions, trace elements are discharged along with particulate matters to the atmosphere. Contaminated scrubber water and sludges also may be problem pollutant sources that will need to be considered in hazardous waste destruction facilities.

In the following case studies, the incremental emissions due to hazardous waste incineration will be examined. Note that there is no information on PIC's.

Case Study No. 1 -- Ethylene Manufacturing Waste (Ref. 47)

The waste obtained was a clear, light-brown solution with no apparent sediment or particulate at room temperature. It had a specific gravity of about 0.9 and a heating value of slightly more than 41.8 MJ/kg (18,000 Btu/lb) The chemical composition was approximately 89 percent carbon and 9 percent hydrogen. Halogens (as chlorine) were about 0.004 percent. Sulfur and nitrogen contents were 1.3 and 0.13 percent, respectively. A viscosity of 1.28 centistokes (at 22°C) allowed this material to be transferred and injected into an incinerator without preheating or other conditioning. Organic components of the waste, summarized in Table 37, were primarily unsaturated cyclic and aromatic compounds. No organic sulfur compound could be identified. Thus, the 1.3 percent sulfur found by elemental analysis was most likely present in an inorganic form. The total inorganic (ash) content of the ethylene waste was less than 0.01 percent. Trace elemental analysis showed no metals present at greater than 1 ppm levels.

The gas composition for the combustion of fuel oil and for ethylene wastes are nearly identical. Even with high excess air, the incinerator produced CO and HC (see Table 38). This indicates that total oxidation of hazardous wastes will not be easy. SO_2 levels for the ethylene waste were high due to the sulfur content in the waste. NO_x levels from the combustion of ethylene waste were comparable to fuel oil combustion. The operation of the incinerator is summarized in Table 39.

TABLE 37. COMPOSITION OF ETHYLENE MANUFACTURING WASTE SAMPLE (REF. 47)

Compound	Approximate concentration (wt percent)
Cyclopentadiene	<0.1
1, 3, 5-hexatriene	<0.1
Methylcyclopentadiene	<0.1
Benzene	30
Cyclohexadiene	20
Toluene	20
Xylene	3
Styrene	20
Methyl styrene	0.8
Indene	5.0
Methylindene and divinyl benzene	1
Naphthalene	3
Ethyl naphthalene	0.2

TABLE 38. TOTAL GAS COMPOSITION IN THE COMBUSTION ZONE BY VOLUME (REF. 47)

Run no. and fuel description	O_2 (%)	CO_2 (%)	N_2 (%)	H_2O (%)	HCl (ppm)	Cl_2 (ppm)	CO (ppm)	NO_x (ppm)	HC (ppm)	SO_2 (ppm)
Fuel oil background										
I	6.3	9.8	73.0	10.9	--	--	12	170	10	15-20
Ethylene waste										
II	7.2	10.3	74.0	8.5	--	--	17	200	5-10	100-200
III	9.4	9.3	73.4	7.9	--	--	17	150	5-15	100-200
IV	4.9	11.5	74.2	9.4	--	--	22	520	5	100-200
C-5,6 + fuel oil										
V	8.1	8.4	69.6	12.7	1.15	0.05	20	145	10-25	5
VI	5.6	9.7	68.8	14.2	1.64	15	17	140	35-65	6
VII	5.5	9.5	67.9	14.7	2.35	56	17	120	30-35	5

TABLE 39. INCINERATION RESULTS SUMMARY (REF. 47)

	Ethylene waste	C-5,6 waste
Combustion temperature (°C)	1,349-1,752	1,348-1,378
Residence time (s)	0.14-0.19	0.17-0.18
Feed rate (kg/min)	1.4-2.0	2.0-2.9[a]
Particulate (mg/m^3)	20-25	36-113[b]
Trace metals (mg/m^3)	0.001-0.003 Pb	0.003-0.006 Pb 0.034-0.065 Mn
Quality of combustion gas		
Total organics (mg/m^3)	13-22	21-27
Waste content (mg/m^3)	Not detected (<0.02)	Not detected (<0.02)
Trace metals (mg/m^3)	0.002-0.003 Pb	0.55-1.40 Pb 0.077-0.19 Mn 0.090-0.030 Co
Quality of scrubber water		
Total organics (mg/liter)	Not significant[c]	Not significant[c]
Trace metals (mg/liter)	0.002 Pb	0.21-0.54 Pb
Quality of solid residue (clinker)		
Waste content (mg/g)	1.8-2.4	Not detected (<0.026)
Destruction efficiency		
Total organics (percent)	99.95-99.96	99.94-99.95
Waste constituents (percent)	>99.999	>99.999

[a]Mixture of C-5,6 and No. 2 oil; feed rate of C-5,6 alone is 0.66-1.5 kg/min.

[b]85-90 percent of stack particulate content was calculated (based upon sodium analysis) to be NaCl produced by the reaction of the caustic scrubber solution with HCl and CO_2.

[c]Levels not significantly higher than background water samples.

Case Study No. 2 -- Hexachlorocyclopentadiene Manufacturing Waste (Ref. 47)

This highly chlorinated waste is produced at a rate of 4.5 Gg annually. It would not sustain combustion without auxiliary fuel; its heating value was only 5.6 MJ/kg (2,400 Btu/lb). The C-5,6 waste was a dark brown liquid with suspended, waxlike particulate matter. Specific gravity was in excess of 1.7, while its viscosity was 11.3 centistokes (at 22°C). The waste was miscible in no. 2 fuel oil which facilitated injection and combustion in the liquid incinerator. The chemical composition of a sample of the C-5,6 waste was approximately 21 percent carbon, 0.67 percent hydrogen, 0.37 percent nitrogen, 0.016 percent sulfur, and 77 percent total halogens (as chlorine). Hexachlorocyclopentadiene (C_5Cl_6) and octachlorocyclopentene (C_5Cl_8) were the major organic waste components found in two separate samples of the C-5,6 waste, as shown in Table 40.

The total inorganic (ash) content of a sample of the waste was less than 0.04 percent. Elemental inorganics found were primarily innocuous metals, such as Fe, Si, Al, Na, and Ca. Lower and trace (ppb) levels were found for numerous elements.

The NO_x levels from C-5,6 wastes were lower than for fuel oil combustion suggesting that thermal NO_x production in C-5,6 waste is lower than fuel oil (see Table 38). The reason may be due to high water content of the waste and consequent lower combustion zone temperatures.

Case Study No. 3 -- Fluidized Bed Incineration of Phenol and Methyl Methacrylate Wastes (Ref. 48)

In Tables 41 and 42, the organic composition and trace metal contents of the phenol waste are noted. Similarly, the organic composition and trace metal contents of the methyl methacrylate wastes are given in Tables 43 and 44 Table 45 summarizes the incineration parameters for the destruction of the above two chemical wastes. Table 46 tabulates the composition of the flue gases produced in the incineration of the phenol and methyl methacrylate wastes. The trace metal concentrations in the stack gases produced in the above incineration process are shown in Table 47.

Analyzing the results of test numbers I and III against test no. II (Table 46), it is noted that NO_x levels in phenol combustion process are four to five times higher than for the fuel oil burning (test no. II).

The presence of CO and HC in the flue gas indicates incomplete combustion of the hydrocarbon elements of the waste. Incomplete combustion may have occurred due to inefficient mixing with air and insufficient residence time at proper temperatures. Incomplete combustion of a hydrocarbon fuel also suggests insufficient heat release from the oxidation process. The presence of CO and HC in the flue gas is not necessarily an indication that inadequate waste destruction has occurred. CO and HC may have originated from carrier chemicals or may have originated in oxidation reactions.

TABLE 40. COMPOSITION OF HEXACHLOROCYCLOPENTADIENE WASTE (REF. 47)

Compound	Approximate concentration (wt percent)
C_5Cl_6 (hexachlorocyclopentadiene)	50-65
C_5Cl_7 (heptachlorocyclopentene)	3-5
C_5Cl_8 (octachlorocyclopentene)	30-45
Other minor constituents	1-2

TABLE 41. ORGANIC COMPOSITION OF PHENOL WASTE REPRESENTATIVE SAMPLE (REF. 48)

Compound	Estimated level in the waste sample (wt percent)[a]
Ethanol	0.06
Acetone	0.07
2-Butanone (MEK)	0.05
Butanol	<0.01
2-Ethoxyethanol	0.07
Toluene	0.3
Xylene (two isomers)	0.6
Isopropyl benzene	0.8
Phenol	3.7
o-Cresol	0.9
m-Cresol	1.9

[a]Sample contained 86 percent water and approximately 5.5 percent nonignitable solids.

TABLE 42. TRACE METALS IN THE PHENOL WASTE (REF. 48)

Element	Approximate concentration in waste (ppm)	Calculated theoretical concentration in combustion gas (mg/m^3)
Ca	310	26
Si	240	20
Al	86	7.3
K	75	6.4
Mg	40	3.4
P	33	2.8
Pt	16	1.4
Ti	15	1.3
Sn	6	0.5
Zn	6	0.5
V[a]	5	0.4
Cr[a]	4	0.3
Cu	4	0.3
Mo	4	0.3
Zr	4	0.3
Mn	3	0.3
Pb[a]	3	0.3
B	1	0.1
Ba[a]	1	0.1
Ce	1	0.1
Nb	1	0.1
Ni	1	0.1
La	0.6	0.05
Cd[a]	0.5	0.04
Sr	0.5	0.04
Li	0.3	0.03
Nd	0.2	0.02
Ag	0.1	0.01
Hg[a,b]	0.1	0.01
Rb	0.1	0.01
Y	0.1	0.01
Sb[a]	0.08	0.007
Se[a]	0.06	0.005
As[a]	0.03	0.003
Be[a]	0.01	0.001
Co[a]	0.01	0.001

[a]Potentially toxic metals -- ACGIH TLV of <1 mg/m^3 for an 8-hour exposure. (Ref. 48)
[b]Determined by atomic fluorescence.

TABLE 43. ORGANIC COMPOSITION OF METHYL METHACRYLATE WASTE REPRESENTATIVE SAMPLE (REF. 48)

Compound	Estimated levels (wt percent)[a]
Methanol	4.9
Acetone	0.7
Methylene chloride	0.7
2-Butanone	1.1
Methyl propanoate	0.4
Methyl methacrylate	33.9
2-Ethoxy ethanol	0.4
Toluene	1.1
Xylene	0.9
2-Ethoxy ethyl acetate	0.4
Phenol	12.6
Cresols	3.4

[a]Sample contains 38 percent water and approximately 1.7 percent nonignitable solids.

TABLE 44. TRACE METALS IN THE MMA WASTE (REF. 48)

Element	Approximate concentration in waste (ppm)	Calculated theoretical concentration in combustion gas (mg/m^3)
Al	160-315	9-18
Ca	160-315	9-18
Pb[a]	160-315	9-18
Si	110	6.4
Mg	75	4.4
P	57	3.3
Cr[a]	38	2.2
Fe	38	2.2
Ba	35	2.0
Zn	26	1.5
Cd[a]	21	1.2
Cu	19	1.1
Mn	10	0.6
Pt	4	0.2
Sn	4	0.2
Bi	2	0.1
K	2	0.1
Mo	2	0.1
Ni	2	0.1
Sb[a]	2	0.1
Sr	2	0.1
Ag	0.4	0.02
Co[a]	0.4	0.02
Zr	0.4	0.02
B	0.2	0.01
Rb	0.2	0.01
V[a]	0.2	0.01
Nb	0.1	0.006
W	0.1	0.006
As[a,b]	0.04	0.002
Hg[a,b]	0.03	0.002
Be[a]	0.006	0.0003
Se[a]	0.004	0.0002

[a]Potentially toxic metals -- ACGIH TLV of $<1\ mg/m^3$ for an 8-hour exposure. (Ref. 48)
[b]Determined by atomic fluorescence.

TABLE 45. INCINERATOR SYSTEM PARAMETERS DATA SUMMARY (REF. 48)

Test no.	I	II	III	IV	V
Waste tested	Phenols	--[a]	Phenols	Methyl metha-crylate	Methyl metha-crylate
Waste/auxiliary fuel ratio	3.0:1[b]	--	2.3:1	2.0:1	2.6:1
Flowrates					
Waste (l/min)	49.9	--	33.6	29.5	36.4
Aux. fuel (l/min)	16.4	12.1	14.8	15.0	14.0
Fluidizing air (m^3/min)	350	435	425	425	425
Overbed air (m^3/min)	72	105	105	105	105
Scrubber quench water (l/min)	2.3	2.3	2.3	2.3	2.3
Scrubber seal water (l/min)	1.6	1.6	1.6	1.6	1.6
Temperatures					
Reactor bed TE-1 (°C)	734	782	760	774	791
Reactor bed TE-2 (°C)	757	779	763	777	791
Reactor bed TE-3 (°C)	721	766	749	771	782
Reactor bed controller (°C)	735	779	754	771	788
Reactor freeboard (°C)	899	793	813	824	843
Reactor duct north end (°C)	888	768	796	796	841
Scrubber inlet (°C)	85	76	80	80	82
Scrubber outlet (°C)	92	82	85	85	91
Pressures					
Waste feed pump (kPa)	270	--	190	140	190
Windbox (cm H_2O)	152	157	168	168	163
Bed differential (cm H_2O)	81	79	86	81	81
Freeboard (cm H_2O)	8.9	10.7	17.8	15.7	14.7
Exhaust oxygen (percent)	9.6[c]	14.8	17.7	14.5	14.1
Calculated residence time (s)	14	12	12	12	12

[a]Background test with auxiliary fuel (No. 2 oil) only.
[b]Volumetric flow ratio -- liters per minute waste/liters per minute auxiliary fuel.
[c]Oxygen content of exhaust gases in stack as measured by site personnel.

TABLE 46. GAS COMPOSITION DATA SUMMARY (REF. 48)

Test No.	O_2 (percent)	CO_2 (percent)	CO (ppm)	NO_x (ppm)	N_2 (percent)	HC (ppm)[a]	EA[b] (percent)
I	10.7	9.4	6-7	56	79.9	5-20	101
II	15.4	5.9	5	10	78.7	10-50	278
III	12.6	7.6	8-26	41-51	79.8	0-33	146
IV	13.4	7.4	8-10	--[c]	79.2	0-40	175
V	11.6	8.0	10-20	--[c]	80.4	0-10	119

[a]As methane.
[b]Excess air.
[c]Instrument was down due to a broken power supply.

TABLE 47. SURVEY FOR TRACE METALS IN THE STACK FILTER DIGESTS BY ICAP (REF. 48)

Element ppm	Test II (background) (ppm)	Test III (phenol) (ppm)	Test V (MMA) (ppm)
Al	≤0.55[a]	≤0.72	≤0.83
Ba	≤0.44	≤0.90	≤0.44
B	≤0.83	≤1.5	≤0.38
Ca	≤0.55	≤1.01	≤1.6
Cd	0.007	0.016	0.15
Cr	0.036	0.061	0.13
Cu	0.035	0.47	0.38
Fe	0.15	4.5	0.56
Pb	0.082	0.40	2.3
Mg	≤0.36	≤0.31	≤0.53
Mn	≤0.004	≤0.015	≤0.013
Ni	0.004	0.012	ND (0.0002)
P	0.075	0.31	0.68
K	ND (0.012)	3.1	2.4
Si	≤0.28	≤0.09	≤0.07
Ag	≤0.004	≤0.001	≤0.013
Na	2.7	89	23
Sr	≤0.015	≤0.017	≤0.018
Ti	≤0.082	≤0.048	≤0.088
V	0.0005	0.0015	0.0008
Zn	≤0.15	≤0.28	≤0.22

[a]"≤", a less than or equal to sign indicates those elements which were detected but not significantly above background levels.

Note
ICAP = Ion coupled argon plasma spectroscopy.

Examining the data of Tables 44 and 47, it is observed that the scrubber was effective in reducing the trace elements from both the waste's effluent gas streams. It is important to note the separated trace elements will find their way to the scrubber water and sludge. Oil and gas burning industrial boilers seldom use any pollution control device; when such boilers are utilized to destroy hazardous wastes for which a scrubber will be required, the presence of trace elements in scrubber water and sludges may be considered as incremental emissions of a hazardous waste thermal destruction process. Sludges that contain high concentrations of toxic trace elements could only be disposed safely in specially designed landfills. Similarly, scrubber waters that contain high concentrations of trace elements and polycyclic organic matter (POM) could not be discharged to city sewers without proper treatment.

Case Study No. 4 -- Emissions from Coal- and Oil-Fired Controlled Industrial Boilers (Ref. 49)

Tables 48 through 52 summarize results of criteria pollutants and trace element emissions from coal-and oil-fired industrial boilers from a comprehensive assessment of emissions (pre and postscrubber) from these units. (Ref. 49)

Comparing the criteria pollutant emissions from the coal-and oil-fired boilers (Tables 48 and 49) with emissions from the incineration of hazardous wastes (case studies 1 and 2), it is noted that all categories of criteria pollutant emissions for coal and oil-fired boilers are many times higher.

When hazardous waste is used as supplemental fuel, conventional fossil fuels are displaced by the equivalent heat input of the hazardous wastes. The net emissions of criteria pollutants from cofiring hazardous wastes and a fossil fuel, may not increase above the emissions level of a simple fossil fuel (coal/oil) fired boiler. For many hazardous wastes with heating values comparable to fossil fuels (coal/oil), the overall net emissions of criteria pollutants may decrease. For many of the aqueous mixtures of hazardous wastes with low heating values, the amount of hazardous wastes for an equivalent heat input of the displaced fossil fuel will be high, but the net emission may still be lower than a single fossil fuel-fired boiler.

Tables 50 and 51 list trace element emissions from coal-and oil-fired boilers. These emissions are many times higher than for typical hazardous wastes. Therefore, by cofiring hazardous wastes with coal or oil as the primary fuel, the level of trace element emissions may not increase above those levels generated from a simple fossil-fuel-fired boiler. It is noted that trace element mass emissions from coal-fired boilers are higher than from oil-fired boilers by a factor of about 80.

Table 52 summarizes the annual emissions resulting from firing coal and oil in industrial boilers. The data show that NO_x, CO, and sulfate emissions from coal-fired boilers exceed those from oil-fuel-fired boilers by more than a factor of 3. Since the emissions of these compounds from hazardous wastes firing are lower than from firing either coal or oil, the combined net

TABLE 48. SUMMARY OF CRITERIA POLLUTANT EMISSIONS -- COAL FIRING (REF. 49)

	ng/J (lb/million Btu)							
Test no.	NO_x (as NO_2)	CO	SO_2	HC (as CH_4)	C_1 to C_{16} Organics	C_7 to C_{16} Organics	Organics higher than C_{16}	Total particulates
200 Inlet	417 (0.97)	20.7 (0.05)	1,009 (2.35)	3.79 (0.01)	<5.49 (<0.01)	0.34 (0.00)	2.28 (0.01)	2,361 (5.49)
200 Outlet	367 (0.85)	18.5 (0.04)	25.4 (0.06)	4.22 (0.01)	<5.49 (<0.01)	0.27 (0.00)	0.33 (0.00)	14.3 (0.03)
201-1 Inlet	491 (1.14)	16.5 (0.04)	1,284 (2.99)	4.35 (0.01)	<5.65 (<0.010)	--	--	3,122 (7.26)
201-1 Outlet	457 (1.06)	15.7 (0.04)	39.0 (0.09)	5.22 (0.01)	<5.65 (<0.01)	--	--	20.9 (0.05)
201-2 Inlet	455 (1.06)	17.7 (0.04)	1,295 (3.01)	0.88 (0.00)	<5.75 (<0.01)	--	--	--
201-2 Outlet	358 (8.33)	15.9 (0.04)	35.5 (0.08)	1.33 (0.00)	<5.75 (<0.01)	--	--	--
201-3 Inlet	330 (0.77)	16.2 (0.04)	1,028 (2.39)	10.9 (0.03)	<5.69 (<0.01)	--	--	--
201-3 Outlet	258 (0.60)	12.7 (0.03)	31.8 (0.07)	10.9 (0.03)	<5.69 (<0.01)	--	--	--
201-4 Inlet	409 (0.95)	8.56 (0.02)	942 (2.19)	8.95 (0.02)	<5.06 (<0.01)	--	--	3,370 (7.84)
201-4 Outlet	420 (0.98)	8.95 (0.02)	49.7 (0.12)	9.73 (0.02)	<5.06 (<0.01)	--	--	20.6 (0.05)
Average inlet	421 (0.98)	15.9 (0.04)	1,112 (2.59)	5.79 (0.01)	<5.53 (<0.01)	0.34 (0.00)	2.28 (0.01)	2,951 (6.86)
Average outlet	372 (0.87)	14.3 (0.03)	36.3 (0.08)	6.29 (0.01)	<5.53 (<0.01)	0.27 (0.00)	0.33 (0.00)	18.6 (0.04)

TABLE 49. SUMMARY OF CRITERIA POLLUTANT EMISSIONS -- OIL FIRING (REF. 49)

	ng/J (lb/million Btu)							
Test no.	NO_x	CO	SO_2	HC[b] (as CH_4)	C_1 to C_{16} Organics	C_7 to C_{16} Organics	Organics higher than C_{16}	Total particulates
200-1 Inlet	175 (0.41)	5.08 (0.01)	938 (2.18)	2.84 (0.01)	<4.63 (<0.01)	--	--	166 (0.39)
200-1 Outlet	166 (0.39)	4.70 (0.01)	32.1 (0.07)	3.20 (0.01)	<4.63 (<0.01)	--	--	20.2 (0.05)
202-2 Inlet	175 (0.41)	5.30 (0.01)	1,075 (2.50)	4.61 (0.01)	<4.79 (<0.01)	--	--	--
202-2 Outlet	165 (0.38)	5.03 (0.01)	29.2 (0.07)	5.01 (0.01)	<4.79 (<0.01)	--	--	--
203-3 Inlet	181 (0.42)	6.22 (0.01)	1,085 (2.52)	1.71 (0.00)	<4.73 (<0.01)	--	--	--
203-3 Outlet	177 (0.41)	5.92 (0.01)	26.7 (0.06)	1.98 (0.01)	<4.73 (<0.01)	--	--	--
203-4 Inlet	141 (0.33)	5.30 (0.01)	874 (2.03)	0.82 (0.00)	<4.14 (<0.01)	--	--	59.0 (0.14)
203-4 Outlet	138 (0.32)	5.60 (0.01)	19.2 (0.04)	0.76 (0.00)	<4.14 (<0.01)	--	--	15.0 (0.03)
203 Inlet[b]	--	--	--	--	--	0.17 (0.00)	2.61 (0.01)	--
203 Outlet[b]	--	--	--	--	--	0.02 (0.00)	0.43 (0.00)	--
Average inlet	168 (0.39)	5.47 (0.01)	993 (2.31)	2.49 (0.01)	<4.57 (<0.01)	0.17 (0.00)	2.61 (0.01)	113 (0.26)
Average outlet	161 (0.37)	5.31 (0.01)	26.8 (0.06)	2.74 (0.01)	<4.57 (<0.01)	0.02 (0.00)	0.43 (0.00)	17.6 (0.04)

[a]Emission factors were computed assuming an O_2 concentration of 5.5 percent, the average concentration for test 202.
[b]Total hydrocarbons as determined by FID.

TABLE 50. EMISSION FACTORS AND MASS EMISSION RATES OF TRACE ELEMENTS DURING COAL FIRING -- TEST 201-1 (REF. 49)

	Emission factor (ng/J)		Emission rate (g/hr)			
Trace element	Scrubber inlet	Scrubber outlet	Scrubber inlet	Scrubber outlet	Efficiency (percent)	Enrichment factor
Be[a]	0.04	0.001	5	0.09	98	3.7
Hg[b]	0.08	0.037	0.50	0.23	55	84
Ca	32	0.015	3,300	1.6	99	0.09
Mg	8.2	0.0046	860	0.48	99	0.11
Sb	1.6	0.010	170	1.1	99	1.2
As	3.4	0.092	350	9.7	97	5.3
B[a]	0.1	0.01	10	1.2	88	2.1
Cd	0.20	0.00042	21	0.044	99	0.4
Cr	1.1	0.054	120	5.7	95	9.5
Co	1.6	0.0050	160	0.53	99	9.6
Cu	4.1	0.0084	430	0.88	99	0.4
Fe	190	1.0	20,000	110	99	0.99
Pb	3.7	0.0088	380	0.92	99	0.5
Mn	0.34	0.0063	35	0.68	98	3.4
Mo	4.3	0.026	450	1.2	99	0.5
Ni	0.60	0.026	61	2.8	95	8.6
V	1.3	0.024	140	2.5	98	3.6
Zn	0.99	0.020	100	2.1	98	3.9
Se	1.4	0.041	140	4.3	97	5.8
Sr	4.7	0.024	500	2.5	99	0.9
Al	210	1.1	22,000	110	99	1.0
Zr	0.69	0.0075	72	0.79	99	2.1
Total	470	2.6	50,000	270	99	

[a]Appropriate values as determined by SSMS. The other values, with the exception of mercury, were determined by ICPOES analysis.
[b]Mercury was determined by cold vapor analysis of SASS train samples taken during test 200.

TABLE 51. EMISSION FACTORS AND MASS EMISSION RATES OF TRACE ELEMENTS DURING OIL FIRING -- TEST 202-4 (REF. 49)

Element	Emission factor (ng/J) Scrubber inlet	Emission factor (ng/J) Scrubber outlet	Emission rate (g/hr) Scrubber inlet	Emission rate (g/hr) Scrubber outlet	Removal efficiency Δ	Enrichment factor
Be[a]	<0.0003	0.00003	<0.04	0.04	Unknown	>11.9
Hg[b]	0.0006	0.0001	0.05	0.006	87	1.48
Ca	0.13	0.022	16	2.7	83	2.03
Mg	0.10	0.0094	12	1.1	91	1.15
Sb	0.02	0.0019	2.5	0.23	91	1.15
As	0.049	0.0094	5.9	1.1	81	2.37
B	0.17	0.012	21	1.5	93	0.87
Cd	0.091	0.021	11	2.5	77	2.80
Cr	0.055	0.0057	6.7	0.69	90	1.26
Co	0.033	0.0038	3.9	0.46	89	1.43
Cu	0.18	0.002	21	0.27	99	0.15
Fe	1.6	0.088	190	11	95	0.69
Pb	0.065	0.0041	7.9	0.50	94	0.77
Mn	0.010	0.0013	1.2	0.15	87	1.58
Mo	0.072	0.0079	8.7	0.95	89	1.35
Ni	0.36	0.063	43	7.7	83	2.16
V	0.88	0.26	110	31	71	3.61
Zn	0.20	0.02	24	2.5	90	1.27
Se	0.016	0.002	2.0	0.23	87	1.43
Sr	0.014	0.0003	1.7	0.038	98	0.28
Al	1.9	0.15	220	18	92	1.0
Zr	0.0049	0.0003	0.59	0.038	94	0.79
Total	6.0	0.78	710	96	87	

[a]Beryllium was determined by SSMS. The other elements, except mercury, were determined by ICPOES.

[b]Mercury was determined by cold vapor analysis of SASS train samples taken during test 203.

TABLE 52. ANNUAL EMISSIONS (REF. 49a)

	Pollutant	Scrubber inlet, kg/year			Scrubber outlet, kg/year		
		Coal firing	Oil firing	Coal/oil	Coal firing	Oil firing	Coal/oil
Gaseous	NO_x (as NO_2)	500,810	164,230	3.05	442,520	157,390	2.81
	SO_2	1,127,300	906,202	1.24	36,800	24,453	1.51
	SO_3	6,184	7,249	0.85	4,157	5,183	0.80
	$SO_4^=$	67,214	20,894	3.22	8,110	8,303	0.98
	CO	16,119	4,991	3.23	14,497	4,845	2.99
	Organics (as CH_4)	5,870	2,272	2.58	6,377	2,500	2.55
	C_1 - C_6[b]	<5,606	<4,164	--	<5,606	<4,164	--
	C_7 - C_{16}	345	155	2.22	274	18	15.2
	C_{16}^+	2,311	2,381	0.97	335	392	0.85
	Total particulates	2,991,700	53,832	55.6	18,856	13,686	1.38
	<1μ	--	--	--	11,691	11,359	1.03
	1 - 3μ	--	--	--	5,657	1,642	3.45
	3 - 10μ	--	--	--	1,320	634	1.93
	>10μ	--	--	--	188	0[c]	--
		m^3 per year					
Liquid	Blowdown/wastewater	~76,000	~76,000	~1	~76,000	~76,000	~1
	Cooling water	~86,000	~86,000	~1	86,000	~86,000	~1
Solid	Bottom ash	~778,600	~7,600	~103	~778,600	~7,600	~103
	Flyash	1,800,000	~15,000	~120	~1,800,000	~15,000	~120
	Scrubber cake	0	0	--	8,054,100	3,011,000	2.67

[a]Assuming 100 percent load, 45 weeks per year (7,560 hr/year).
[b]These values represent the detection limit of the instrument used.
[c]These values represent oil-firing particulate with a minimum of coal ash contamination.

emissions from hazardous wastes cofiring may be equal to or less than from coal or oil firing.

The data on Tables 48 through 52 illustrate the removal efficiency of the scrubber for each of the criteria pollutants. It is interesting to note that the efficiency of the scrubber for coal particulate removal is over 99 percent, but for particulates from oil firing, the removal efficiency is only approximately 75 percent. The lower removal efficiency achieved during oil firing is due to the increased fraction of the particulates smaller than the 3 micron size range. The same characteristics of the scrubber are exhibited for trace elements from the flue gas (Tables 50 and 51) of coal-and oil-fired boilers.

Particle size distributions obtained during liquid hazardous wastes firing should be the same or smaller than the size of the oil-fired particulates. Increased emissions of smaller particulates may therefore be expected in the cofiring of hazardous wastes with fuel oil or coal. Under such circumstances, a two-stage particulate control system may be desirable.

8.1.2 Fugitive Emissions

Possible sources of fugitive emissions from a hazardous waste thermal destruction facility are summarized below:

1. Waste handling
    - Receiving site (transfer from truck trailers and railroad cars)
    - Transfer to storage tanks and to processing stations
2. Waste processing
    - Mixing or blending
    - Heating
    - Stirring
    - Shredding or crushing
    - Drying
3. Waste feeding to boiler
    - Leakage from pump seals and valves
    - Leakage from blower seals and dampers
    - Leakage from transport pipes, ducts and conveyors
    - Leakage from burner assembly, fittings and recirculation lines
4. Waste evaporation/emission
    - Fumes from spilled liquid waste collection pit
    - Fine particle emissions from waste handling, processing and feeding stations
    - Fumes from gas cooling/cleaning system

5. Waste burning flue gas system

   - Leakage from boiler and accessories
   - Leakage from pollution control equipment

6. Waste residues

   - Scrubber wash water
   - Scrubber sludges
   - Hopper ash from the particulate boiler
   - Boiler bottom ash
   - Recycle -- water treatment facility

Facility system design to decrease the occurrences of fugitive emissions would include:

- Prevention designs for leakages from ducts, pipes, mechanical handling equipment
- Safety feature designs for pumps, blowers and fans
- Safe collection and containment of spills and leaks from equipment system
- Safe disposal method for spills, vacuum cleaners or line catches
- Provision for containment room for a waste processing station
- Use of mechanical handling equipment in transport of wastes

Even with the best engineering designs, fugitive emissions may occur in the plant facility. The following steps may reduce the occurrences.

Good Operating and Maintenance

Adoption of good operating and maintenance practices are essential to ensure control of fugitives. These include:

- Regular cleanup schedules for the operating area
- Immediate cleanup of small spills
- Scheduled cleanup procedures to remove dust, spills and accumulation of hazardous wastes from outside surfaces of the machines, equipment, ducts, pipes and fittings, and other plant facility fixtures
- Mandatory use of appropriate protective clothing and adherence to the safety regulations in the work area
- Complete recordkeeping for any mishaps, spills and work stoppage due to dangerous work environment
- Periodic refresher training programs for maintaining healthy work environment and safe work practices
- Periodic health check up programs for the workers

Monitoring and Measurement of Fugitive Emissions

A methodology to monitor the fugitive emissions from the hazardous waste destruction facility should be identified. Some of the monitoring steps to be considered include:

- Routine source leak test programs
- Scheduled measurements of dust concentrations for the work area (high-volume sampler tests)
- Analysis of dust constituents and particle size distribution measurements
- Scheduled measurement programs to detect emissions of pollutant gases in the work area
- Continuous instrumentation monitoring of pollutants from high emission risk areas
- Regular checkups of pump and blower seals
- Periodic measurements of emissions from bag filter system
- Accurate records of workers' health
- Periodic measurements of ambient air quality outside the plant area
- Overview recordkeeping of the health data of the neighboring residents
- Effective routine and preventive maintenance work plans for the equipment systems

8.2 IMPACT OF CHANGES IN BOILER OPERATION

The boiler characteristic curves presented in Section 5 are based on mean residence times and bulk temperatures. One variable, boiler load, is addressed. Other effects have been addressed in the model by the use of conservative temperatures and residence times. However, the normal fluctuations in boiler operation and maintenance can impact the final destruction efficiency of the waste.

The liquid waste being destroyed must be heated, then vaporized, before it can be burned. This is normally done by atomizing the waste into very small droplets. Typical oil atomizers achieve this through steam or air atomization, or by high-pressure mechanical atomization. A representative mean atomized droplet diameter is 80 µm. An upper bound of 200 µm is typical. A haphazard approach to waste atomization or poor atomizer maintenance would result in much larger droplet sizes. The results of poor atomization may be extrapolated from the consideration of the 80 and 200 µm droplets below.

In Section 6, estimates of the time required to vaporize droplets of dichlorobenzene were developed. For an 80-µm droplet the time requirement was approximately 0.05 sec. The 200 µm droplet required approximately 0.12 sec to vaporize. In Figure 36, the time required for vaporization is superimposed on the boiler full load characteristic curves. The important thing to note is that for the firetube boiler, the peak temperature has been passed before the droplet is vaporized. Going one step further, and considering that the residence time will vary from one half to twice the mean

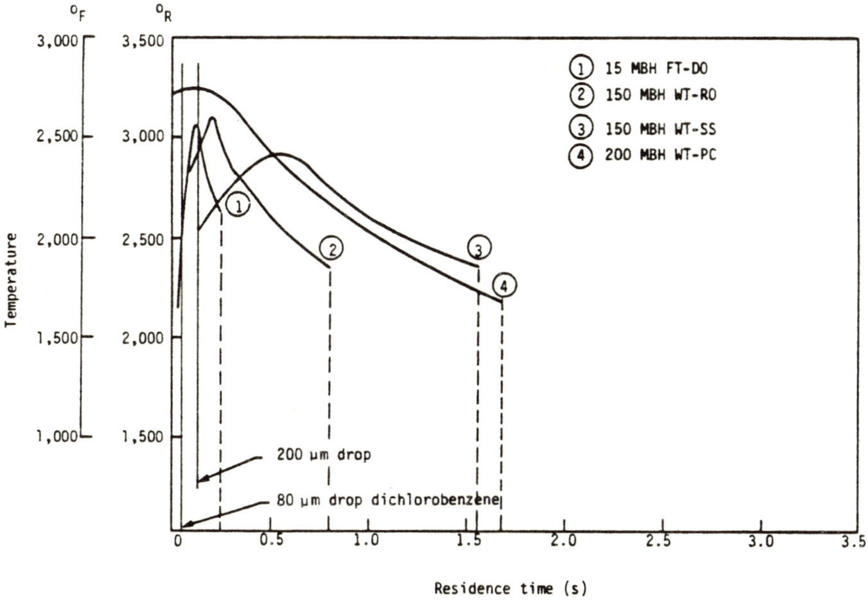

Figure 36. Effect of droplet size on vaporization times at full load.

residence time, waste may be leaving the furnace without being fully vaporized. With the rapid temperature drop after leaving the furnace, unreacted waste may be leaving the boiler. At full load, the residence time is approximately half the residence time at half load, as shown in Figure 37.

Other boiler operating and maintenance procedures may have similar impacts. Therefore, all procedures that would impact combustion conditions must be identified and procedures adopted to minimize the impacts.

Unfortunately, the effect of boiler operation at other than design conditions may result in the formation of products of incomplete combustion. The nature of these compounds may cause the bottom or flyash collected to be hazardous, or cause the generation of toxic air emissions.

8.3 BOILER RETROFIT TO ACCEPT HAZARDOUS WASTES COFIRED WITH A PRIMARY FUEL

If it is determined that the waste and boiler are theoretically compatible and that fugitive and secondary emissions are controllable, the final major area of concern is the retrofit requirements of the boiler itself. It is expected that this activity will represent a major economic impact. The following discussion outlines the major items of consideration.

8.3.1 Handling and Pretreatment of Waste

Combustion conditions, such as the temperture-time profile in industrial boiler furnaces discussed in Section 5, are of crucial importance for the effective disposal of hazardous waste in a boiler. However, the successful retrofit of an industrial boiler to accept and efficiently destroy hazardous wastes depends on a number of other major considerations. These considerations can be grouped into the following categories:

- Handling and pretreatment of waste
- Feeding and burning requirements
- Handling and posttreatment of residues, and
- Process control and inventory instrumentation

Handling and Pretreatment

Proper handling, storage, and feeding of hazardous wastes to a retrofitted industrial boiler is vital to reduce or eliminate exposure, spill, and contamination with adjoining surroundings (people and equipment systems) thereby ensuring the health and safety of the workers. Careful design of the facility including selection of equipment and process technology is important to meet the basic obligation of safe handling and proper disposal of hazardous wastes.

Handling and pretreatment of hazardous wastes consist of storage, transport, and preprocessing for injection into a boiler. The main environmental concerns in the storage and transport of hazardous wastes, and preprocessing equipment, is the danger posed by accidental spills, leaks from

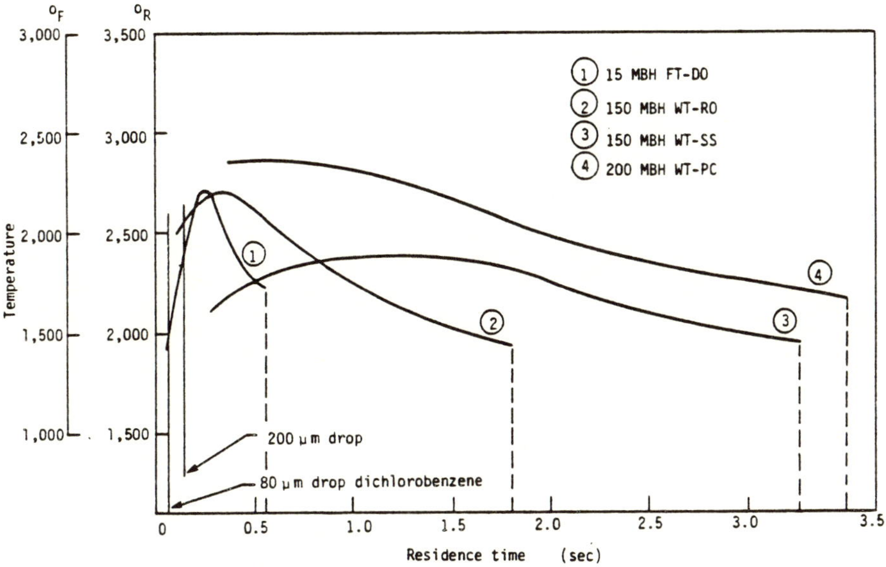

Figure 37. Effect of droplet size on vaporization times at half load.

piping, valves and joints, and possible breakdown of the material handling systems.

Preprocessing

Depending upon the nature of the hazardous waste and the type of industrial boiler in which the waste will be burned, it may be necessary that the "as-received" waste be processed before injection into the boiler. This preprocessing step for the pumpable wastes may involve heating, blending, screening, settling and other processing methods. Preprocessing of nonpumpable wastes, such as, sludge, solids, and containerized wastes may involve grinding, screening and shredding.

Heating

For highly viscous hazardous waste liquids, controlled steam heating with tank coils or in-line heaters have been used. Before the decision to heat the liquid is made, the flash point of the waste should be determined, and the heating device should be fitted with a set point temperature controller.

Blending

If the liquid cannot be heated, miscible liquid may be added to the viscous waste and blended to bring down the viscosity to pumpable or atomizing condition. However, the combination of two or more types of wastes in the mixture may produce sometimes uncontrolled and undesirable reactions, resulting in the formation of:

- Flammable substances
- Toxic fumes
- Explosive mixtures
- Complicated polymers
- Compounds of greater toxicity
- Compounds that may cause pressurization in closed vessels
- Shock and friction sensitive compounds

Where blending warrants extreme safety, the blending vessel may be blanketed with an inert gas and pneumatically-driven diaphragm feed pumps may be used to transport the liquid from storage tank to blender, and blender to boiler burner.

8.3.2 Feeding and Burning Requirements

The injection and combustion of hazardous waste in existing industrial boilers requires careful consideration of the boiler design, existing burners, potential retrofit of additional burners, as well as the typical operating cycle of the boiler.

The following types of atomization techniques may be used for liquid hazardous wastes:

- Rotating cup burner
- Mechanical or oil pressure air atomizing burner
- Steam or high-pressure air atomizing burner
- Low-pressure-air atomizing burner

For highly viscous hazardous wastes containing various amounts of noncombustible materials, steam or high-pressure air atomizing burners are preferred. For nonviscous highly organic wastes, mechanical low-pressure air or rotary cup burners may be used. Solid waste material can be injected with a pulverized fuel or dropped onto stoker grates.

Burner

The primary function for a properly designed burner is to transform the homogeneous liquid fuel to a finely atomized mist spray and to initiate turbulent mixing of combustion air with the finely atomized liquid fuel mist.

Boiler

The primary function of an industrial boiler is to supply steam flow as required by the plant process needs. One important consideration in a retrofit design to allow hazardous waste burning is that the steam supply rate of the boiler match the plant's steam requirements. From the fuel properties of the primary fossil fuel(s) and the quantity of available waste fuel, the portions of the total heat input to the boiler by the hazardous waste fuel and the auxiliary burner can be calculated.

The sensible heat loss in drying, heating, and bringing the hazardous waste to ignition point must be added to the heat input requirements for steam generation. Some of the hazardous wastes may contain 30 to 70 percent moisture, and thermal dissociation of the wastes may be required before the chemical heat from the hazardous waste fuel can be released. Appropriate proportioning of primary to waste fuels, together with minimum excess air supply and satisfactory burner design are key to achieving the desired thermal destruction of hazardous waste.

8.3.3 Posttreatment of Residues

Noncombustible residues discharged from an industrial boiler cofired with hazardous waste are associated with liquid and solid streams from pollution control devices and from the discharges from bottom and flyash hoppers.

Regulatory agencies will require the proper selection, operation and maintenance of pollution control devices by the industrial boiler operator burning hazardous wastes. Because of the heterogeneous characteristics of certain hazardous wastes, systems may be required to control both particulate and gaseous pollutants.

Proper selection of pollution control devices will require the following data of the flue gas stream:

- Flue gas volume, temperature, and particulate loading
- Estimated particle size distribution
- Composition of the flue gas
- Emission regulations

Particulate matter can be removed from the flue gas through the use of mechanical separators, electrostatic precipitators, or filtration devices. Gaseous pollutants can be removed by venturi, packed bed tower (or other gas-liquid contacting device), and mist eliminator assemblies, or by adsorption systems such as activated carbon. In some installations, a properly designed high energy scrubber and mist eliminator system may be adequate for controlling both particulate and gaseous pollutants. Monitoring the effluent gas stream discharged to the atmosphere is conducted to assure that the allowable pollutant concentrations are being maintained during normal operation.

8.3.4 Process Control and Monitoring Instrumentation

When an industrial boiler facility is retrofit to burn hazardous wastes, close monitoring of the entire facility may be required to continuously assure complete (99.99 percent) destruction. Such a monitoring system may cover all operations beginning with the transfer of wastes from storage tanks to valving, pumping, heating, agitating and feeding of the wastes to the boiler burners.

An important measurement in the boiler that is burning hazardous waste is the oxygen concentration in the flue gas. Stable combustion of a fuel generally occurs within a range of air-to-fuel ratios extending from the lean and rich limits of complete combustion and flame stability. An oxygen analyzer can be utilized to set the feed ratio to a level that will assure complete thermal destruction of the waste and efficient operation of the boiler.

References

1. Harrison, J. W., J. B. White, and W. J. King. Assess Hazard Potential from Combustion of Wastes in Industrial Boilers. Draft Report, Research Triangle Institute, Research Triangle Park, NC, October 1980.

2. United States Environmental Protection Agency. Background Document Resource Conservation and Recovery Act, Subtitle C -- Identification and Listing of Hazardous Waste Section 261.31 and 261.32. May 2, 1980, Draft Document.

3. United States Environmental Protection Agency. Preliminary Final Economic Impact Analysis (Regulatory Analysis Supplement for Subtitle C Resource Conservation and Recovery Act (RCRA) -- 1976). April 30, 1980.

4. Key J. A., et al. Identification of Potential Secondary Sources of VOC from Organic Chemical Manufacturing Facilities. Draft Document. IT Enviroscience, Inc., Knoxville, TX, July 1980.

5. Kensicki, R., et al. Hazardous Waste Generation and Commercial Hazardous Waste Management Capacity: An Assessment. EPA Prepublication Draft Report, EPA-SW-894, 1980.

6. Rosenberg, D. G. et al., Assessment of Hazardous Waste Practices in the Petroleum Refining Industry. Environmental Series No. SW-129c, June 1976.

7. McCandless, L. C., et al., Assessment of Industrial Hazardous Waste Practices, Storage, and Primary Battery Industries. EPA-530-SW-1026, January 1975.

8. Gruber, I., Assessment of Industrial Hazardous Waste Practices, Organic Chemicals, Pesticides and Explosives Industries. EPA-530/SW-118c, April 1975.

9. Lannotti, J. E., et al. An Inventory of Industrial Hazardous Waste Generation in New York State. New York State Department of Environmental Conservation, Albany, New York, June 1979.

10. Fennelly, P. F. Surveying Massachusetts' Hazardous Wastes. Environmental Science and Technology, 11:762, 1977.

11. The Impact of Hazardous Waste Generation in Minnesota. Minnesota Pollution Control Agency Report, Roseville, MN, October 1977.

12. Stradley, M. W., et al., An Evaluation of the Status of Hazardous Waste Management in Region X. EPA 910/9-76-024, December 1975.

13. Van Noordwyk, et al., Quantification of Municipal Disposal Methods for Industrially Generated Hazardous Wastes. EPA-600/2-79-135, August 1979.

14. Coxe, E., et al., Energy Conversion Potential of Industrial Waste Streams. Reynolds, Smith and Hills, Inc., Report No. 68-01-3569, September 1977.

15. Assessment of Industrial Hazardous Waste Management Petroleum Re-refining Industry. EPA Environmental Series No. SW-144c, 1977.

16. Manson, L., and S. Unger. Hazardous Material Incinerator Design Criteria. EPA-600/2-79-198, October 1979.

17. Rappe, C., et al., Formation of Polychlorinated Dibenzo-p-dioxins and Dibenzofurans by Burning or Heating Chlorophenates. Chemosphere No. 3, Pergamon Press, London, 1978, p. 269-281.

18. Hess, L. Y. Reprocessing and Disposal of Waste Petroleum Oils. Noyes Data Corporation, Park Ridge, New Jersey, 1979.

19. Development Document for Effluent Limitations Guidelines and New Source Performance Standards for the Unbleached Kraft and Semichemical Pulp Segment of the Pulp, Paper, and Paperboard Mills Point Source Category. EPA-440/1-74-025a, May 1974.

20. EPA. Development Document for Effluent Limitations Guidelines and Standards for the Petroleum Refining Point Source Category. EPA-440/1-79/014-b, December 1979.

21. Yates, J. L., et al., Chaudhry, and J. A. Dewey. An Inventory of Used and Byproducts Hydrocarbon Streams. Final Report, EPA Contract 68-03-2728, March 1980.

22. Ostojio, N. End Use of Solvents Containing Volatile Organic Compounds. EPA-450/3-79-032, May 1979.

23. National Petroleum Council Reports on U.S. Refinery Flexibility. Oil & Gas Journal, December 15, 1980, p. 34.

24. Wicks, P. H., et al. Hazardous Waste Management Planning-1972-73. State of Oregon, Department of Environmental Quality Report, Portland, OR, March 1974.

25. Manuel, R. J., and S. C. Vinson. Special Waste Survey Report -- August 1974-September 1975. Mississippi State Board of Health, Division of Solid Waste Management and Vector Control Report, 1975.

26. Private communication, Gene Crumpler, Office of Solid Waste, EPA, Washington, D.C., and H. K. Willard, Acurex Corporation, Mountain View, California, November 13, 1980.

27. Subtitle C, Resource Conservation and Recovery Act of 1976, Final Environmental Impact Statement-Part 1 & 2, Appendices, April 1980.

28. Locklin, D. W., et al. Design Trends and Operating Problems in Combustion Modification of Industrial Boilers. EPA-650/2-74-032, April 1975.

29. Putnam, A. A., et al. Evaluation of National Boiler Inventory. EPA-600/2-75-067, October 1975.

30. Devitt, T., et al. Population and Characteristics of Industrial/Commercial Boilers in the U.S. EPA-600/7-79-178a, August 1979.

31. Summary Report Compliance Status of Industrial Boilers. PEDCo Environmental, Cincinnati, Ohio. Report under EPA Contract 68-02-2603, Work Assignment 25, Task 3, November 1978.

32. Cato, G. A., et al. Field Testing: Application of Combustion Modifications to Control Pollutant Emissions from Industrial Boilers -- Phase I. EPA-600/2-74-087a, October 1974.

33. Cato, G. A., et al. Field Testing: Application of Combustion Modifications to Control Pollutant Emissions from Industrial Boilers -- Phase II. EPA-600/2-76-086a, April 1976.

34. Private communication. Weir, R., Riley Stoker Corp., Worcester, Massachusetts, with C. Castaldini, Acurex Corporation, Mountain View, California, November 12, 1980.

35. Private communication. Farrell, R. N., Trane Co., LaCrosse, Wisconsin, with C. Castaldini, Acurex Corporation, Mountain View, California, November 12, 1980.

36. Private communication. Bouquet, D. N., Nebraska Boiler Co., Lincoln, Nebraska, with C. Castaldini, Acurex Corporation, Mountain View, California, November 12, 1980.

37. Private communication. Wilcox, J. C., Babcock & Wilcox Co., Industrial and Marine Division, North Canton, Ohio, with C. Castaldini, Acurex Corporation, Mountain View, California, November 12, 1980.

38. Hottel, H. C., and Cohen, E. S. Radiant Heat Transfer in a Gas Filled Enclosure: Allowance for Nonuniformity of Gas Temperature. American Institute of Chemical Engineering Journal, 4:3, 1958.

39. Field M. A., et al. Combustion of Pulverized Coal. The British Coal Utilization Research Association, 1967.

40. Johnson, T. R., and Beer, J. M. The Zone Method Analysis of Radiant Heat Transfer: A Model for Bituminous Radiation. Fourth Symposium on Flames and Industry. Predictive Methods for Industrial Flames. Proceedings of the British Flame Research Committee. The Institute of Fuel, September 1972.

41. Foster, P. J., Calculation of the Optical Properties of Dispersed Phases. Combustion and Flame, 7:277, 1963.

42. Beer, J. M., Methods for Calculating Radiative Heat Transfer from Flames in Combustors and Furnaces. Heat Transfer in Flames. J. Wiley & Sons, New York, 2:29, 1974.

43. Radiation from the Flame Over an Overfeed Coal Bed. Proceedings of the Residential Conference on Science in the Use of Coal. The Institute of Fuel, April 1958.

44. Lucas, D. M., and Toth, H. E. The Calculation of Heat Transfer in the Fire Tubes of Shell Boilers. Journal of the Institute of Fuel, October 1972.

45. Benson, S. W. The Foundations of Chemical Kinetics, McGraw Hill, 1960, Chapters 10 and 11.

46. Lee, K. C., et al., Predictive Model of the Time-Temperature Requirements for Thermal Destruction of Dilute Organic Vapors. 72nd Annual Meeting of the Air Pollution Control Association, June 24-29, 1979.

47. Mathews, B., et al. Destroying Chemical Wastes in Commercial-Scale Incinerators. Phase II Summary Report No. SW-115C, TRW Defense & Space Systems Group and Arthur D. Little, Inc., prepared for Environmental Protection Agency, Office of Solid Waste Management, Contract 68-01-2966.

48. Ackerman, et al. Destroying Chemical Wastes in Commercial Scale Incinerators. Facility Report No. 3, Systems Technology, TRW Defense & Space Systems Group, prepared for Environmental Protection Agency, Office of Solid Waste Management, NTIS No. PB-265-540.

49. Leavitt, C., et al. Environmental Assessment of Coal-and Oil-Firing in a Controlled Industrial Boiler --Comparative Assessment, Volume II. TRW Inc., prepared for Industrial Environmental Research Laboratory, NTIS No. PB-289-981, August 1978.

50. Duvall, D. S., W. A. Rubey, and J. A. Mescher. High Temperature Decomposition of Organic Hazardous Waste. In Proceedings of 6th Annual Research Symposium on the Treatment of Hazardous Waste. U.S. EPA Publication No. EPA-600/9-80-011, March, 1980.

Appendix A
Schematics of Industrial Boiler Types Showing Isothermal Zones

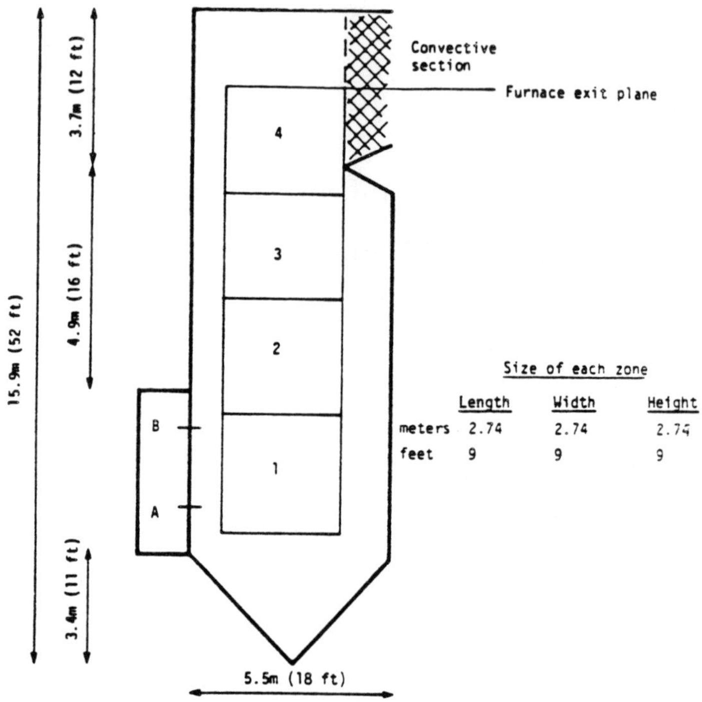

Figure A-1. Schematic of pulverized-coal-fired boiler, 58.6 MW (200 million Btu/hr) -- side view.

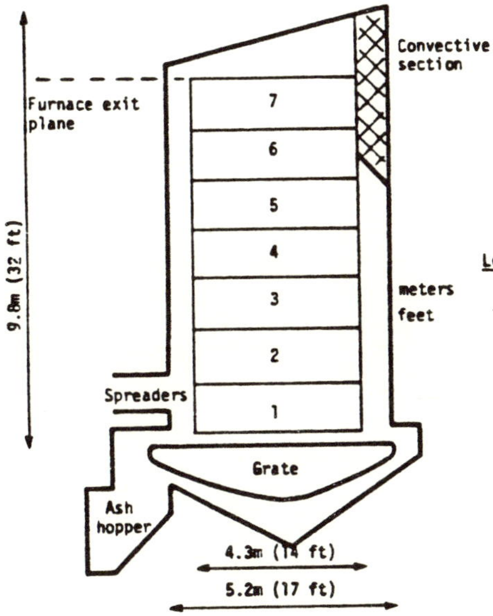

Size of each zone

	Length	Width	Height
meters	4.3	4.3	1.07
feet	14	14	3.5

Figure A-2. Schematic of coal-fired spreader stoker boiler, 44 MW (150 million Btu/hr) -- side view.

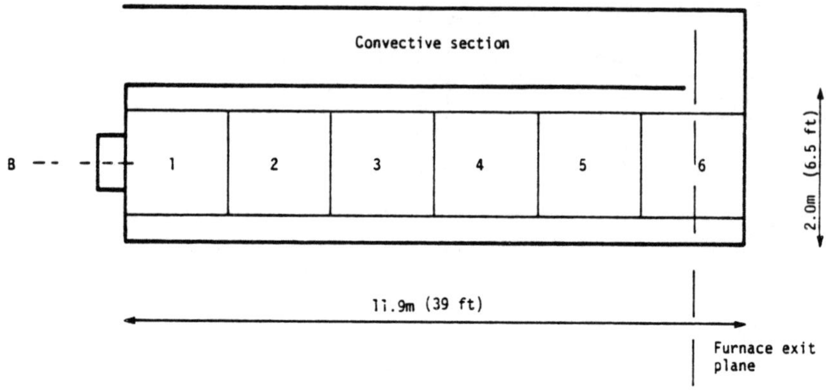

Size of each zone

	Length	Width	Height
meters	1.83	1.83	1.83
feet	6	6	6

Figure A-3. Schematic of residual-oil-, distillate-oil-, and natural-gas-fired package watertube boiler, 44 MW (150 million Btu/hr) -- plan view

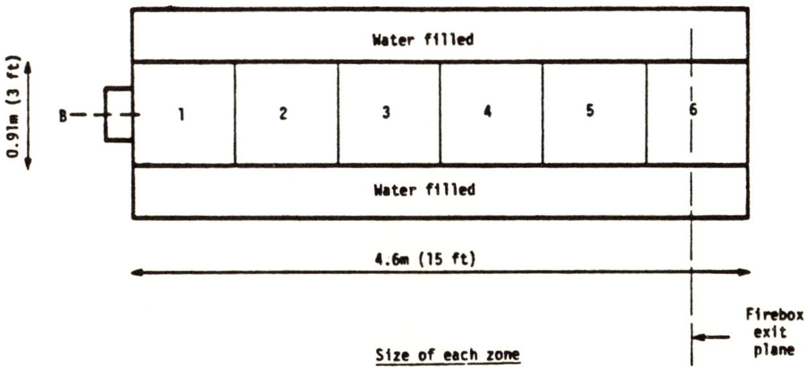

	Diameter	Length
meter	0.91	0.91
feet	3	3

Figure A-4. Schematic of distillate-oil- and natural-gas-fired firetube boiler, 4.4 MW (15 million Btu/hr) -- plan view

Appendix B

Heat Transfer Results—Computer Listings

TABLE B-1. HEAT TRANSFER RESULTS -- WATERTUBE BOILER BURNING PULVERIZED COAL (FULL LOAD)

ZONE	1	2	3	4
ABSORPTION COEFFICIENT (1/FT)	.610000-01	.610000-01	.610000-01	.610000-01
INLET AIR TEMP (DEG R)	860.000			
TOTAL GAS FLOWRATE (LB/HR)	199140.			
HEAT GAIN FROM CHEMICAL REACTION (BTU/HR)	.166456+09	.184951+08	.000000	.000000
GAS RECIRCULATION FROM 1				
2	59742.0			
3		.000000		
4			.000000	
MEAN GAS TEMPERATURE (DEG R)	3227.41	3172.80	2907.03	2690.11
NET RADIANT TRANSFER (BTU/HR)	.232898+08	.207604+08	.147808+08	.115009+08
NET CONVECTIVE TRANSFER (BTU/HR)	.283960+07	.275768+07	.241798+07	.215269+07
TOTAL HEAT INPUT (BTU/HR)	.186411+09	.225099+09	.155259+09	.138060+09
SENSIBLE HEAT GAIN (BTU/HR)	.160282+09	.201581+09	.138060+09	.124406+09
VELOCITY OF GASES (FT/SEC)	22.4983	22.1177	20.2650	18.7528
MEAN RESIDENCE TIME (SEC)	.400030	.406914	.444115	.479927
TOTAL RESIDENCE TIME (SEC)	1.73099			

"FIN

TABLE B-2. HEAT TRANSFER RESULTS -- WATERTUBE BOILER BURNING PULVERIZED COAL (HALF LOAD)

ZONE	1	2	3	4
ABSORPTION COEFFICIENT (1/FT)	.610000-01	.610000-01	.610000-01	.610000-01
INLET AIR TEMP (DEG R)	860.000			
TOTAL GAS FLOWRATE (LB/HR)	113770.			
HEAT GAIN FROM CHEMICAL REACTION (BTU/HR)	.832280+08	.924756+07	.000000	.000000
GAS RECIRCULATION FROM 1				
2	34131.0			
3		.000000		
4			.000000	
MEAN GAS TEMPERATURE (DEG R)	2850.93	2780.96	2475.95	2250.98
NET RADIANT TRANSFER (BTU/HR)	.142339+08	.118241+08	.787121+07	.586190+07
NET CONVECTIVE TRANSFER (BTU/HR)	.234470+07	.221399+07	.186847+07	.160823+07
TOTAL HEAT INPUT (BTU/HR)	.945744+08	.109122+09	.732053+08	.634656+08
SENSIBLE HEAT GAIN (BTU/HR)	.779958+08	.950842+08	.634656+08	.559955+08
VELOCITY OF GASES (FT/SEC)	11.3541	10.9560	9.86069	8.99661
MEAN RESIDENCE TIME (SEC)	.792666	.821471	.912715	1.00838
TOTAL RESIDENCE TIME (SEC)	3.52723			

*FIN

TABLE B-3. HEAT TRANSFER RESULTS -- SPREADER STOKER (FULL LOAD)

ZONE	1	2	3	4	5	6	7
ABSORPTION COEFFICIENT (1/FT)	.500000-01	.900000-01	.500000-01	.400000-01	.300000-01	.200000-01	.200000-01
INLET AIR TEMP (DEG R)	810.000						
TOTAL GAS FLOWRATE (LB/HR)	157700.						
HEAT GAIN FROM CHEMICAL REACTION (BTU/HR)	.769758+08	.419868+08	.209954+08	.000000	.000000	.000000	.000000
GAS RECIRCULATION FROM 1							
2	55195.0						
3		.000000					
4			.000000				
5				.000000			
6					.000000		
7						.000000	
MEAN GAS TEMPERATURE (DEG R)	2487.11	2804.73	2911.45	2703.06	2550.33	2437.62	2335.20
NET RADIANT TRANSFER (BTU/HR)	.545516+07	.171055+08	.117721+08	.706206+07	.454404+07	.270740+07	.239454+07
NET CONVECTIVE TRANSFER (BTU/HR)	.291748+07	.357690+07	.380272+07	.337961+07	.305791+07	.201553+07	.250645+07
TOTAL HEAT INPUT (BTU/HR)	.913012+08	.161443+09	.125226+09	.109652+09	.992099+08	.916080+08	.868800+08
SENSIBLE HEAT GAIN (BTU/HR)	.829286+08	.140760+09	.109652+09	.992099+08	.916080+08	.868800+08	.818990+08
VELOCITY OF GASES (FT/SEC)	15.7290	15.4832	16.0724	14.9264	14.0788	13.4566	12.0912
MEAN RESIDENCE TIME (SEC)	.254920	.226051	.217765	.234484	.248608	.260095	.271502
TOTAL RESIDENCE TIME (SEC)	1.71342						

*FIN

TABLE B-4. HEAT TRANSFER RESULTS -- SPREADER STOKER (HALF LOAD)

ZONE	1	2	3	4	5	6	7
ABSORPTION COEFFICIENT (1/FT)	.500000-01	.900000-01	.300000-01	.400000-01	.300000-01	.200000-01	.200000-01
INLET AIR TEMP (DEG R)	810.000						
TOTAL GAS FLOWRATE (LB/HR)	91100.0						
HEAT GAIN FROM CHEMICAL REACTION (BTU/HR)	.349890+08	.209934+08	.699780+07	.699780+07	.000000	.000000	.000000
GAS RECIRCULATION FROM 1							
2	31885.0						
3		.000000					
4			.000000				
5				.000000			
6					.000000		
7						.000000	
MEAN GAS TEMPERATURE (DEG R)	2105.80	2367.73	2386.82	2362.91	2189.65	2062.08	1947.48
NET RADIANT TRANSFER (BTU/HR)	.289604+07	.867418+07	.460035+07	.416828+07	.244900+07	.130308+07	.120359+07
NET CONVECTIVE TRANSFER (BTU/HR)	.215466+07	.267403+07	.264161+07	.268356+07	.230813+07	.205947+07	.184318+07
TOTAL HEAT INPUT (BTU/HR)	.431050+08	.750170+08	.547773+08	.544531+08	.476818+08	.428792+08	.394317+08
SENSIBLE HEAT GAIN (BTU/HR)	.381343+08	.644695+08	.474553+08	.476818+08	.428792+08	.394317+08	.368049+08
VELOCITY OF GASES (FT/SEC)	6.71448	7.55074	7.51498	7.53837	6.98284	6.57608	6.21806
MEAN RESIDENCE TIME (SEC)	.521262	.463531	.465787	.464476	.501229	.532289	.563506
TOTAL RESIDENCE TIME (SEC)	3.51203						

"FIN

TABLE B-5. HEAT TRANSFER RESULTS -- FIRETUBE BOILER BURNING DISTILLATE OIL (FULL LOAD)

ZONE	1	2	3	4	5	6
ABSORPTION COEFFICIENT (1/FT)	.650000	.550000	.350000	.250000	.200000	.150000
INLET AIR TEMP (DEG R)	640.000					
TOTAL GAS FLOWRATE (LB/HR)	14420.0					
HEAT GAIN FROM CHEMICAL REACTION (BTU/HR)	.600240+07	.600240+07	.300120+07	.000000	.000000	.000000
GAS RECIRCULATION FROM 1						
2	4526.00					
3		.000000				
4			.000000			
5				.000000		
6					.000000	
MEAN GAS TEMPERATURE (DEG R)	2150.37	2832.29	3077.51	2877.14	2732.02	2622.99
NET RADIANT TRANSFER (BTU/HR)	522623.	.160203+07	.166628+07	954902.	658815.	470380.
NET CONVECTIVE TRANSFER (BTU/HR)	45000.8	72872.1	82630.7	74420.4	68773.6	64844.0
TOTAL HEAT INPUT (BTU/HR)	.673997+07	.153283+08	.135010+08	.117521+08	.107228+08	.999522+07
SENSIBLE HEAT GAIN (BTU/HR)	.617234+07	.136554+08	.117521+08	.107228+08	.999522+07	.946000+07
VELOCITY OF GASES (FT/SEC)	43.3502	57.0710	62.0123	57.9748	55.0507	52.8536
MEAN RESIDENCE TIME (SEC)	.576964-01	.438051-01	.403146-01	.431222-01	.454127-01	.473005-01
TOTAL RESIDENCE TIME (SEC)	.277651					

*FIN

TABLE B-6. HEAT TRANSFER RESULTS -- FIRETUBE BOILER BURNING DISTILLATE OIL (HALF LOAD)

ZONE	1	2	3	4	5	6
ABSORPTION COEFFICIENT (1/FT)	.650000	.550000	.350000	.250000	.200000	.150000
INLET AIR TEMP (DEG R)	640.000					
TOTAL GAS FLOWRATE (LB/HR)	7635.00					
HEAT GAIN FROM CHEMICAL REACTION (BTU/HR)	.300120+07	.300120+07	.150060+07	.000000	.000000	.000000
GAS RECIRCULATION FROM 1						
2	763.500					
3		.000000				
4			.000000			
5				.000000		
6					.000000	
MEAN GAS TEMPERATURE (DEG R)	1908.38	2575.35	2730.10	2499.97	2340.42	2224.69
NET RADIANT TRANSFER (BTU/HR)	315240.	.109824+07	.102626+07	542208.	354290.	241785.
NET CONVECTIVE TRANSFER (BTU/HR)	35087.7	62163.5	68404.2	59115.3	52839.2	48415.4
TOTAL HEAT INPUT (BTU/HR)	.338842+07	.652657+07	.637949+07	.528483+07	.468350+07	.427638+07
SENSIBLE HEAT GAIN (BTU/HR)	.303809+07	.536616+07	.528483+07	.468350+07	.427638+07	.398617+07
VELOCITY OF GASES (FT/SEC)	20.3604	27.4763	29.1274	26.6720	24.9699	23.7351
MEAN RESIDENCE TIME (SEC)	.122787	.909874-01	.858299-01	.937311-01	.100121	.105329
TOTAL RESIDENCE TIME (SEC)	.598786					

*FIN

TABLE B-7. HEAT TRANSFER RESULTS -- FIRETUBE BOILER BURNING NATURAL GAS (FULL LOAD)

ZONE	1	2	3	4	5	6
ABSORPTION COEFFICIENT (1/FT)	.280000	.250000	.200000	.180000	.100000+00	.800000-01
INLET AIR TEMP (DEG R)	640.000					
TOTAL GAS FLOWRATE (LB/HR)	15255.0					
HEAT GAIN FROM CHEMICAL REACTION (BTU/HR)	.749975+07	.599980+07	.149995+07	.000000	.000000	.000000
GAS RECIRCULATION FROM 1						
2	3976.50					
3		1988.25				
4			.000000			
5				.000000		
6					.000000	
MEAN GAS TEMPERATURE (DEG R)	2569.62	3262.77	3267.90	3051.01	2931.58	2830.87
NET RADIANT TRANSFER (BTU/HR)	639188.	.168032+07	.137661+07	962740.	496775.	367861.
NET CONVECTIVE TRANSFER (BTU/HR)	61850.7	90445.8	90720.1	81599.1	76364.3	72107.9
TOTAL HEAT INPUT (BTU/HR)	.816424+07	.169760+08	.149536+08	.117249+08	.106805+08	.101074+08
SENSIBLE HEAT GAIN (BTU/HR)	.746320+07	.152053+08	.134863+08	.106805+08	.101074+08	.966738+07
VELOCITY OF GASES (FT/SEC)	47.5951	60.4330	60.5280	56.5262	54.2994	52.5768
MEAN RESIDENCE TIME (SEC)	.525264-01	.413676-01	.413027-01	.442273-01	.460410-01	.475497-01
TOTAL RESIDENCE TIME (SEC)	.273815					

*FIN

TABLE B-8. HEAT TRANSFER RESULTS -- FIRETUBE BOILER BURNING NATURAL GAS (HALF LOAD)

ZONE	1	2	3	4	5	6
ABSORPTION COEFFICIENT (1/FT)	.280000	.250000	.200000	.180000	.100000+00	.800000-01
INLET AIR TEMP (DEG R)	640.000					
TOTAL GAS FLOWRATE (LB/HR)	7021.50					
HEAT GAIN FROM CHEMICAL REACTION (BTU/HR)	.374987+07	.299990+07	749975.	.000000	.000000	.000000
GAS RECIRCULATION FROM 1						
2	2106.45					
3		1053.22				
4			.000000			
5				.000000		
6					.000000	
MEAN GAS TEMPERATURE (DEG R)	2367.32	2934.74	2888.47	2687.41	2504.07	2402.26
NET RADIANT TRANSFER (BTU/HR)	469306.	.111158+07	829950.	532252.	262338.	187028.
NET CONVECTIVE TRANSFER (BTU/HR)	53697.8	77304.4	74924.7	64615.1	80871.0	84439.8
TOTAL HEAT INPUT (BTU/HR)	.409921+07	.819237+07	.692361+07	.525232+07	.463846+07	.431425+07
SENSIBLE HEAT GAIN (BTU/HR)	.357621+07	.700349+07	.601873+07	.463546+07	.431425+07	.407198+07
VELOCITY OF GASES (FT/SEC)	23.2273	28.7947	28.2623	25.8774	24.3691	23.5702
MEAN RESIDENCE TIME (SEC)	.107682	.868214-01	.884872-01	.966096-01	.101784	.106066
TOTAL RESIDENCE TIME (SEC)	.587840					

*FIN

TABLE B-9. HEAT TRANSFER RESULTS -- WATERTUBE BOILER BURNING RESIDUAL OIL (FULL LOAD)

ZONE	1	2	3	4	5	6
ABSORPTION COEFFICIENT (1/FT)	.350000	.300000	.200000	.150000	.120000	.120000
INLET AIR TEMP (DEG R)	810.000					
TOTAL GAS FLOWRATE (LB/HR)	156580.					
HEAT GAIN FROM CHEMICAL REACTION (BTU/HR)	.965692+08	.413868+08	.000000	.000000	.000000	.000000
GAS RECIRCULATION FROM 1						
2	54632.0					
3		54632.0				
4			.000000			
5				.000000		
6					.000000	
MEAN GAS TEMPERATURE (DEG R)	2833.50	3057.30	2856.88	2666.76	2524.96	2397.87
NET RADIANT TRANSFER (BTU/HR)	.147711+08	.207108+08	.106590+08	.674032+07	.480818+07	.394407+07
NET CONVECTIVE TRANSFER (BTU/HR)	.207276+07	.228244+07	.210424+07	.193272+07	.180350+07	.168781+07
TOTAL HEAT INPUT (BTU/HR)	.108815+09	.176443+09	.149662+09	.976013+08	.889282+08	.825867+08
SENSIBLE HEAT GAIN (BTU/HR)	.919709+08	.153449+09	.136899+09	.889282+08	.825867+08	.769048+08
VELOCITY OF GASES (FT/SEC)	45.3901	48.9739	45.7634	42.7179	40.4804	38.4027
MEAN RESIDENCE TIME (SEC)	.132108	.122514	.131109	.140456	.148374	.156289
TOTAL RESIDENCE TIME (SEC)						.830850

*FIN

TABLE B-10. HEAT TRANSFER RESULTS -- WATERTUBE BOILER BURNING RESIDUAL OIL (HALF LOAD)

ZONE	1	2	3	4	5	6
ABSORPTION COEFFICIENT (1/FT)	.350000	.300000	.200000	.150000	.120000	.120000
INLET AIR TEMP (DEG R)	810.000					
TOTAL GAS FLOWRATE (LB/HR)	72290.0					
HEAT GAIN FROM CHEMICAL REACTION (BTU/HR)	.482846+08	.206934+08	.000000	.000000	.000000	.000000
GAS RECIRCULATION FROM 1						
2	28916.0					
3		28916.0				
4			.000000			
5				.000000		
6					.000000	
MEAN GAS TEMPERATURE (DEG R)	2556.86	2601.56	2440.31	2237.09	2002.68	1950.66
NET RADIANT TRANSFER (BTU/HR)	.987122+07	.123078+08	.577930+07	.334109+07	.210783+07	.160182+07
NET CONVECTIVE TRANSFER (BTU/HR)	.182015+07	.195660+07	.175249+07	.156006+07	.141601+07	.120988+07
TOTAL HEAT INPUT (BTU/HR)	.547448+08	.829076+08	.664290+08	.419425+08	.370413+08	.335175+08
SENSIBLE HEAT GAIN (BTU/HR)	.430454+08	.686431+08	.588971+08	.370413+08	.335175+08	.305463+08
VELOCITY OF GASES (FT/SEC)	21.6702	22.7330	20.7579	18.9670	17.6575	16.5385
MEAN RESIDENCE TIME (SEC)	.276776	.263924	.289046	.316338	.339799	.362789
TOTAL RESIDENCE TIME (SEC)						1.84867

*F!N

TABLE B-11. HEAT TRANSFER RESULTS -- WATERTUBE BOILER BURNING NATURAL GAS (FULL LOAD)

/CM	1	2	3	4	5	6
ABSORPTION COEFFICIENT (1/FT)	.200000	.150000	.120000	.100000+00	.400000-01	.300000-01
INLET AIR TEMP (DEG R)	810.000					
TOTAL GAS FLOWRATE (LB/HR)	125120.					
HEAT GAIN FROM CHEMICAL REACTION (BTU/HR)	.700740+08	.490518+08	.210222+08	.000000	.000000	.000000
GAS RECIRCULATION FROM 1						
2	37536.0					
3		37536.0				
4			25024.0			
5				.000000		
6					.000000	
MEAN GAS TEMPERATURE (DEG R)	2605.56	3205.57	3271.96	3059.98	2889.10	2735.11
NET RADIANT TRANSFER (BTU/HR)	.740502+07	.145404+08	.131164+08	.870307+07	.510576+07	.440699+07
NET CONVECTIVE TRANSFER (BTU/HR)	.186985+07	.239343+07	.244557+07	.226556+07	.212122+07	.199430+07
TOTAL HEAT INPUT (BTU/HR)	.806012+08	.151626+09	.156494+09	.128472+09	.979339+08	.907069+08
SENSIBLE HEAT GAIN (BTU/HR)	.713264+08	.134692+09	.140930+09	.117503+09	.907069+08	.843056+08
VELOCITY OF GASES (FT/SEC)	38.2555	47.0403	48.0147	44.9040	42.3964	40.1366
MEAN RESIDENCE TIME (SEC)	.156922	.127550	.124962	.133619	.141522	.149490
TOTAL RESIDENCE TIME (SEC)						.834064

*FIN

TABLE B-12. HEAT TRANSFER RESULTS -- WATERTUBE BOILER BURNING NATURAL GAS (HALF LAOD)

ZONE	1	2	3	4	5	6
ABSORPTION COEFFICIENT (1/FT)	.200000	.150000	.120000	.100000+00	.800000-01	.800000-01
INLET AIR TEMP (DEG R)	810.000					
TOTAL GAS FLOWRATE (LB/HR)	68060.0					
HEAT GAIN FROM CHEMICAL REACTION (BTU/HR)	.350370+08	.245259+08	.105111+08	.000000	.000000	.000000
GAS RECIRCULATION FROM 1						
2	20418.0					
3		20410.0				
4			13612.0			
5				.000000		
6					.000000	
MEAN GAS TEMPERATURE (DEG R)	2347.39	2819.61	2831.05	2604.56	2418.93	2256.29
NET RADIANT TRANSFER (BTU/HR)	.502166+07	.875050+07	.719472+07	.437970+07	.239674+07	.195482+07
NET CONVECTIVE TRANSFER (BTU/HR)	.165076+07	.205883+07	.205039+07	.184922+07	.169130+07	.155761+07
TOTAL HEAT INPUT (BTU/HR)	.407645+08	.729487+08	.725844+08	.575468+08	.428458+08	.387577+08
SENSIBLE HEAT GAIN (BTU/HR)	.340921+08	.621394+08	.633393+08	.513179+08	.387577+08	.352453+08
VELOCITY OF GASES (FT/SEC)	18.7376	22.5071	22.5985	20.7905	19.5088	18.0105
MEAN RESIDENCE TIME (SEC)	.320211	.266582	.265505	.288593	.310740	.333140
TOTAL RESIDENCE TIME (SEC)						1.78477

FIN

TABLE B-13. HEAT TRANSFER RESULTS -- WATERTUBE BOILER BURNING DISTILLATE OIL (FULL LOAD)

ZONE	1	2	3	4	5	6
ABSORPTION COEFFICIENT (1/FT)	.350000	.300000	.150000	.120000	.100000+00	.100000+00
INLET AIR TEMP (DEG R)	840.000					
TOTAL GAS FLOWRATE (LB/HR)	132540.					
HEAT GAIN FROM CHEMICAL REACTION (BTU/HR)	.965874+08	.413946+08	.000000	.000000	.000000	.000000
GAS RECIRCULATION FROM 1						
2	53016.0					
3		53016.0				
4			.000000			
5				.000000		
6					.000000	
MEAN GAS TEMPERATURE (DEG R)	2864.62	3089.95	2919.47	2740.86	2601.06	2476.50
NET RADIANT TRANSFER (BTU/HR)	.157166+08	.216018+08	.854461+07	.596830+07	.428074+07	.367238+07
NET CONVECTIVE TRANSFER (BTU/HR)	.204725+07	.230580+07	.215157+07	.198993+07	.186161+07	.174552+07
TOTAL HEAT INPUT (BTU/HR)	.108488+09	.174366+09	.147266+09	.974654+08	.895071+08	.835648+08
SENSIBLE HEAT GAIN (BTU/HR)	.906745+08	.150459+09	.136570+09	.895071+08	.835648+08	.779471+08
VELOCITY OF GASES (FT/SEC)	44.5500	48.0328	45.3826	42.6062	40.4330	38.4936
MEAN RESIDENCE TIME (SEC)	.134741	.124915	.132209	.140825	.148594	.155870
TOTAL RESIDENCE TIME (SEC)						.836955

*FIN

TABLE B-14. HEAT TRANSFER RESULTS -- WATERTUBE BOILER BURNING DISTILLATE OIL (HALF LOAD)

ZONE	1	2	3	4	5	6
ABSORPTION COEFFICIENT (1/FT)	.350000	.300000	.150000	.120000	.100000+00	.100000+00
INLET AIR TEMP (DEG R)	810.000					
TOTAL GAS FLOWRATE (LB/HR)	71770.0					
HEAT GAIN FROM CHEMICAL REACTION (BTU/HR)	.452937+08	.206973+08	.000000	.000000	.000000	.000000
GAS RECIRCULATION FROM 1						
2	28708.0					
3		28708.0				
4			.000000			
5				.000000		
6					.000000	
MEAN GAS TEMPERATURE (DEG R)	2554.86	2706.57	2518.89	2318.91	2162.14	2023.63
NET RADIANT TRANSFER (BTU/HR)	.100657+08	.120741+08	.451770+07	.298504+07	.205495+07	.170749+07
NET CONVECTIVE TRANSFER (BTU/HR)	.182573+07	.194299+07	.178196+07	.161413+07	.148479+07	.137250+07
TOTAL HEAT INPUT (BTU/HR)	.547081+08	.823959+08	.666438+08	.431972+08	.385981+08	.350583+08
SENSIBLE HEAT GAIN (BTU/HR)	.428167+08	.683789+08	.603442+08	.385981+08	.350583+08	.319783+08
VELOCITY OF GASES (FT/SEC)	21.5054	22.7625	21.2027	19.5193	18.1998	17.0338
MEAN RESIDENCE TIME (SEC)	.278999	.263361	.282783	.307387	.329675	.352240
TOTAL RESIDENCE TIME (SEC)						1.81465

*FIN

Appendix C

Model Application—An Example Case

This appendix demonstrates the application of the model developed in Section 7. The example case is a test burn of PCB's in an industrial boiler owned and operated by General Motors Corporation at their Bay City, Michigan facility. The test program was carried out by GCA/Technology Division under U.S. Environmental Protection Agency Contract No. 68-02-3168, Work Assignment No. 9. The work was completed October 31, 1980. The test report is "Evaluation of PCB Destruction Efficiency in an Industrial Boiler," EPA-600/2-81-055a.

The boiler was a 60 million Btu/hr package water tube unit burning no. 6 distillate oil. The waste stream was waste oil containing approximately 500 ppm Arochlor 1254. The waste stream and oil were mixed in a ratio of 1:10 prior to injection. The boiler was operated at half load (i.e., the fuel injection rate was half the maximum rate).

The time and temperature profile within the fire box was generated with the Acurex Zonal Heat Balance Program. Kinetic data were derived from thermal destruction unit data on 2, 2', 4, 5, 5'-pentachlorobiphenyl generated by Duvall at the University of Dayton Research Institute (Ref. 50).

The model input data is summarized in Table 1, and the predictive model diagram is shown in Figure 1. The model predicts that for half load and 2.5 percent O_2 content the boiler should achieve greater than 99.99 percent destruction. The test results are given in Table 2 and confirm this prediction.

Table C-1. INPUT PARAMETERS FOR EXAMPLE CASE MODEL APPLICATION

Compound (2,2', 4, 5, 5' - Pentachlorobiphenyl)			
	Pyrolysis	Oxidation (2.5% O_2)	Oxidation (21% O_2)
E (Kcal/g-mole)	53.6	66.7 ± 0.5	70.0
A (sec^{-1})	7.44×10^9	2.6×10^{12}	1.10×10^{16}
Boiler			
T_m	1700°K ±50°	(Flame)	
T_x	1144°K ±50°	(Back wall)	
t_m	0.4 sec	Acurex boiler zonal heat balance	
t_x	1.8 sec		
(O_2% in flue gas	6+%)		

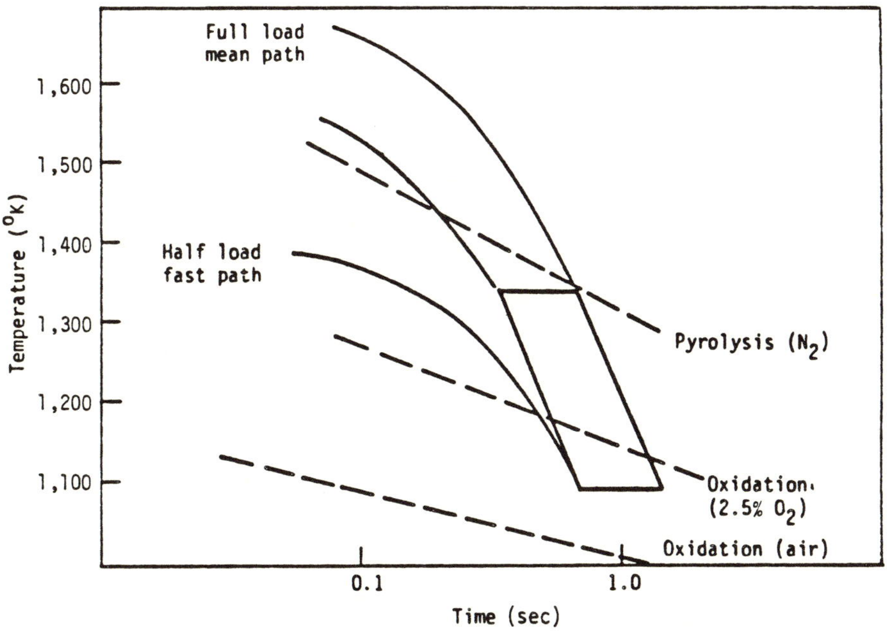

Figure C-1. Predicted boiler operational zone and destruction efficiency curves for example case.

Table C-2. REPORTED DESTRUCTION EFFICIENCY OF PCB's IN AN INDUSTRIAL BOILER

Run number	PCB in fuel[a] concentration range (mg/kg)	PCB[b] in (mg/min)	PCB out (mg/min)	PCB[c] destruction efficiency (%)
PCB-2	34-72	480-1,000	5.8×10^{-2}	≥99.99[d]
PCB-3	34-76	480-1,100	5.6×10^{-2}	≥99.99[d]
PCB-4	34-76	480-1,100	5.6×10^{-2}	≥99.99[d]

Note: Density of 1:10 Dilution of Waste Oil: No. 6 fuel oil is 3.4 kg/gal.

[a]Based on GM and GCA reported results.

[b]Fuel combustion rate: 3.43 kg/gal x 4 gal/min fuel flow = 14 kg/min.

[c]Percent destruction = $100 \; \frac{\text{PCB in} - \text{PCB out}}{\text{PCB in}}$

[d]This assumes 100 percent sample collection efficiency. A validated test sample collection efficiency was not conducted as part of this verification burn.

PART II

FURNACES, KILNS, AND COMBUSTORS

The information in Part II is from *Evaluation of the Feasibility of Incinerating Hazardous Waste in High-Temperature Industrial Processes,* prepared by F.D. Hall, W.F. Kemner, G. Annamraju, R. Krishnan, M.A. Taft-Frank, and D.N. Albrinck of PEDCo Environmental, Inc. for the U.S. Environmental Protection Agency, April 1983.

Acknowledgment

This report was prepared for the U.S. Environmental Protection Agency by PEDCo Environmental, Inc. The EPA Work Assignment Managers were Ms. Laurel J. Staley and Mr. Robert Mournighan. The PEDCo Project Director was Mr. William F. Kemner, and the Project Manager was Mr. Fred D. Hall. The principal authors were Messrs. Kemner, Hall, Gopal Annamraju, and Radha Krishnan and Ms. Mary A. Taft-Frank and Diane N. Albrinck.

Executive Summary

In the search for disposal alternatives, the U.S. Environmental Protection Agency is evaluating the potential use of high-temperature processes for the incineration of hazardous wastes. Many kinds of waste have already been destroyed in boilers and cement kilns; this report considers the many other potential processes, such as metallurgical furnaces, brick and lime kilns, glass furnaces, and sewage sludge combustors. Each process is examined against such criteria as time/temperature profile, geographical location, product quality considerations, institutional factors, and environmental impacts. Promising alternatives are identified.

The first task was to select common industries using high-temperature processes. Table 1 summarizes the screening criteria against which each process was evaluated.

TABLE 1. SCREENING CRITERIA FOR POTENTIAL HIGH-TEMPERATURE PROCESSES FOR THE DESTRUCTION OF HAZARDOUS WASTE

- Compatibility of process
  - Thermal destruction conditions: temperature and residence time (turbulence not considered)
  - Product quality
  - Potential for fugitive emissions
- Number of facilities
- Geographic matching of industry location and waste generators

The initial list of processes was based on operating temperature; any major process with an operating temperature of greater than 650°C was included. The initial listing of approximately 100 processes included the metallurgical, chemical, and mineral industries and sludge and waste incinerators. When the other criteria in Table 1 were applied, several of these processes were eliminated from consideration

Table 2 lists those industries and processes that remained after several processes had been eliminated from further consideration by the application of the screening criteria. These processes were evaluated in more detail by use of a hazardous waste destruction model that compares temperature and residence times in the process with those required for complete (99.99%) destruction of various hazardous wastes.

Based on kinetic data on the constituents of the wastes, characteristic curves showing the various time-temperature combinations for destroying the wastes were obtained, as shown in Figure 1. For the industrial process, such information as the furnace volume, volumetric flow rate of gases, and temperature range of the furnace made it possible to plot "process curves" showing the residence time at which the furnace gas is above a certain temperature within the furnace. Laying the "process curves" over the hazardous waste characteristic curves made it possible to predict the waste categories destructible at various temperatures. Figure 2 shows this for a short rotary kiln. The required 99.99 percent destruction is predicted for those wastes whose lines are below and to the left of the process curves. Table 3 summarizes those processes evaluated in detail and the waste categories that are predicted to be 99.99 percent destroyable in each process.

Based on the process evaluation and discussions with many persons in industry and government, it is concluded that both technical and institutional factors are major obstacles to the use of high-temperature processes. The economics of each case have not been evaluated in detail,

TABLE 2. INDUSTRIES AND PROCESSES SELECTED FOR FURTHER HAZARDOUS WASTE DESTRUCTION EVALUATION

Industry	Process or furnace	Exit and maximum temperature, °C	Residence time (seconds, unless otherwise noted)[a]
Brick	Tunnel kiln	260-1180	4.3
Carbon black	Oil furnace process	870-1400	1.1
Primary copper	Reverberatory furnace	1300-1400	2.2
Primary lead	Blast furnace	700-1200	5.9
Primary zinc	Multiple-hearth roaster Suspension roaster Fluidized-bed roaster	200-980 930-1010 950-1740	>10 >10 >10
Iron and steel	Blast furnace Open-hearth furnace	1100[b]-1870 1200-1800	1.1[c] 2.0
Lime	Long rotary kiln Short rotary kiln with stone preheater	680-1900 1160-1840	8.3 7.6
Lightweight aggregate	Rotary kiln	370-1150	3.6
Glass	Melting furnace	620-1480	4.1
Sewage sludge	Fluid-bed furnace Multiple-hearth furnace	760-870 480-980	1.4 0.5
Hazardous waste incineration[d]	Rotary kiln Liquid injection	820-1600 700-1650	2.0 s to 2 h[e] 0.5 to 2

[a]Residence time above the greater of the exit temperature or 760°C.

[b]Represents estimated temperature at the top of the combustion zone in the blast furnace.

[c]Residence time has been calculated using 33% of the total furnace volume.

[d]Included for comparison only.

[e]The residence time for solids and sludges can be adjusted.

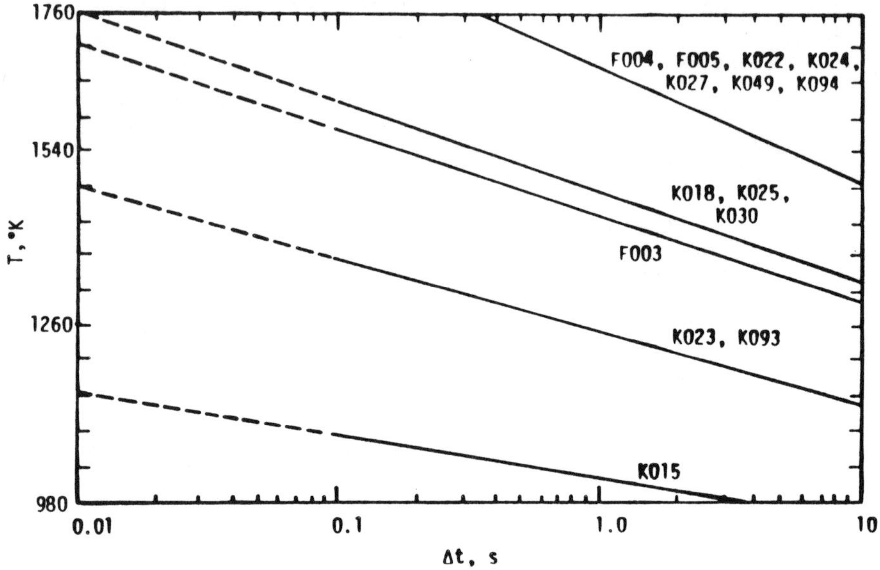

Figure 1. Plots of time-temperature requirements for 99.99 percent destruction of selected hazardous waste categories.

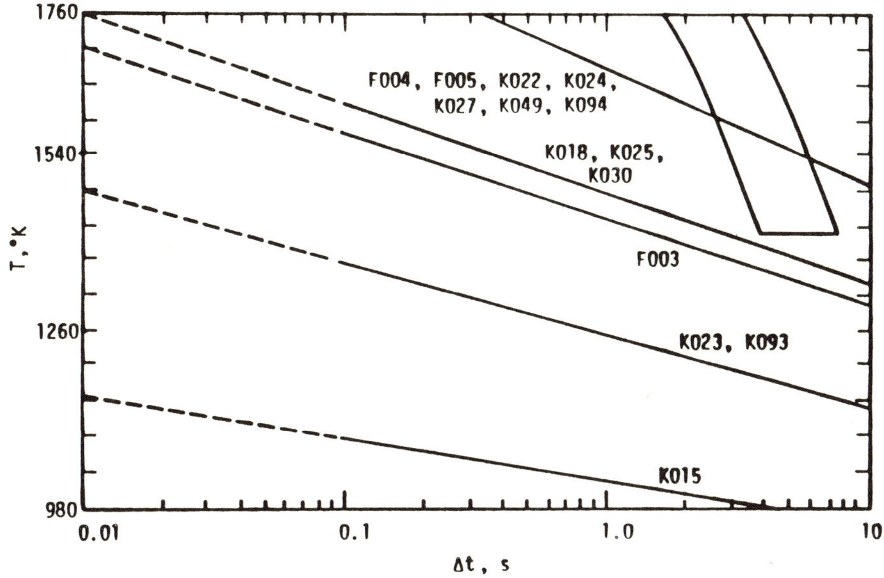

Figure 2. Hazardous waste destructibility for the short rotary lime kiln.

TABLE 3. HAZARDOUS WASTE CATEGORIES THAT ARE PREDICTED DESTRUCTIBLE (99.99%) IN SELECTED INDUSTRIAL PROCESSES

Industry	Process or furnace	Waste category						
		K015	K023, K093	F003	K018 K030	K025	F004, F005	K022, K024, K027 K049, K094
Brick	Tunnel kiln	Yes	Yes	No	No	No	No	No
Carbon black	Oil furnace process	Yes	Yes	Maybe	No	No	No	No
Glass	Melting furnace	Yes	Yes	Yes	Yes	Yes	No	No
Iron and steel	Blast furnace	No	Yes	Yes	No	No	Yes	No
	Open-hearth furnace	No	Yes	Yes	No	Yes	Yes	No
Lightweight aggregate	Rotary kiln	Yes	Yes	No	No	No	No	No
Lime	Long and short rotary kilns	Yes	Yes	Yes	Yes	Yes	Yes	Yes
Primary copper	Reverberatory furnace	No	Yes	Yes	No	Yes	No	No
Primary lead	Blast furnace	Yes	Yes	No	No	No	No	No
Primary zinc	Multiple-hearth roaster	Yes	No	No	No	No	No	No
	Suspension roaster	Yes	Yes	No	No	No	No	No
	Fluid-bed roaster	Yes	Yes	No	No	No	No	No
Sewage sludge	Fluid-bed furnace	Yes	No	No	No	No	No	No
	Multiple-hearth furance	Yes	No	No	No	No	No	No

but they will clearly be proportional to energy cost factors and the degree of hazard associated with the waste. The driving force to utilize high-temperature processes must come from the companies involved. The availability of low- or zero-cost energy combined with possible disposal fees from the generators must be sufficiently attractive to justify the effort required to conduct test burns and make the necessary process modifications.

1. Introduction

Approximately 36 billion kilograms (40 million short tons) of hazardous waste is generated annually in the United States by some 270,000 industrial plants and other facilities. Most of this waste is now landfilled, but alternative waste disposal methods are under consideration that allow for the recovery of heat (through incineration) or valuable constituents (through recycling). The rising costs of transportation and landfilling and the increasingly stringent RCRA regulations governing land disposal of organic liquids will undoubtedly lead to more interest in thermally destroying hazardous waste or reducing their quantity through recycling.

High-temperature industrial processes are being considered as an alternative means of thermal destruction. The Incineration Research Branch of the U.S. Environmental Protection Agency (EPA) is involved in the investigation, evaluation, and development of thermal destruction techniques for the disposal of hazardous waste. Two high-temperature processes (industrial boilers and cement kilns) are already being evaluated as possible thermal destruction candidates, and many kinds of wastes have already been destroyed in these processes. The Agency's interest in other possible high-temperature industrial processes for this purpose prompted the initiation of this feasibility screening study.

The objectives of this study were to locate processes that offer suitable combustion conditions, to provide uniform criteria for the selection of processes that appear to have the greatest potential for waste destruction, and to examine the selected processes for overall consideration.

In the selection of possible processes, the first task was to find common industries using high-temperature processes. Typical operating parameters of various processes were obtained from the literature, trade associations, and industry contacts. Any process with an operating temperature greater than 650°C (1200°F) was included. This temperature is believed to be the minimum temperature allowable for the destruction of the wastes under consideration. The initial listing contained approximately 100 processes from the metallurgical, chemical, and mineral industries, and sludge and waste incinerators.

The processes on this list were then evaluated against the following screening criteria for their potential use in the destruction of hazardous wastes:

- Compatibility of process
  - Thermal destruction conditions (temperature and residence time; turbulence not considered)
- Number of facilities
- Geographic matching of industrial location and waste generators.

As a result of this screening effort, the list was narrowed down to processes in the following industries, each of which is addressed in this report:

Brick and tile	Lime
Carbon black	Primary copper
Glass	Primary lead
Iron and steel	Primary zinc
Light-weight aggregate	Sewage sludge incinerators

Section 2 of the report discusses the characteristics and sources of hazardous wastes considered for thermal destruction.

Section 3 presents the hazardous waste destruction model developed to identify those processes most applicable for the destruction of hazardous wastes. Section 4, which composes the major part of the report, presents an assessment of each of the high-temperature industrial processes selected as the most promising candidates for hazardous waste destruction. Appendices A and B present backup data for the selection process.

2. Waste Selection and Characterization

2.1 SELECTION AND WASTE COMPONENT CHARACTERIZATION

In this assessment, the selection of specific hazardous wastes as candidates for thermal destruction in an industrial process is of primary importance in examining this method as a means of destroying these wastes. The type of waste selected and its characteristics play a major role in determining the process operating conditions (temperature and residence time) required for complete destruction. This information also aids in assessing the impact on the process hardware and the emission control system and ultimately determines the compatibility of the wastes with a proposed process in an industrial facility.

Initially, all hazardous waste categories from specific and nonspecific sources (as found in Parts 261.31 and 261.32, 46 FR 27476) were considered potential candidates for thermal destruction. Certain waste categories, however, were eliminated during preliminary review because of their low combustibility or their explosive nature. The waste types produced by the following industries were eliminated from consideration: inorganic pigments, inorganic chemicals, explosives, iron and steel, and secondary lead. Most of these industries generate waste streams that are highly inorganic in nature and hence not suitable for combustion.

After the preliminary review, more detailed information gathered and compiled on the remaining waste categories led to the elimination of additional wastes because of the low combustibility of their components. Appendix B presents background data (physical state, composition, present disposal practices, geographic distribution, and annual generation rate) for those waste categories included in this initial screening. Background information collected on all waste categories was obtained from EPA's Resource Conservation and Recovery Act, Subtitle C, Identification and Listing of Hazardous Waste, Background Document.[1]

The hazardous waste categories that were ultimately selected for further evaluation were those in which the wastes appeared to exhibit high combustibility or incinerability. The definition of incinerability here is considered to be the relative ease with which flame incineration can completely oxidize a constituent to its final products of combustion.[2] The ease of incinerability can be related to the heat of combustion of each constituent.[2] The heat of combustion was estimated for each waste category by use of the heat of combustion for each of the principal organic hazardous constituents. These data were then used to predict incinerability. This predicted incinerability was, in turn, used as a basis for the selection of potential hazardous waste candidates for thermal destruction in high-temperature industrial processes.

Table 2-1 presents the selected waste candidates, the type of generating industry and its SIC code, and a description of the waste. The Organic Chemicals Industry generates most of the wastes selected for consideration.

Table 2-2 presents a characterization of the hazardous waste types and their components. Most wastes are composed of several constituents. In general, wastes from nonspecific sources (those in the F series) vary considerably in composition among the different processes, whereas wastes from specific sources (those in the K series) tend to be more consistent in the type and percentage of each waste component.

2.2 MATHEMATICAL MODEL FOR RATE OF THERMAL DESTRUCTION OF HAZARDOUS WASTES

2.2.1 Theory

The chemical reaction sequence involved in the destruction of organic constituents present in hazardous wastes is a

TABLE 2-1. DESCRIPTION OF SELECTED HAZARDOUS WASTE STREAMS

Hazardous waste No.	Waste generating industry		Description of waste
	Type	SIC	
F003	Nonspecific sources		Spent nonhalogenated solvents and still bottoms from the recovery of these solvents
F004	Nonspecific sources		Spent nonhalogenated solvents and still bottoms from the recovery of these solvents
F005	Nonspecific sources		Spent nonhalogenated solvents and still bottoms from the recovery of these solvents
K015	Organic chemicals	2865	Still bottoms from the distillation of benzyl chloride
K018	Organic chemicals	2869	Heavy ends from the fractionation column in ethyl chloride production
K022	Organic chemicals	2869	Distillation bottom tars from the production of phenol/acetone from cumene
K023	Organic chemicals	2869	Distillation light ends from the production of phthalic anhydride from naphthalene
K024	Organic chemicals	2869	Distillation bottoms from the production of phthalic anhydride from naphthalene
K025	Organic chemicals	2865	Distillation bottoms from the production of nitrobenzene by the nitration of benzene
K027	Organic chemicals	2869	Centrifuge and distillation residues from toluene diisocyanate production
K030	Organic chemicals	2869	Column bottoms or heavy ends from the combined production of trichloroethylene and perchloroethylene
K049	Petroleum refining	2911	Slop oil emulsion solids from the petroleum refining industry
K093	Organic chemicals	2869	Distillation light ends from the production of phthalic anhydride from ortho-xylene
K094	Organic chemicals	2869	Distillation bottoms from the production of phthalic anhydride from ortho-xylene

TABLE 2-2. PHYSICAL DATA FOR ORGANIC CONSTITUENTS OF HAZARDOUS WASTE CATEGORIES AND WASTE GENERATION RATES

Hazardous waste number (physical state)	Chemical compound[a] (% of total)	Melting point,[b] °C	Boiling point,[b] °C	Specific gravity[b]	Heat of combustion,[c] kcal/g	Annual generation rate,[a,d] Mg
F003 (organic liquid and sludges)	o-Xylene	-25.0	144	0.880		489,900[e] (all xylenes)
	m-Xylene	-48.0	139	0.860		
	p-Xylene	13.0	138	0.861		
	Ethylbenzene	-95.0	136	0.867		54,430[e]
	Cyclohexanone	-45.0	155	0.947		9,072[e]
	Acetone	-95.4	56.2	0.792	7.36	86,200[e]
	Ethylacetate	-83.6	77	0.894	6.12	69,900[e]
	Ethylether	-116	34.5	0.715		54,430[e]
	Methyl isobutyl ketone	-85.0	116	0.804		78,000[e]
	n-Butyl alcohol	-89.0	118	0.811	8.62	45,360[e]
	Methanol	-97.8	64.5	0.792	5.70	317,500[e]
F004 (organic liquid and sludges)	O-Cresol	31.0	191	1.03	8.18	11,800[e] (cresols and cresylic acid)
	m-Cresol	11.0	201	1.03	8.18	
	p-Cresol	35.0	220-250	1.02-1.07	8.09	
	Cresylic acid	20-76	220-250	1.02-1.07	8.09	
	Nitrobenzene	5.7	211	1.20	6.01	9,072[e]
F005 (organic liquid and sludges)	Toluene	-94.5	111	0.866	10.14	317,500[e]
	Pyridine	-42	115	0.982	7.83	907[e]
	Methylethyl ketone	-85.9	79.6	0.805	8.07	
	Carbon disulfide	-109	46.3	1.26		77,100[e]
	Isobutanol	-108	107	0.806		18,600[e]
K015 (organic sludge)	Benzylchloride	-39	179	1.10	6.18	5,300
	Trichlorobenzene	17-63.5	209-219	1.57	3.40	
	Tetrachlorobenzene	47-138	246-254	1.73	2.61	
	Pentachlorobenzene	86	277	1.83	2.05	
	Hexachlorobenzene	228	326	2.04	1.79	
	Benzotrichloride	-5	220	1.38	3.90	
	Chlorobenzene	-45.2	131	1.11	6.60	
	Toluene	-94.5	111	0.866	10.14	
K018 (organic liquid)	Ethylchloride (3%)	-139	13.0	0.917	3.00	35,000
	1,2-dirchloroethane (22%)	-35.3	83.7	1.26	3.00	
	Hexachlorobenzene	228	326	2.04	1.79	
	1,1,1-Trichloroethane	-38	75	1.32	1.99	
	Hexachlorobutadiene	-22	210-220	1.68	2.12	
	Trichloroethylene (32%)	-73	86.7	1.46	1.74	
K022 (organic sludge)	Acetone	-94.3	56.2	0.792	7.36	45,000-100,000
	Phenol	42.5	182	1.07	7.78	
	Cumene	-96	152	0.862		
	PAH[f] (90%)					
K023 (organic liquid)	Phthalic anhydride (90)	131	285	1.53	5.29	950
	Maleic anhydride (10%)	53	200	0.934	3.40	
K024 (organic sludge)	Phthalic anhydride (50%)	131	285	1.53	5.29	2,300
	1,4-Naphthoquinone (50%)	123-126	Sublimes		6.97	

(continued)

TABLE 2-2 (continued)

Hazardous waste number (physical state)	Chemical compound[a] (% of total)	Melting point,[b] °C	Boiling point,[b] °C	Specific gravity[b]	Heat of combustion[c] kcal/g	Annual generation rate,[a,d] Mg
K025 (organic sludge)	m-Dinitrobenzene (94%)	89.9	303	1.55	4.15	170-200
	2,4 Dinitrotoluene (6%)	70.5	300	1.32	4.68	
	Nitrobenzene (trace)	5.7	211	1.20	6.01	
K027[g] (organic sludge)	Toluene diisocyanate (<3%)	19-22	251	1.22	5.92	6,000
	Toluene diamines (<3%)	99	280		8.24	
	Polymers and tarlike materials (90%)				5-9	
K030 (organic liquid)	1,1,2,2-Tetrachloroethane (23%)	-44	146	1.60	1.39	15,000
	Hexachlorobutadiene (34%)	-22	210-220	1.68	2.12	
	Hexachlorobenzene (20%)	228	326	2.04	1.79	
	1,1,1,2-Tetrachloroethane (6%)	68	130	1.55	1.39	
	Ethylene dichloride (4%)	-36	84	1.25	1.90	
	Hexachloroethane	186	186	2.09	0.46	
	Pentachloroethane (3%)	-29	162	1.68	0.53	
	Perchloroethylene (4%)	-22	121	1.63	1.19	
	1,1,2-Trichloroethane (4%)	-36.4	114	1.44	1.99	
K049[h] (sludge)	H_2O (40%)	0	100	1.00		2,000-10,000
	Oil (34-43%)			0.83-0.87	10-11	
	Phenol (14 ppm)	42.5	182	1.07	7.78	
	Sand, silt (12%)					
K093	Phthalic anhydride (>90%)	131	285	1.53	5.29	2,300
	Maleic anhydride (trace)	53	200	0.934	3.40	
K094 (organic sludge)	Phthalic anhydride (>90%)	131	285	1.53	5.29	410

[a] Ref. 1.

[b] Ref. 5.

[c] Ref. 2.

[d] Data is based on generation rate in 1979 and 1980.

[e] Quantities for F003, F004, and F005 are based on solvent usage and may be larger if including other applications. The generation rate is amount generated which is not an exact indication of the amount to be incinerated since some of these solvents are recovered and reused. Depending on the recovery process the still bottoms contain 5-50% of the original solvent. Most U.S. recovery processes result in an average of 25% solvent in still bottoms.

[f] Polycyclic aromatic hydrocarbons.

[g] Highly reactive. Also contains 6% ferric chloride.

[h] Also contains diatomaceous earth and trace metal contaminants, especially chromium and lead.

complicated process consisting of a series of decomposition, polymerization, and free radical reactions. The destructive processes of significance in incineration are pyrolysis (thermolytic cracking) and oxidation.[3] In general, at sufficiently high temperatures, the overall reaction rate can be conservatively described by first order kinetics.[4]

Because the processes under consideration are all above 760°C (1400°F), the hazardous wastes included in the study should go from a liquid phase to a gas phase soon after they enter the high-temperature processes. Based on the assumption that the hazardous waste is in the gas phase before any decomposition or oxidation occurs, the rate of disappearance of any organic hazardous waste constituent can be described by first order kinetics as:

$$\frac{dC_A}{dt} = -kC_A \qquad (1)$$

where C_A = the concentration of hazardous constituent A at time t

k = the reaction rate constant

The solution to Equation 1 can be expressed as:

$$\ln \frac{C_A}{C_{AO}} = -k\ \Delta t \qquad (2)$$

where C_{AO} = initial concentration of hazardous constituent A

Δt = time required to reach final concentration C_A

The temperature dependence of the reaction rate constant k can be correlated by the Arrhenius equation:

$$k = A \exp(-E/RT) \qquad (3)$$

where A = Arrhenius pre-exponent frequency factor (S^{-1})
E = Energy of activation (J/kg-mole)
R = Universal gas constant (8314 J/kg-mole-°K)
T = Absolute temperature (°K)

For 99.99 percent destruction of a hazardous waste constituent, the left-hand concentration term in Equation 2 is:

$$\ln \frac{C_A}{C_{AO}} = -9.21 \qquad (4)$$

Substituting Equation 4 for $\ln \frac{C_A}{C_{AO}}$ and Equation 3 for k in Equation 2, the time Δt required for 99.99 percent destruction of a compound can be represented as:

$$\ln \Delta t = \ln \left(\frac{9.21}{A}\right) + \frac{E}{RT} \qquad (5)$$

2.2.2 Time-Temperature Curves for Destruction of Selected Waste Categories

A literature search was conducted to obtain kinetic data (A and E values) for organic compounds associated with the selected hazardous waste categories. Table 2-3 presents the results. As the data column indicates, kinetic data were not available for many of the organic constituents in the wastes. Most of the available data were on pyrolytic destruction as opposed to oxidative destruction. The pyrolysis data represent either decomposition into stable molecules or production of free radicals by cleavage of a specific bond in the compound. In some cases where pyrolysis data were not available for a specific compound, values of A and E are presented for the limiting compound in the destruction of the particular complex (Table 2-4 indicates the limiting compound in the destruction of several organic complexes).

Several assumptions were involved in the choice of appropriate A and E values for the hazardous waste categories from the kinetic data for the organic waste constituents. A conservative approach was taken by representing the hazardous waste by A and E values corresponding to the worst-case compound (most difficult to destroy) in that category. Pyrolysis data instead of oxidation data were used whenever they were available because they are the most conservative.[3] Table 2-5 presents the relative

TABLE 2-3. KINETIC DATA FOR COMPOUNDS PRESENT IN CANDIDATE HAZARDOUS WASTES

Hazardous waste No.	Chemical compound	Kinetic parameters		Type of data[a]
		A (s^{-1})	E MJ/kg-mole (Btu/lb mole)	
F003	o-xylene	4.8×10^{12}	314 (1.35×10^5)	P
	m-xylene	4.2×10^{12}	325 (1.40×10^5)	P
	p-xylene	4.8×10^{12}	318 (1.37×10^5)	P
	Ethyl benzene	1.0×10^{13}	263 (1.13×10^5)	P
	Cyclohexanone	-[b]	- -[b]	
	Acetone	-	- -	
	Ethyl acetate	3.2×10^{12}	200 (8.60×10^4)	P
	Ethyl ether	-	- -	
	Methyl isobutyl ketone	-	- -	
	n-butyl alcohol	4.8×10^{14}	277 (1.19×10^5)	P
F004	Cresols	-	- -	
	Cresylic acid	-	- -	
	Nitrobenzene	2.1×10^{12}	325 (1.40×10^5)	P
F005	Toluene	2.1×10^{12}	325 (1.40×10^5)	P
	Methyl ethyl ketone	-	- -	
	Pyridine	-	- -	
	Carbon disulfide	-	- -	
	Isobutanol	4.8×10^{14}	277 (1.19×10^5)	P
K015	Benzyl chloride	-	- -	
	Trichlorobenzene	-	- -	
	Tetrachlorobenzene	-	- -	
	Pentachlorobenzene	-	- -	
	Hexachlorobenzene	1.9×10^{16}	304 (1.31×10^5)	O
	Chlorobenzene	-	- -	
K018	Ethyl chloride	4.0×10^{14}	284 (1.22×10^5)	P
	1,2-dichloroethane	1.0×10^{13}	293 (1.26×10^5)	P
	Hexachlorobenzene	1.9×10^{16}	304 (1.31×10^5)	O
	1,1,1-trichloroethane	3.2×10^{12}	200 (8.62×10^4)	P
	Hexachlorobutadiene	2.1×10^{12}	325 (1.40×10^5)	P
	Trichloroethylene	-	- -	

(continued)

TABLE 2-3 (continued)

Hazardous waste No.	Chemical compound	Kinetic parameters: A (s^{-1})	Kinetic parameters: E MJ/kg-mole	(Btu/lb mole)	Type of data[a]
K022	Acetone	-	-	-	
	Phenol	-	-	-	
	Cumene	3.2×10^{13}	242	(1.04×10^{5})	P
	Tars	-	-	-	
K023	Phthalic anhydride	-	-	-	
	Maleic anhydride	1.0×10^{13}	293	(1.26×10^{5})	P
K024	Phthalic anhydride	-	-	-	
	1,4-naphthaquinone	-	-	-	
K025	m-dinitrobenzene	-	-	-	
	2,4-dinitrotoluene	-	-	-	
	Nitrobenzene	2.1×10^{12}	325	(1.40×10^{5})	P
K027	Toluene diisocyanate	-	-	-	
	Toluene diamines	-	-	-	
	Polymer tarlike material	-	-	-	
K030	1,1,2,2-tetrachloroethane	1.0×10^{13}	293	(1.26×10^{5})	P
	Hexachlorobutadiene	2.1×10^{12}	325	(1.40×10^{5})	P
	Hexachlorobenzene	1.9×10^{16}	304	(1.31×10^{5})	O
	1,1,2,2-tetrachloroethane	-	-	-	
	Ethyl chloride	4.0×10^{14}	253	(1.09×10^{5})	P
	Trichloroethylene	-	-	-	
	1,1,2-trichloroethane	3.2×10^{12}	200	(8.62×10^{4})	P
	Pentachloroethane	1.0×10^{13}	293	(1.26×10^{5})	P
K049	Phenol	-	-	-	
	Oil	-	-	-	
K093	Phthalic anhydride	-	-	-	
	Maleic anhydride	1.0×10^{13}	293	(1.26×10^{5})	P
K094	Phthalic anhydride	-	-	-	

[a]P - Pyrolysis.
O - Oxidation.

[b]Blanks indicate no data were available.

TABLE 2-4. WORST CASE COMPOUND FOR DESTRUCTION BY STRUCTURAL CLASS[4]

Compound	Limiting compound	Reasoning[a]
Methyl chloride	Methane	UCT
Methylene chloride	Methane	UCT
Chloroform	Methane	Analogy
Carbon tetrachloride	Methane	Analogy
Carbon disulfude	Dichloroproprane	UCT
Trichloroethanes	Propane	UCT
Tetrachloroethanes	Propane	Analogy
Pentachloroethanes	Propane	Analogy
Hexachloroethanes	Propane	Analogy
Trichlorotrifluoroethane	Propane	Analogy
Vinylidene chloride	Vinyl chloride	UCT
Trichloroethylene	Propane	UCT
Perchloroethylene	Propane	Analogy
Bis(chloromethyl)ether	Dichloroethane	Analogy with bis(2-chloromethyl) ether
Chloropropane	Propane	TP
Dichloropropane	Propane	UCT
Trichloropropane	Dichloroethanes	UCT
Dichloropropanols	Propane	Analogy
Epichlorohydrin	Dichloroethane	UCT
Isobutanol	Propane	UCT
Chlorobutadiene	Toluene	Analogy
Hexachlorobutadiene	Toluene	UCT
Maleic anhydride	Dichloroethane	UCT
Bis(2-chloroethyl)ether	Dichloroethane	UCT
Methyl ethyl ketone	Dichloroethane	UCT
Methyl isobutyl ketone	Dichloroethane	UCT
Chlorobenzene	Hexachlorobenzene	UCT
Dichlorobenzene	Hexachlorobenzene	UCT
Trichlorobenzene	Hexachlorobenzene	UCT
Tetrachlorobenzene	Hexachlorobenzene	Analogy
Pentachlorobenzene	Hexachlorobenzene	Analogy
Pyridine	Methane	UCT

(continued)

TABLE 2-4 (continued)

Compound	Limiting compound	Reasoning[a]
Nitrobenzene	Toluene	UCT
Phenol	Methane	UCT
Chlorophenol	Methane	Analogy
Dichlorophenol	Methane	Analogy
Trichlorophenol	Methane	Analogy
Tetrachlorophenol	Methane	Analogy
Pentachlorophenol	Methane	Analogy
Nitrophenols	Methane	Analogy
Benzotrichloride	Dichloroethane	UCT
Cresols	Benzene	UCT
Cresylic acid	Benzene	Analogy
Nitrocresols	Benzene	Analogy
Dinitrocresols	Benzene	Analogy
Aceteophenone	Toluene	UCT
Phthalic anhydride	Propane	UCT
Dimethyl phenols	Methane	Analogy
Cumyl phenol	Methane	Analogy
Naphthalene	Propane	UCT
Naphthoquinone	Propane	Analogy

[a]UCT - Union Carbide predicted temperature.
TP - Thermal pyrolysis data.
Analogy - Structural correlation and speculation.

TABLE 2-5. ORDER OF INCREASING DESTRUCTIBILITY BASED ON TEMPERATURE REQUIRED FOR 99.99 PERCENT DESTRUCTION AT 1-SECOND RESIDENCE TIME[3]

Order	Compound	Temperature[a] °C	(°F)
1	Methyl chloride	844	(1550)
2	Methane	836	(1535)
3	Phenol	807	(1484)
4	Methylene chloride	797	(1466)
5	Pyridine	791	(1455)
6	Chlorobenzene	768	(1413)
7	Dichlorobenzene	765	(1408)
8	Hexachlorobenzene	763	(1405)
9	Ethane	761	(1401)
10	Vinyl chloride	746	(1374)
11	Ethyl chloride	737	(1358)
12	Benzene	737	(1358)
13	Cresol	722	(1331)
14	Ethylene	721	(1329)
15	Toluene	720	(1327)
16	Nitrobenzene	720	(1327)
17	Hexachlorobutadiene	719	(1325)
18	Trichlorobenzene	716	(1320)
19	Vinylidene chloride	712	(1313)
20	Acetophenol	711	(1310)
21	Propane	708	(1305)
22	1,2,2-trichloro-1,1,2-trifluoroethane	692	(1277)
23	Dichloropropane	686	(1266)
24	Trichloroethane	683	(1260)
25	Phthalic anhydride	672	(1240)
26	Isobutanol	666	(1229)
27	Trichloroethylene	662	(1222)
28	Naphthalene	661	(1221)
29	Methyl ethyl ketone	657	(1213)
30	Dichloroethane	656	(1212)
31	Epichlorohydrin	618	(1143)
32	Maleic anhydride	607	(1123)
33	Bis(2-chloroethyl) ether	602	(1115)
34	Methyl isobutyl ketone	599	(1109)
35	Trichloropropane	571	(1060)
36	Benzotrichloride	528	(982)
37	Carbon disulfide	400	(751)

[a]Temperature required for 99.99 percent destruction at a residence time of 1 second.

destructibilities of various organic compounds based on their autoignition temperatures. This listing was used to determine the worst-case compound for several hazardous waste categories. If the compounds did not appear on this table, heat of combustion was used to determine the worst-case compound, i.e., the lower the heat of combustion, the more difficult the combustion of the compound is believed to be.

In cases where kinetic data were not available for a compound, an arbitrary conservative A value of 1.0×10^{10} s^{-1} and E value of 293 MJ/kg-mole (1.26×10^{5} Btu/lb-mole) were assigned.[3] It should be noted that low A values ($<1.0 \times 10^{10}$) and high E values ($>1.26 \times 10^{5}$) do not usually occur with the same compound.

Table 2-6 presents the composite kinetic parameters for the hazardous waste categories. Based on the relationship derived for estimating time-temperature requirements for 99.99 percent destruction (Equation 5, Section 2.2.1), plots of time-temperature relationships were constructed for the selected hazardous waste categories. Figure 2-1 depicts the predicted linear gas phase time-temperature plots. Because the kinetic behavior of the waste categories is uncertain at times less than 0.1 second, dotted lines are used to extend the time-temperature relationships below 0.1 second.

2.3 SUMMARY OF WASTE CHARACTERISTICS

Table 2-7 summarizes the reported characteristics of each selected waste. The approximate waste heating values were obtained from the literature.[2,6] When specific data were not available, estimates were based on an approximate waste density and an assumed average waste heating value of 33.5 MJ/liter (120,000 Btu/gal).[3] The kinetic data were obtained from tables in the preceding subsection, and the remaining data were taken from Table 2-2.

TABLE 2-6. COMPOSITE KINETIC PARAMETERS FOR HAZARDOUS WASTE CATEGORIES

		Kinetic parameters	
Hazardous waste No.	Worst case compound	A, s^{-1}	E, MJ/kg-mol (Btu/lb-mol)
F003	Xylene	4.2×10^{12}	325 (1.40×10^{5})
F004	Cresols	1.0×10^{10}[a]	293 (1.26×10^{5}[a])
F005	Pyridine	1.0×10^{10}[a]	293 (1.26×10^{5}[a])
K015	Hexachlorobenzene	1.9×10^{16}	304 (1.31×10^{5})
K018	Hexachlorobutadiene	2.1×10^{12}	325 (1.40×10^{5})
K022	Phenol	1.0×10^{10}[a]	293 (1.26×10^{5}[a])
K023	Maleic anhydride	1.0×10^{13}	293 (1.26×10^{5})
K024	Phthalic anhydride	1.0×10^{10}[a]	293 (1.26×10^{5}[a])
K025	Nitrobenzene	2.1×10^{12}	325 (1.40×10^{5})
K027	Toluene diisocyanate	1.0×10^{10}[a]	293 (1.26×10^{5}[a])
K030	Hexachlorobutadiene	2.1×10^{12}	325 (1.40×10^{5})
K049	Phenol	1.0×10^{10}[a]	293 (1.26×10^{5}[a])
K093	Maleic anhydride	1.0×10^{13}	293 (1.26×10^{5})
K094	Phthalic anhydride	1.0×10^{10}[a]	293 (1.26×10^{5}[a])

[a]Conservative values were assigned where kinetic data were not available.

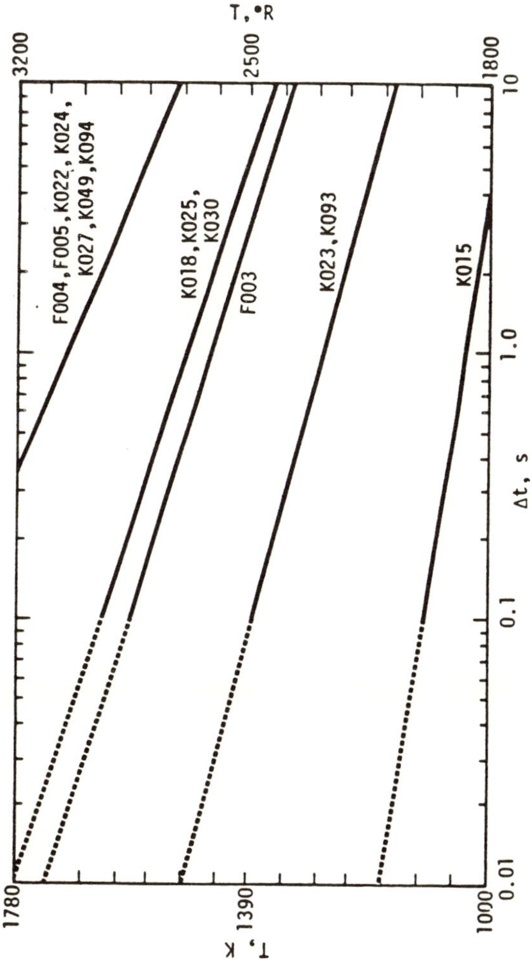

Figure 2-1. Plots of time - temperature requirements for 99.99 percent destruction of selected hazardous waste categories.

TABLE 2-7. SUMMARY OF DATA NEEDED TO EVALUATE THERMAL DESTRUCTIBILITY OF SELECTED HAZARDOUS WASTES

Waste	Approximate heat value,[a] kcal/g	Most conservative compound for calculating kinetic data[b]	A,[c,d] sec^{-1}	E,[c,d] kcal/mol	Physical state[e]	Predominant organic compound,[e] % of total	Geographic location[f]
F003	8.3	Xylene	4.2×10^{12}	77.5	Organic liquid and sludge	Nonhalogenated solvent (25%)	Widespread
F004	6.7	Cresols	(1.0×10^{10})[g]	(70)[g]	Organic liquid and sludge	Nonhalogenated solvent (25%)	Widespread
F005	6.0	Pyridine	(1.0×10^{10})[g]	(70)[g]	Organic liquid and sludge	Nonhalogenated solvent (25%)	Widespread
K015	4.2	Benzyl chloride	1.9×10^{16}	72.6	Organic sludge	Benzyl chloride (2-10%)	
K018	1.8	Hexachlorobutadiene	2.1×10^{12}	77.5	Organic liquid	Heavy chlorinated organics (43%) (hexachlorobenzene, etc.)	Tex., La., N.J., Calif.
K022	7.2	Phenol	(1.0×10^{10})[g]	(70)[g]	Organic sludge	Tars (90%)	Widespread
K023	5.2	Maleic anhydride	1.0×10^{13}	70	Organic liquid	Maleic anhydride (90%)	Pa., Ill., N.J., Calif., La., Tex.
K024	5.2	Phthalic anhydride	(1.0×10^{10})[g]	(70)[g]	Organic sludge	Phthalic anhydride (<50%)	Pa., Ill., N.J., Calif., La., Tex.
K025	6.0	Nitrobenzene	2.1×10^{12}	77.5	Organic sludge	m-Dinitro Benzene (94%)	Miss., Tex., N.J., W.Va., La.
K027	6.0	Toluene diisocyanate	(1.0×10^{10})[g]	(70)[g]	Organic sludge	Polymers and Tars (90%)	Tex., La., Ohio, W.Va.
K030	2.2	Hexachlorobutadiene	2.1×10^{12}	77.5	Organic liquid	Hexachlorobutadiene (34%)	La., Tex., Ky., Kans., Calif.
K049	3.3	Oil	(1.0×10^{10})[g]	(70)[g]	Aqueous sludge	H_2O (40%); Oil (34-43%)	Widespread
K093	5.2	Maleic anhydride	1.0×10^{13}	70	Organic liquid	Phthalic anhydride (<100%)	Ill., Tex., N.J., Puerto Rico, Calif.
K094	5.2	Phthalic anhydride	(1.0×10^{10})[g]	(70)[g]	Organic sludge	Phthalic anhydride (<100%)	Ill., Tex., N.J., Puerto Rico, Calif.

[a] Ref. 2, 3, 6.

[b] Table 3-6.

[c] Table 3-6.

[d] A = Arrhenius pre-exponent frequency factor; E = energy of activation for bond rupture.

[e] Ref. 1.

[f] Ref. 1, 2.

[g] Parenthesis indicate no A, E data were available and worst case conditions were used.

3. Hazardous Waste Destruction Model for Identifying Waste Categories Destructible in Industrial Processes

The destruction of hazardous wastes in an industrial process is primarily dependent on the temperature and residence time profiles of the waste combustion products in the high-temperature furnace. Hence, a knowledge of the temperature-residence time relationship throughout the furnace is essential for estimating the destruction potential of any hazardous waste. A comparison of the time-temperature characteristics of the wastes (developed in Section 2) and the process temperature-residence time profiles should identify the potentially destructible (99.99 percent) hazardous waste.

3.1 PROCESS TEMPERATURE-RESIDENCE TIME PROFILES

3.1.1 Bulk Residence Time Calculations

The computation of bulk residence times for the various processes requires that a flow model be postulated for the gases in the furnace. The plug flow model modified by a degree of mixing serves as a good approximation. This model assumes that there is no longitudinal mixing of the gaseous combustion products along the flow path.[7] Consequently, the temperature, composition, and velocity of the gases are uniform at any chosen cross section of the furnace.

For an isothermal plug flow furnace, the mean residence time (t) is given by the furnace volume (V) divided by the actual gas volumetric flow rate (Q'):

$$t = \frac{V}{Q'} \cdot 60 \qquad (1)$$

The ideal gas law should closely describe the temperature and volume relationship of the gases in the furnace since process operating pressures are low (close to atmospheric) and temperatures are high (significantly higher than the boiling point of the compounds). Any slight error introduced by use of the ideal gas law should be small in relation to the distribution of residence times. Using the ideal gas law, the actual volumetric flow rate (Q') can be computed from the standard volumetric flow rate (Q_s) by correcting for temperature:

$$Q' = Q_s \cdot \frac{T}{273} \tag{2}$$

where T is in °K

Combining (1) and (2), the mean residence time of an isothermal furnace can be expressed as:

$$t = \frac{16,400\ V}{Q_s T} \tag{3}$$

where t = mean residence time, s

V = furnace volume, m^3

Q_s = standard volumetric flow rate, Nm^3/min (273°K or 0°C)

T = temperature, K

In reality, however, some longitudinal mixing and gas recirculation often occur in the furnaces and result in deviations from total plug flow behavior. Thus, a portion of the gas may follow a faster path and exit the furnace at a time less than t. Experimental measurements have shown that the minimum residence time is approximately 0.5 t.[3] This time is defined as the fast-path residence time ($\bar{t}$).

3.1.2 Incorporation of Temperature-Residence Time Profile

The temperature-time relationship developed in Section 3.1.1 is applicable for isothermal processes only. In an actual furnace, the gas exits at a temperature that is lower than the maximum reaction zone temperature, i.e., the furnace operation is not isothermal. Hence, it is necessary to incorporate the temperature-axial distance profile of the furnace to determine the temperature-residence time relationship for various temperatures in the furnace.

A differential analysis appears to be suitable.[8] The mean residence time dt in an element of volume dV can be expressed as:

$$dt = \frac{dV}{Q'} \tag{4}$$

Equation 4 is the appropriate differential form of Equation 1.

Further, if A is the cross-sectional area of the furnace (assumed constant), and dx represents an element of axial length in the furnace,

$$dV = A\ dx \tag{5}$$

Combining Equations 4 and 5,

$$dt = \frac{A\ dx}{Q'} \tag{6}$$

Noting that Q' is a function of temperature as given by Equation 2,

$$dt = \frac{16{,}400\ A\ dx}{Q_s \cdot T} \tag{7}$$

The temperature T in the above equation is a function of axial distance. Because detailed temperature-axial distance profiles are rarely available for furnaces used by industrial processes, an approximate solution to Equation 7 can be obtained by assuming a linear gas temperature-axial distance profile between the point of maximum temperature (T_m at distance x_m) and the furnace exit (T_e at distance x_e). The linear profile can be mathematically expressed as:

$$T = \frac{(T_m - T_e)x + T_e x_m - T_m x_e}{(x_m - x_e)} \tag{8}$$

Substitution of Equation 8 for T in Equation 7 and subsequent integration gives

$$(t - t_m) = \frac{16{,}400\ A}{Q_s} \cdot \frac{(x_m - x_e)}{T_m - T_e} \ln \frac{T}{T_m} \tag{9}$$

If we further assume that the maximum temperature (T_m) in the furnace occurs very near the entrance to the furnace, i.e., $x_m << x_e$, and noting that Ax_e is the furnace volume, V,

$$(t - t_m) = \frac{16,400\ V}{Q_s\ (T_e - T_m)} \ln \frac{T}{T_m} \tag{10}$$

where $t - t_m$ is in seconds

V is in m^3

Q is in Nm^3/min

T, T_e, T_m are in K

Equation 10 is a simplified expression for the mean residence time of the gas above a given temperature, T, in the furnace. According to Equation 10, the mean residence time of the gas above the furnace exit temperature T_e is:

$$(t_e - t_m) = \frac{16,400\ V}{Q\ (T_e - T_m)} \ln \frac{T_e}{T_m} \tag{11}$$

The fast-path residence time $(\bar{t} - \bar{t}_m)$ of the gas above a certain temperature T can be calculated from the mean residence time determined from Equation 10, according to the following equation:

$$(\bar{t} - \bar{t}_m) = 0.5\ (t - t_m) \tag{12}$$

Use of Equation 10 requires a knowledge of the following parameters for the industrial process/furnace:

- Furnace volume
- Volumetric flow rate of the gases
- Maximum temperature attained in the reaction zone
- Exit temperature of gas from the furnace

An extensive search was conducted to obtain representative data for the parameters for each of the candidate industrial processes. When data on volumetric flow of furnace exit gases were not available, inlet flow rates were used to compute Q, based on the assumption that no major reactions occur in the furnace that might drastically change the volumetric flow rate of the gases.

After the necessary data were obtained, process curves were plotted for each of the candidate industrial processes. These curves comprise plots of $(t - t_m)$ vs. T and $(\bar{t} - \bar{t}_m)$ vs. T, respectively called the mean-path and fast-path process curves.

3.2 PREDICTIONS OF DESTRUCTIBLE HAZARDOUS WASTE CATEGORIES

A prediction of destructible hazardous waste categories can now be made by overlaying the temperature-residence time process curves for each candidate process on the hazardous waste temperature-time plots developed earlier in Section 2. It is assumed that all the hazardous wastes are in the gaseous state by the time (t_m) the maximum temperature (T_m) in the furnace is reached. This assumption has been shown to be valid for the types of hazardous wastes being considered in this study.[3]

The approach towards predicting destructible hazardous wastes in the industrial process is composed of 1) an initial conservative estimate, and 2) an extended conservative estimate.

3.2.1 Initial Conservative Estimate

Essentially, the initial conservative estimate of destructible hazardous waste categories considers only the lowest temperature profile of the process, i.e., T_e corresponding to residence time above T_e. Equation 11 $(t_e - t_m)$ allows computation of the mean residence time above the exit temperature. The fast-path residence time above the exit temperature can then be calculated by use of equation 12 $(\bar{t}_e - \bar{t}_m)$. A horizontal line connecting the points $(t_e - t_m, T_e)$ and $(\bar{t}e - \bar{t}_m, T_e)$ can be drawn on the process curve plots for each process considered. This line is defined as the most conservative line. By overlaying this line on the hazardous waste time-temperature plots, one can make an initial conservative estimate of destructible wastes according to the following guidelines:

- If the hazardous waste plot lies completely below the most conservative line, 99.99 percent destruction of the waste will be achieved.
- If the hazardous waste plot lies completely above the most conservative line, 99.99 percent destruction of the waste will not be reached.
- If the hazardous waste plot lies only partially above or below the most conservative line, 99.99 percent destruction of the waste may only be marginally achieved.

An initial conservative estimate of destructible wastes is not evaluated for industrial processes with a furnace exit gas temperature below 1800°R (the lowest temperature on the hazardous waste plots).

3.2.2 Extended Conservative Estimate

In the industrial processes, the hazardous wastes will be subjected to temperatures above the furnace exit gas temperature, and the rate of destruction will be greater at these higher temperatures. The extended conservative estimate of incinerable hazardous waste categories is developed to account for these higher destruction rates.

The approach to the extended conservative estimate consists of comparing the process curves (plots of $t - t_m$ vs. T and $\bar{t} - \bar{t}_m$ vs. T) with the temperature and times required for thermal destruction of the hazardous wastes. An overlay of the process curves for a candidate industrial process on the hazardous waste plots results in an extended estimate of destructible hazardous waste categories, based on the same guidelines stipulated in Section 3.2.1, but substituting process curves for most conservative line.

Both the initial and the extended conservative estimates can be used to predict those hazardous waste categories that are potentially destructible. This prediction reduces the number of waste categories for which a more extensive evaluation must be undertaken.

3.2.3 Illustrative Example

The following example illustrates the approaches for determining destructible hazardous waste categories in an industrial process. The oil furnace process in the manufacture of carbon black is the candidate process selected.

3.2.3.1 Process Temperature-Time Profiles--

Process parameters:

Furnace volume (V)	= 9.63 m^3 (340 ft^3)
Volumetric flow rate of gases (Q)	= 105 Nm^3/min (4000 scfm)

Note: Nm^3/min is based on 0°C or 273°K whereas scfm is based on 70°F or 530°R.

Furnace temperatures:

Maximum reaction zone temperature (T_m)	= 1670 K (3010°R)
Exit gas temperature (T_e)	= 1030 K (1860°R)

Temperature-residence time profiles--Equation 10 provides a way to calculate the mean residence time above any temperature (T) in the furnace, whereas Equation 12 relates the fast-path residence time to the mean residence time. Table 3-1 presents calculated mean and fast-path residence times above a given furnace temperature.

Figure 3-1 shows plots of the mean path and fast-path process curves, i.e., Δt vs. T and $\Delta\bar{t}$ vs. T, overlaid on the hazardous waste time-temperature plots. It also shows the most conservative line connecting the mean-path and fast-path residence times corresponding to exit gas temperature (T_e).

Predictions of destructible hazardous waste categories--By use of the guidelines presented earlier in this section, predictions regarding the potential of the oil furnace process to achieve 99.99 percent destruction of hazardous wastes can be made, based on Figure 3-1. Table 3-2 presents the results. Waste type F003 is shown as a marginal case because it appears to be destructible only under mean-path conditions.

TABLE 3-1. TIME-TEMPERATURE CALCULATIONS FOR THE OIL FURNACE PROCESS IN CARBON BLACK MANUFACTURE

Temperature (T), K (°R)	Mean residence time above T $\Delta t = (t - t_m)$, s	Fast-path residence time above T $\Delta\bar{t} = (\bar{t} - \bar{t}_m)$, s
1670 (3010) (T_m)		
1610 (2900)	0.088	0.044
1560 (2800)	0.17	0.085
1450 (2600)	0.34	0.17
1330 (2400)	0.53	0.27
1220 (2200)	0.74	0.37
1110 (2000)	0.96	0.48
1030 (1860) (T_e)	1.13	0.56

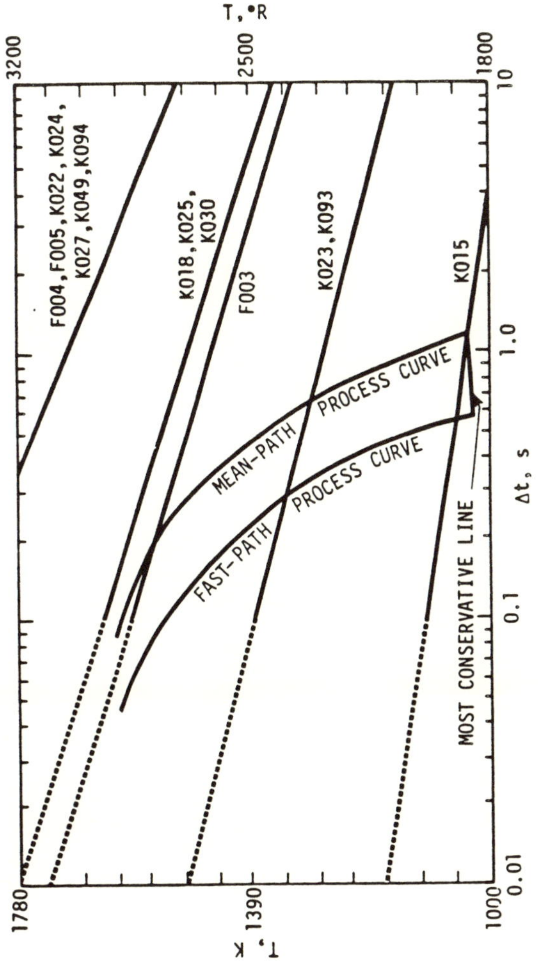

Figure 3-1. Hazardous waste destructibility predictions for the oil furnace process in carbon black manufacture.

TABLE 3-2. FEASIBILITY OF OIL FURNACE PROCESS (CARBON BLACK) FOR THERMAL DESTRUCTION OF HAZARDOUS WASTES

Type of conservative estimate	Hazardous waste type													
	F003	F004	F005	K015	K018	K022	K023	K024	K025	K027	K030	K049	K093	K094
Initial	No	No	No	No	No	No	No	No	No	No	No	No	No	No
Extended	Maybe	No	No	Yes	No	No	Yes	No	No	No	No	No	Yes	No

4. Process Assessment

All of the major industrial processes in the United States were reviewed, from both a technical and availability standpoint, for selection of the most promising candidates for hazardous waste destruction.

The initial listing of approximately 100 processes was based on operating temperature and included any major process with an operating temperature of greater than 649°C (1200°F). (See Appendix C.) It includes processes from the metallurgical, chemical, and mineral industries and sludge and waste incinerators. Each of these industries was then screened against the criteria listed in Table 4-1 to eliminate from further consideration those industries that show little potential for effective hazardous waste disposal.

TABLE 4-1. SCREENING CRITERIA FOR POTENTIAL HIGH-TEMPERATURE PROCESSES FOR THE DESTRUCTION OF HAZARDOUS WASTE

- Compatibility of process
  - Thermal destruction conditions: temperature and residence time (turbulence not considered)
  - Product quality
  - Potential for fugitive emissions
- Number of facilities
- Geographic matching of industry location and waste generators

Gas residence time within the process during high-temperature conditions was an important consideration; the lower the process temperature, the longer is the residence time required for 99.99 percent thermal destruction. Turbulence, the other condition necessary for combustion, was not considered because information on this parameter is generally sparse. Residence time and product quality considerations eliminated organic and miscellaneous chemical manufacturing processes from consideration because high temperatures are often attained in the reactors, but residence times are quite short.

Product quality, both actual and perceived, was another important consideration. Destroying hazardous waste in a steel furnace, for example, could increase trace elements in the steel and affect its physical properties, and in all cases, hazardous waste disposal was considered secondary to product quality. This criterion eliminated processes such as basic oxygen and electric arc furnaces in the steel industry because these processes produce a high-quality steel that cannot tolerate residues from hazardous waste destruction. This criterion also affected chemical manufacturing, food processing (human and animal consumption), fertilizer manufacturing, and petroleum refining processes.

The more potential a process has for fugitive emissions, the less likely it is to be suitable for hazardous waste disposal. Raw materials handling procedures also affect the potential for a process to generate fugitive emissions and its suitability for hazardous waste disposal. In a sintering process at an iron and steel mill, for example, open conveyors are used to transport raw materials to the pug mill (mixing device) and to the sinter strand, which is also open. The major modifications to the materials-handling system that would be required to correct this high potential for fugitive emissions and the short residence time at the maximum temperature of about 1210°C (2200°F) resulted in the elimination of the iron and steel sintering process from further consideration. The fugitive emissions potential and short residence times, coupled with their location in highly populated areas, also eliminated municipal solid waste incinerators from the list of possible processes.

The number of facilities available in areas of high hazardous waste generation was another major consideration. For example, specialty processes that are located in only a few widely scattered places or developing processes that are currently operational at only a few sites were not considered. The processes selected for detailed analysis were those with the potential for disposing of significant quantities of hazardous wastes.

Table 4-2 lists those industries and processes that remained after the screening criteria had been applied. These processes were evaluated in more detail by use of the hazardous waste destruction model discussed in Section 3. The process assessments are presented alphabetically by industry in Subsections 4.1 through 4.10.

4.1 BRICK AND TILE INDUSTRY

The brick and tile industry includes those establishments engaged primarily in the manufacture of common brick, refractory brick, and structural clay tiles. The manufacturing steps involved are blending of the raw material, grinding, screening, forming, and in some cases, drying. The final step consists of heating the ware (brick or tile) in either a tunnel kiln or a periodic kiln. This process is also known as firing or burning of the ware. Figure 4.1-1 is a generalized flow diagram of a brick and tile plant. Several changes occur during the firing process, including the liberation of both free and combined water, the combustion of organic matter, the decomposition of impurities, and the bonding of solid clay particles.[9] Table 4.1-1 presents common clay impurities and their respective breakdown temperatures. Corresponding changes in physical properties accompany the alterations in the composition and

TABLE 4-2. INDUSTRIES AND PROCESSES SELECTED FOR FURTHER HAZARDOUS WASTE DESTRUCTION EVALUATION

Industry	Process or furnace	Exit and maximum temperatures, °C	Residence time (seconds, unless otherwise noted)[a]
Brick	Tunnel Kiln	260-1180	4.3
Carbon black	Oil furnace process	870-1400	1.1
Primary copper	Reverberatory furnace	1300-1400	2.2
Primary lead	Blast furnace	700-1200	5.9
Primary zinc	Multiple-hearth roaster	200-980	>10
	Suspension roaster	930-1010	>10
	Fluidized-bed roaster	950-1740	>10
Iron and steel	Blast furnace	1100[b]-1870	1.1[c]
	Open hearth furnace	1200-1800	2.0
Lime	Long rotary kiln	680-1900	8.3
	Short rotary kiln with stone preheater	1160-1840	7.6
Lightweight aggregate	Rotary kiln		
Glass	Melting furnace	620-1480	4.1
Sewage sludge	Fluid-bed furnace	760-870	1.4
	Multiple-hearth furnace	480-980	0.5
Hazardous waste incineration[d]	Rotary kiln	820-1600	2.0 s to 2 h[e]
	Liquid injection	700-1650	0.5 to 2

[a]Residence time above the greater of the exit temperature or 760°C.

[b]Represents estimated temperature at the top of the combustion zone in the blast furnace.

[c]Residence time has been calculated using 33 percent of the total furnace volume.

[d]Included for comparison only.

[e]The residence time for solids and sludges can be adjusted.

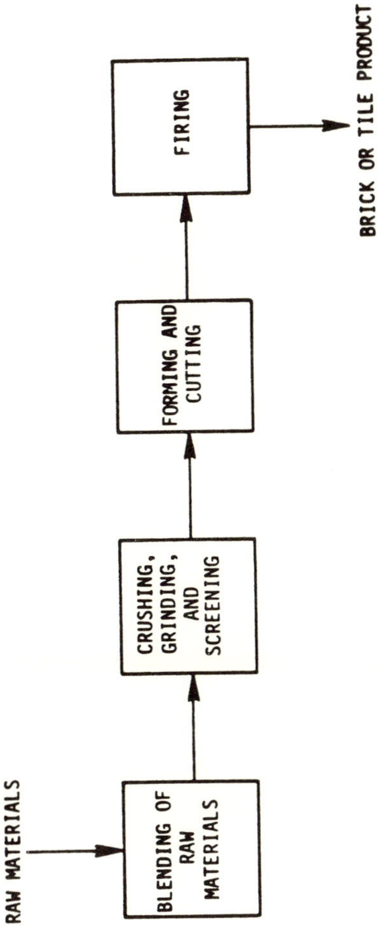

Figure 4.1-1. Generalized flow diagram of a brick and tile plant.

TABLE 4.1-1. BREAKDOWN TEMPERATURES OF CLAY IMPURITIES[9]

	Temperature, °C	°F
$FeS_2 + O_2 \rightarrow FeS + SO_2$	350-450	660-840
$4FeS + 7O_2 \rightarrow 2Fe_2O_3 + 4SO_2$	500-800	930-1470
$Fe_2(SO_4)_3 \rightarrow Fe_2O_3 + 3SO_3$	560-775	1040-1430
$C + O_2 \rightarrow CO_2$	350	660
$S + O_2 \rightarrow SO_2$	250-920	480-1690
$CaCO_3 \rightarrow CaO + CO_2$	600-1050	1110-1920
$MgCO_3 \rightarrow MgO + CO_2$	400-900	750-1650
$FeCO_3 + 3O_2 \rightarrow 2Fe_2O_3 + 4CO_2$	800	1470
$CaSO_4 \rightarrow CaO + SO_3$	1250-1300	2280-2370

mineralogical constitution of the clay product. Most notable are true specific gravity, porosity, volume, strength, and hardness.[10]

4.1.1 Process Descriptions

4.1.1.1 Tunnel Kilns--

Tunnel kilns are used in most modern brick and tile plants. Figure 4.1-2 presents a typical configuration. The product is loaded onto a car and charged into the left end of the kiln. As the car progresses along the length of the kiln, it passes through zones of increasing temperature. In most cases, the kiln has two sets of burners. The main (or upper) fuel-fired burners are located along the middle section of the kiln and provide the bulk of heat necessary for firing.[11] Secondary (or lower) burners are located along the length of the kiln between the preheating zone and the firing zones. Air is passed through the kiln countercurrently to the movement of the ware.[11] The exact amount of excess air varies with the desired product specifications.[10] As the ware leaves the combustion zone, cold air forced in from the right end of the kiln passes over the charge cooling it by countercurrent heat exchange. Some of this air is withdrawn for use as preheated primary combustion air; the remaining air continues to the left, enters the combustion zone, mixes with the combustion gases, and then passes over the incoming charge, where it loses much of its heat. Air locks located at both ends of the kiln prohibit flow conditions from being disturbed by the entrance or exit of cars.[11]

Processing is continuous, and the average tunnel kiln produces 100,000 bricks per day.[12] The inside dimensions of a typical kiln are 98 m (321 feet) long by 3.4 m (11.2 feet) wide by 1.5 m (4.9 feet) high.[12] Historically, tunnel kilns were fired with natural gas; however, oil, coal, or sawdust are often used today. These kilns generally consume approximately 2326 to 2791 kJ/kg (1000 to 1200 Btu/lb) of brick; newer kilns are

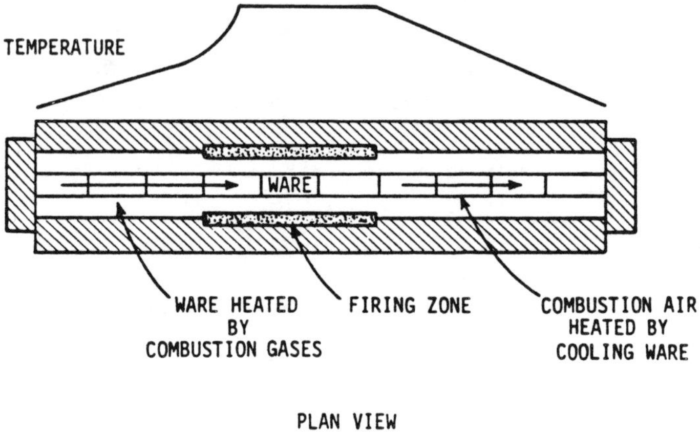

Figure 4.1-2. Diagram of a tunnel kiln.[11]

designed to consume 1860 kJ/kg (800 Btu/lb). An average brick weighs 2.2 kg (5 lb).[12]

4.1.1.2 Periodic Kilns--

Because of its high energy and labor requirements, the periodic (or beehive) kiln is in the process of being phased out.[12] The downdraft periodic kiln is a permanent brick structure with several fireholes through which fuel is fired.[13] The hot combustion gases are drawn over and through the bricks by underground flues. A 2-week cycle is required to fire, cool, and discharge the product.[10] Fuel is burned approximately 50 percent of the time. Although fuel efficiency is not as high as in a tunnel kiln, the uniform temperature distribution throughout the periodic kiln leads to a high-quality product. In most cases it is used for small-scale production.[11]

4.1.2 Process Assessment for Hazardous Destruction

Because of its wide application, relatively long residence time, and moderately high-temperature range, the tunnel kiln appears to offer an alternative for disposing of hazardous wastes. The periodic kiln, on the other hand, is in the process of being phased out and therefore would not provide a long-term solution to this problem.

4.1.2.1 Technical Factors--

Compatibility of hazardous wastes with process fuels--Most new tunnel kilns are being designed to burn more than one fuel type. This dual fuel design would be advantageous if hazardous wastes were to be introduced. Kilns designed to fire only solid or gaseous fuels would have to be retrofitted to accept a liquid waste. Installation of a liquid burner/atomizer system with the appropriate pump and piping would suffice. The waste burners should be located along the upper row near the burner firing the main fuel. Appendix A provides a more detailed discussion on the requirements for equipment and handling procedures necessary to fire hazardous wastes.

Graphic determination of suitable waste categories--The bulk residence time (8.0 seconds, based on a maximum temperature) and temperature range [260° to 1180°C (500° to 2150°F)] of a typical tunnel kiln indicate that destruction of hazardous wastes in this furnace is feasible. A typical brick and tile tunnel kiln processes 100,000 bricks or an equivalent tonnage per day.[12] Inasmuch as burners are only located between the preheating and firing zones, a "working" furnace volume had to be calculated. This working volume more accurately reflects the area where hazardous waste destruction would take place. Based on a typical temperature profile within the kiln, an average working volume of 335 m^3 (11,833 ft^3) was calculated.[12] In addition, the kiln has an average volumetric flow rate of 473 Nm^3/min (18,000 scfm).[12]

Figure 4.1-3 plots the change in residence time for mean (Δt), and fast ($\Delta \bar{t}$) path process curves of a tunnel kiln. These plots are overlaid on the hazardous waste curves to determine which hazardous waste categories are destructible, given the kiln's time-temperature relationship. Based on data from this graph, Table 4.1-2 provides predictions regarding the ability of a typical tunnel kiln to destroy (99.99%) the selected hazardous wastes. Because the exit gas temperature falls below the lowest temperature in the graph (Figure 4.1-2), the conservative process line is not shown. Based on the extended case, it appears that K015, K023, and K093 could be 99.99 percent destroyed.

TABLE 4.1-2. APPLICABILITY OF THE BRICK AND TILE TUNNEL KILN FOR THERMAL DESTRUCTION OF HAZARDOUS WASTES

K015	K023, K093	F003	K018, K025 K030	F004, F005, K022, K024 K027, K049, K094
Yes	Yes	No	No	No

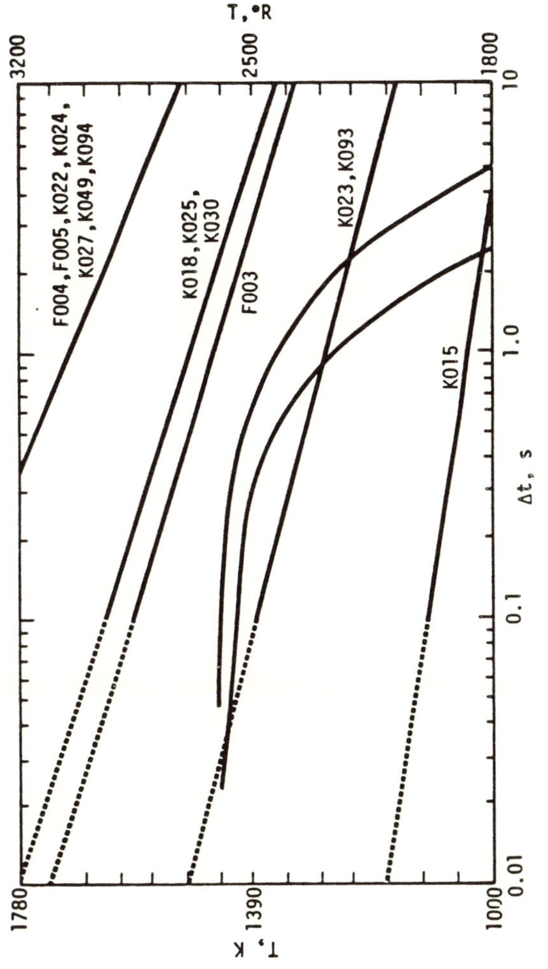

Figure 4.1-3. Hazardous waste destructibility for the brick and tile tunnel kiln.

Effects on product quality and product application--Almost all (97 percent) of the output from the brick and tile industry is purchased for construction.[14] Because of the bulk and weight of brick and tile products, markets tend to be regional in character. Intra-product competition is based as much on specialty features, such as style, finishes, and color, as on price. Any effect on these features resulting from the destruction of hazardous waste would be a major detriment.

Studies are being conducted on the impact of substituting municipal and industrial sludges for a portion of the clay prior to firing the bricks. The sludge burns during firing and releases heat; this leaves tiny bubbles in the bricks which may enhance their insulating capability. Though not as strong as clay bricks, these bricks made with sludge reportedly meet the standards of strength set by the American Society for Testing and Materials.[15] Whereas studies indicate that destroying hazardous organic wastes in the tunnel kiln might have some impact on the brick and tile, it may not seriously degrade product quality.[15]

Changes in the kiln's atmosphere resulting from the incineration of hazardous organic wastes could discolor the brick and tile. Color is dependent on the atmosphere of the kiln (oxidizing and reducing), kiln temperature, and the impurities within the raw materials. The pink color often associated with brick used for residential construction comes from iron impurities.[10] Normally, the iron impurities present in the clay (whether oxides, carbonates, sulfides, or sulfates) are oxidized to ferric oxide (Fe_2O_3), which generally gives rise to various shades of yellow or red.[10] Occasionally the atmosphere within the kiln may not be oxidizing, however, in which case lower oxides such as magnetite (Fe_3O_4) or ferrous oxide (FeO) may be formed and give rise to black or blue-black colors. The introduction and subsequent oxidation of hazardous wastes could alter the atmosphere of the kiln in the immediate area of the ware, and a lack of excess oxygen or a reducing atmosphere may cause this darker color. Although this problem could be solved by providing additional excess air, the threat of potential product discoloration might inhibit the use of kilns to destroy hazardous wastes.

Impact on process hardware--The presence of trace quantities of metals and the formation of acids (such as HCl) could cause duct corrosion and possibly interfere with the operation of air pollution control equipment. Inasmuch as the addition of an appropriate pollution control device system will be necessary at brick and tile plants which decide to use waste fuels, the selection of corrosion-resistant materials of construction could prevent this problem.

Unlike lime or cement kilns, the tunnel kiln does not develop a protective coating; therefore, it is more susceptible to abrasion and thermal shock, and the addition of abrasive or corrosive wastes could increase normal wear and tear on the refractory.

Acceptable amounts of hazardous waste feed--Based on nominal operating parameters and the assumption that 10 percent of the required heat input can be supplied by the use of hazardous waste fuels without affecting equipment or product quality, the heat input of hazardous wastes can be estimated as follows:

$$0.10 \times 2560 \text{ kJ/kg of brick} \times 2.3 \text{ kg/brick} = 590 \text{ kJ/brick}$$

If the extended estimate is applied, this kiln could be capable of thermally destroying hazardous waste types K015, K023, and K093. Based on the assumption that the average heating value of the destroyable waste categories is 23,300 kJ/kg, the quantity of hazardous waste feed can be estimated as follows:

$$590 \text{ kJ/brick} \times \frac{1}{23{,}300 \text{ kJ/kg}} = 0.0253 \text{ kg waste/brick}$$

Based on a continuous output of 100,000 bricks/day in a typical tunnel kiln, up to 2530 kg (5600 lb) of hazardous wastes could be destroyed per day. Assuming 260 days of operation per year, a tunnel kiln's annual hazardous waste disposal rate can be estimated as follows:

2490 kg/day x 260 days/yr x Mg/1000 kg
= 647 Mg/yr

Trends in process fuel consumptions--Newer kilns are being designed to accept more than one fuel type; for example, either sawdust and/or oil. These new kilns are also significantly more fuel-efficient; they consume about 1860 kJ/kg (800 Btu/lb) of brick compared with the current average of 2790 kJ/kg (1200 Btu/lb).[12] The trend toward multiple fuels and higher fuel efficiency is likely to continue in future design modifications and should increase the kiln's potential for destroying hazardous waste.

4.1.2.2 Geographic Matching of Industry Locations and Waste Generators--

Figure 4.1-4 shows the distribution of domestic brick and tile plants by state. It should be noted that the majority of these plants use kilns of the small, batch type, which are unsuitable for destroying hazardous wastes. No firm numbers were available, but it is estimated that approximately 130 of the total 466 brick and tile plants utilize tunnel kilns.

Table 4.1-3 presents the SIC code and geographic distribution of each destructible waste category. The states specifically identified as generating these wastes (California, New Jersey, Louisiana, Pennsylvania, Illinois, Puerto Rico, and Texas) contain 30 percent of the total operating brick and tile plants.

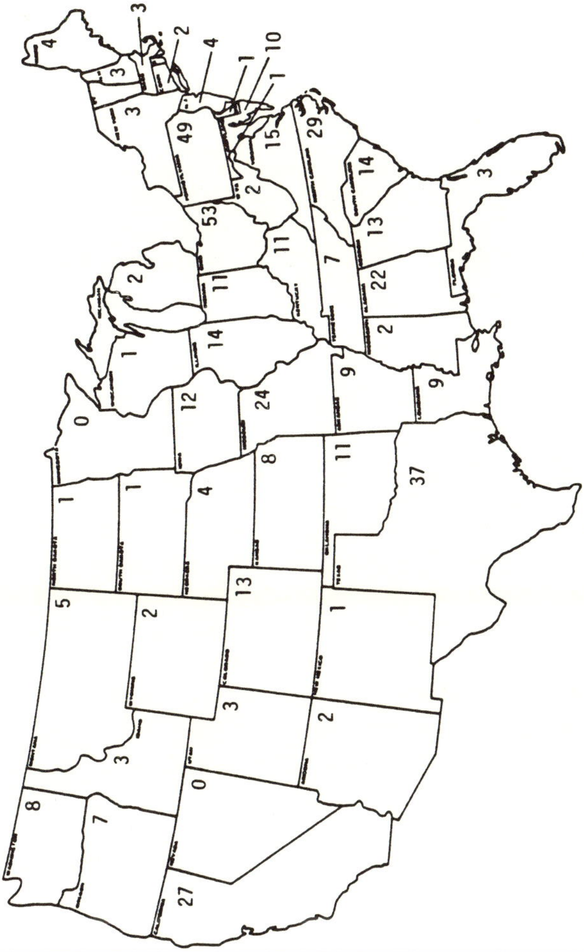

Figure 4.1-4. Geographic distribution of U.S. structural clay products plants (1973.)[16,17,18]

TABLE 4.1-3. SIC CODES AND GEOGRAPHIC DISTRIBUTION OF INDUSTRIES GENERATING HAZARDOUS WASTE DESTRUCTIBLE IN BRICK AND TILE TUNNEL KILNS[1,19]

Category	SIC code of generating industry	Geographic location
K015		Two plants (locations not known)
K023	2869	Richmond and El Sequendo, California; Kearney and Bridgeport, New Jersey; Baton Rouge, Louisiana; Bridgeville and Neville Island, Pennsylvania; Cicero and Millsdale, Illinois; Texas City, Texas
K093	2869	Illinois, Texas, New Jersey, Puerto Rico, California

4.1.2.3 Environmental Factors--

Impact on air pollution control devices--Atmospheric emissions from tunnel kilns consist of total suspended particulates (TSP), sulfur dioxide (SO_2), carbon monoxide (CO), nitrogen oxides (NO_x), hydrocarbons (HC), and in some cases, fluorides (HF), all of which can be detrimental to one's health. The TSP, SO_2, CO, and NO_x are primarily products of combustion and thus a function of the fuel used.[13,14] Some raw materials also contain a relatively high quantity of organic or sulfur-bearing impurities, which could increase these emissions. Kiln temperatures must reach 1370°C (2500°F) or greater for sulfur dioxide to be emitted from impurities within the brick.[13]

Certain clays contain fluorine as an impurity. One of the most common fluorine-containing impurities in clay is fluorite or fluorspar (CaF_2).[9] During firing, this impurity reacts to form hydrogen fluoride (HF), which can hydrolize at flue gas temperatures to form hydrofluoric acid mist. Table 4.1-4 shows emissions factors for tunnel kilns that fire oil, coal, or natural gas.

TABLE 4.1-4. EMISSION FACTORS FOR A TUNNEL KILN[13]
kg/Mg (lb/ton)

	TSP	SO_2	CO	HC	NO_x	HF
Gas-fired	0.02 (0.04)	Neg[a]	0.02 (0.04)	0.01 (0.02)	0.08 (0.15)	0.5 (1.0)
Oil-fired	0.3 (0.6)	2.0 (4.0)	Neg[a]	0.05 (0.1)	0.55 (1.1)	0.5 (1.0)
Coal-fired	0.5 (1.0)	3.6 (7.2)	0.95 (1.9)	0.3 (0.6)	0.45 (0.9)	0.5 (1.0)

[a] Negligible.

Air pollution control devices are generally tailored to the specific contaminants generated from impurities in the clay or fuel. Most kilns do not have pollution control equipment because they use a clay with few impurities or fire with natural gas. At the opposite end of the spectrum, a few plants may use a wet scrubber to control HF and SO_2 emissions and an afterburner to eliminate hydrocarbon emissions. If currently pending New Source Performance Standards are promulgated, most plants built after 1982 will be required to incorporate these control devices into their design.

Plants considering the destruction of hazardous wastes will require both a high-energy wet scrubber to eliminate gaseous emissions and an afterburner to eliminate products of incomplete combustion. Depending on the increases in pollutant loading and extent of existing control equipment, an adjustment in control devices may be all that some plants require, whereas others may need a complete retrofitting of suitable controls.

A potential exists for trace quantities of incomplete combustion products to escape into the stack gases. The afterburner should effectively destroy these compounds, but the actual behavior of hazardous wastes and their associated products of incomplete combustion need to be investigated further. This could be accomplished by conducting a test burn to determine the exact environmental impact associated with thermally destroying specific hazardous wastes in a tunnel kiln.

4.1.2.4 Institutional Factors--

Although the literature does not actually address the destruction of hazardous wastes in these kilns, private companies have been conducting their own studies. Preliminary results from one such study show that tunnel kilns are capable of destroying certain types of organic hazardous waste.[20] Because of the confidential nature of these tests, the specific wastes destroyed and other test data were not available. The fact that these studies have been made, however, indicates an interest on the part of industry. Energy requirements in the brick and tile industry are quite high, and there is a prevailing interest in reducing their energy-related expenditures, either directly by a reduction in the amount of conventional fuel required, or indirectly by offsetting these costs by a disposal fee.

Because many brick and tile plants are located near metropolitan areas, the impact of a waste spill or process failure could be great. Consequently, these kilns must be proven to be an effective and environmentally sound method for disposing of these wastes to prevent adverse impacts on the surrounding community.

4.2 CARBON BLACK INDUSTRY

4.2.1 Process Descriptions

Carbon black can be manufactured by various processes ranging from primitive ones to continuous, high-capacity systems with sophisticated designs and control equipment.[21] The basic reaction by which it is produced may be represented by the following equation:[22]

$$C_nH_m \rightarrow n\ C + \frac{m}{2} H_2 \pm H_R \text{ (heat of reaction)}$$

The type of hydrocarbon feedstock used varies with the process employed, and this feedstock determines whether the reaction is endothermic or exothermic.

About 94 percent of domestic carbon black production is produced by the oil furnace process. The thermal process is next in popularity. Table 4.2-1 presents the latest estimate of the capacities of U.S. carbon black plants.

4.2.1.1 Oil Furnace Process--

The oil furnace process involves the incomplete combustion and cracking of a liquid hydrocarbon feedstock at temperatures of approximately 1370°C (2500°F) and 20 to 50 percent excess air to produce an aerosol of carbon particles.[23] Figure 4.2-1 presents a detailed flow diagram of a typical process.

The input materials to the reactor include a preheated oil feedstock, preheated air, and natural gas for heat and product quality control. The preferred feedstocks are high in aromaticity and contain low levels of asphaltenes, sulfur (<3%), and alkali metals. Refinery byproduct residual tars and decant oil from gasoline production are commonly used as feeds. Because the decomposition of heavy aromatic feedstocks in the furnace is slightly endothermic, a considerable amount of energy is required to vaporize and heat the feedstock to reaction temperatures.

Primary quenching with water reduces the temperature of the gaseous effluent from the reactor to approximately 538°C (1000°F). Secondary quenching further reduces the temperature of the gas to less than 260°C (500°F) before it enters the fabric filters. Carbon black, in light fluffy form, is separated in the filters and conveyed to micropulverizers that discharge into a surge tank. The off-gases are burned in a CO boiler, incinerator, or

TABLE 4.2-1. UNITED STATES CARBON BLACK CAPACITY, 1976[21]

Manufacturer	Capacity, 10^3 Mg (10^3 tons)	
	Oil furnace process	Thermal process
Ashland Chemical	304 (335)	
Arkansas Pass, Texas	68 (75)	
Baldwin, Louisiana	116 (128)	
Belpre, Ohio	45 (50)	
Mojave, California	27 (30)	
Shamrock, Texas	48 (53)	
Cabot Corporation	425 (468)	
Big Spring, Texas	113 (125)	
Franklin, Louisiana	91 (100)	
Pampa, Texas	29 (32)	
Parkersburg, West Virginia	74 (82)	
Ville Platte, Louisiana	118 (130)	
Cities Service	343 (378)	
Conroe, Texas	44 (48)	
El Dorado, Arkansas	37 (41)	
Eola, Louisiana	32 (35)	
Franklin, Louisiana	71 (78)	25 (28)
Hickok, Kansas	23 (25)	
Mojave, California	24 (26)	
Moundsville, West Virginia	71 (78)	
Seagraves, Texas	41 (45)	
Continental Oil	194 (214)	
Bakersfield, California	35 (39)	
Ponca City, Oklahoma	61 (67)	
Sunray, Texas	43 (47)	
Westlake, Louisiana	54 (60)	
J. M. Huber Corporation	179 (197)	19 (21)
Baytown, Texas	117 (129)	
Borger, Texas	62 (68)	
Phillips Petroleum	219 (241)	
Borger, Texas	130 (143)	
Orange, Texas	52 (57)	
Toledo, Ohio	36 (40)	
Richardson Company	95 (105)	
Addis, Louisiana	45 (50)	
Big Spring, Texas	50 (55)	
Thermatomic Carbon		
Sterlington, Louisiana[a]		59 (65)
Total	1938	114

[a]Plant closed in 1977.

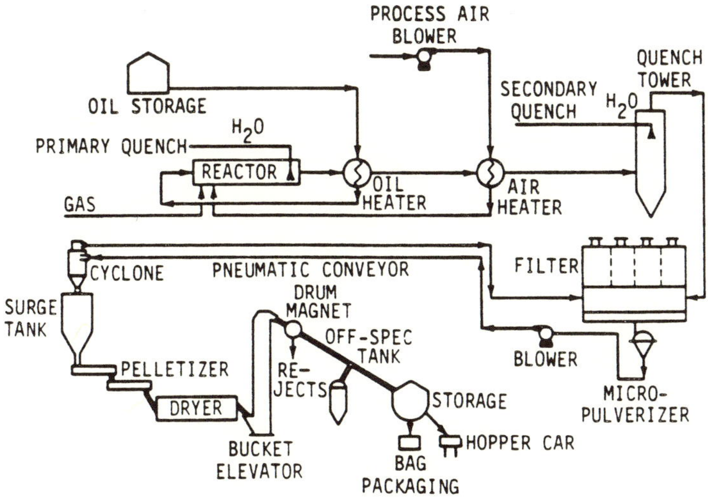

Figure 4.2-1. Flow diagram of oil furnace process.[21]

flare to reduce emissions of CO, hydrocarbons, and H_2S. If a boiler is used, heat is recovered as steam.

As shown in Table 4.2-2, different grades of carbon black can be produced by changing reactor temperatures and residence times.

TABLE 4.2-2. TIME-TEMPERATURE CONDITIONS IN CARBON BLACK REACTORS

Black grade	Residence time, s	Temperature, °C (°F)
Thermal (N990)	10	1210-1340 (2200-2450)
SRF (N700 series)	0.9	1400 (2550)
HAF (N300 series)	0.031	1540 (2800)
SAF (N100 series)	0.008	1790 (3250)

4.2.1.2 Thermal Process--

The thermal process involves decomposition of a gaseous hydrocarbon feed by thermal cracking at 1320° to 1540°C (2400° to 2800°F) in the absence of air to yield carbon and hydrogen. Natural gas is used as the feedstock. The product stream from the cracking chamber is filtered to remove carbon black, and the exit gas is burned to generate heat for additional cracking.

4.2.2 Process Assessment for Hazardous Waste Destruction

Of the two processes discussed, the oil furnace process was selected for further evaluation for thermal disposal of hazardous wastes. The thermal decomposition process was not investigated further mainly because of the limited number of these facilities.

4.2.2.1 Technical Factors--

Compatibility of hazardous wastes with process fuel--Most of the waste types being considered for this study are aromatic in nature and hence excellent candidates for disposal by the oil furnace process.

Graphical determination of suitable waste categories--Residence times and temperatures cited for SRF and thermal grades of carbon black appear to be feasible for hazardous waste destruction, whereas the residence times for HAF and SAF grades are too short for waste destruction purposes.

For example, based on the conditions for manufacture of SRF grade carbon black presented in Table 4.2-2:

Residence time at maximum temperature of 1420°C (2550°F) = 0.9 s

Therefore, a gas flow rate of 105 Nm^3/min (4000 $scfm^3$) will require a furnace with a volume of 9.63 m^3 (340 ft^3).

Based on computations of t (mean residence time) and $\bar{t}$ (fast-path residence time) for several temperatures between the maximum temperature [1400°C (2550°F)] and the exit gas temperature [760°C (1400°F)], values are obtained for Δt and $\Delta\bar{t}$. Figure 4.2-2 shows plots of Δt and $\Delta\bar{t}$ versus T (mean- and fast-path process curves) overlaid on the hazardous waste plots to determine destructible hazardous waste categories.

Table 4.2-3 presents predictions regarding the potential of the oil furnace process to achieve 99.99 percent destruction of the hazardous wastes. It appears that the wastes in categories F003, K015, K023, and K093 can be destroyed at the conditions prevailing in the furnace. Waste category F003 is shown as being marginal because these wastes appear to be destructible only under mean-path conditions. High turbulence in the furnace would enhance the destruction of these wastes.

TABLE 4.2-3. FEASIBILITY OF OIL FURNACE PROCESS (CARBON BLACK) FOR THE THERMAL DESTRUCTION OF HAZARDOUS WASTES

Hazardous waste type													
F003	F004	F005	K015	K018	K022	K023	K024	K025	K027	K030	K049	K093	K094
Maybe	No	No	Yes	No	No	Yes	No	No	No	No	No	Yes	No

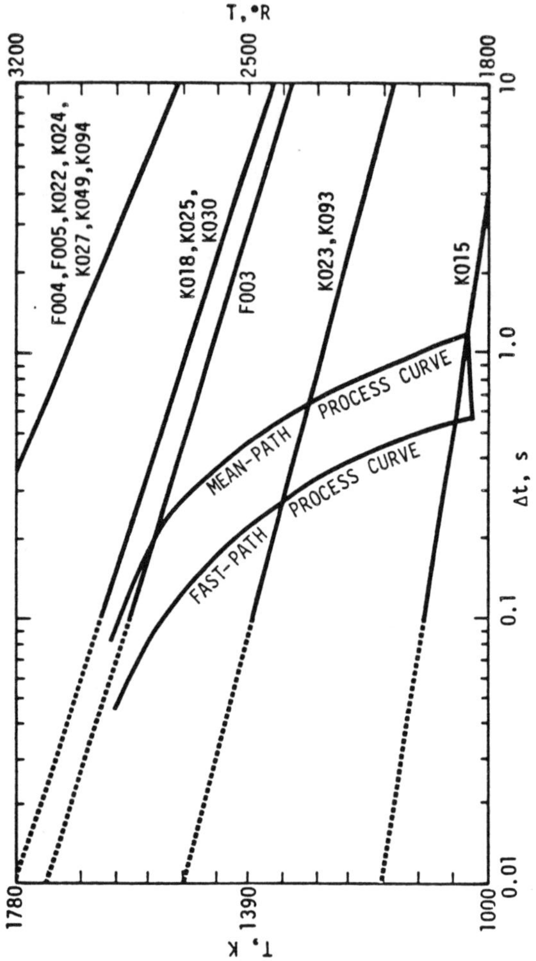

Figure 4.2-2. Hazardous waste destructibilty predictions for the oil furnace process in carbon black manufacture.

Effects on product quality and product application--Because carbon black has a potential for adsorbing gaseous hydrocarbons,[24] some of the hazardous components in the waste may be adsorbed on the surface of the carbon black particles. It is difficult to predict the extent of adsorption of various hazardous components, and test burns would be necessary to determine levels of toxic substances that could be associated with the product.

About 93 percent of the carbon black produced in the United States is consumed by the rubber industry, primarily for use as a reinforcing agent in rubber tires.[24] Contamination of the product should not prevent the use of the carbon black process for destroying hazardous wastes. Tests would be required, however, to ensure that any product contamination would not affect the tensile strength or elasticity and thus the integrity of the product. Secondary uses of carbon black are in inks and paints, and waste destruction is not expected to have any detrimental effect on the quality of these products, either.

Impact on process hardware--The design and construction features of reactors employed for carbon black manufacture are determined by economics and product specifications.[21] Furnace reactors have undergone substantial design changes during the last decade to give higher turbulence, rapid mixing of reactant gases and feedstock, and increased capacity. Refractory-lined metal construction appears to be a notable feature in the latest reactor designs. Inasmuch as these furnaces are designed to accept a fairly wide variety of oil feedstocks (e.g., byproduct residual tars and decant oil from gasoline production), combustion of compatible liquid wastes should not have any deleterious effect on process hardware. Destruction of highly halogenated hazardous wastes, however, may decrease the useful life of the refractory lining.

Acceptable amounts of hazardous waste feed--Operating parameters for furnace processes vary from plant to plant and, within a given plant, depend on the grade of product manufactured.

Typical reactor feed rates and carbon black production rates are as follows:[23]

Rate of feed oil:	757 liters/h (200 gal/h)
Rate of air supplied:	6660 m^3/h (235,000 ft^3/h
Rate of natural gas supplied:	623 m^3/h (22,000 ft^3/h)
Rate of carbon black production:	390 kg/h or 0.52 kg/liter (860 lb/h or 4.3 lb/gal oil)
Yield of carbon black (lb of product per 100 lb of carbon in feedstock:	60 percent

Based on the assumption that the hazardous wastes will represent 10 percent of the volumetric flow rate of feed oil to the furnace, the hazardous waste feed rate will be 76 liters/h. Since the wastes have an average density of 0.95 kg/liter (7.92 lb/gal),[5] the feed rate amounts to about 73 kg/h (160 lb/h) of hazardous waste feed.

Assuming 24 liters/day of process operation 260 days/yr, the amount of disposable hazardous wastes that could be processed in a typical reactor is calculated as follows:

73 kg/h x 24 h/day x 260 days/yr x 1 Mg/1000 kg
= 456 Mg/yr

Provisions for waste introduction--The physical compatibility of the wastes and the oil feedstock appears to be good; therefore, the wastes could be blended with the oil and then preheated and atomized before they enter the furnace. No retrofit problems are anticipated.

Projected trends in process fuel--Recent trends in carbon black production include more emphasis on the furnace process as a result of advances in furnace technology that allow a wide range of carbon blacks to be produced.[24] Hence, it appears that oil (typically highly aromatic petroleum fractions) will be the predominant fuel in the carbon black industry in the future.

4.2.2.2 Geographic Matching of Industry Locations and Waste Generators--

Categories of wastes that were shown to be destructible (in the previous section) are F003, K015, K023, and K093. Table 4.2-4 identifies the SIC code and geographic distribution of the industries generating these destructible wastes. Figure 4.2-3 shows the distribution of carbon black manufacturing plants in the United States. No SIC code is associated with F003 because this waste is generated by nonspecific sources. Two of the selected destructible waste categories (K023, K093) are generated primarily in Texas, California, and New Jersey. Table 4.2-1 (presented earlier) and Figure 4.2-3 reveal that the states of Texas and California account for a substantial fraction of total domestic carbon black production. Hence, the geographic matching of the industry and waste generator locations for waste types K023 and K093 appears to be excellent. This makes the carbon black industry even more attractive for thermal disposal of the candidate hazardous wastes because the shorter distances between waste generators and industry location reduces transportation costs and lessens possible hazards resulting from transport.

4.2.2.3 Environmental Factors--

Impact on air pollution control devices--Suspended carbon black particles usually are collected in control devices such as electrostatic precipitators, wet scrubbers, cyclones, or fabric filters.[24] Electrostatic precipitators and cyclones were widely utilized in older plants. The current trend, however, is to use an agglomerating device or cyclone followed by a fabric filter system. These collection systems are estimated to remove more than 99 percent of the carbon black from the combustion gases. Particulate emissions from carbon black plants are generally low for economic reasons and because of environmental control requirements.[23]

Combustion of hazardous waste is not believed to have any harmful effect on air pollution control devices. Wastes in one

TABLE 4.2-4. SIC CODES AND GEOGRAPHIC DISTRIBUTION OF INDUSTRIES GENERATING HAZARDOUS WASTE DESTRUCTIBLE IN THE OIL FURNACE CARBON BLACK PROCESS[19]

Category	SIC code of generating industry	Geographic location
F003	-	Widespread
K015	2865	Two plants (locations not known)
K023	2869	Richmond and El Sequendo, California; Kearney and Bridgeport, New Jersey; Baton Rouge, Louisiana; Bridgeville and Neville Island, Pennsylvania; Cicero and Millsdale, Illinois; Texas City, Texas
K093	2869	Texas, California, New Jersey

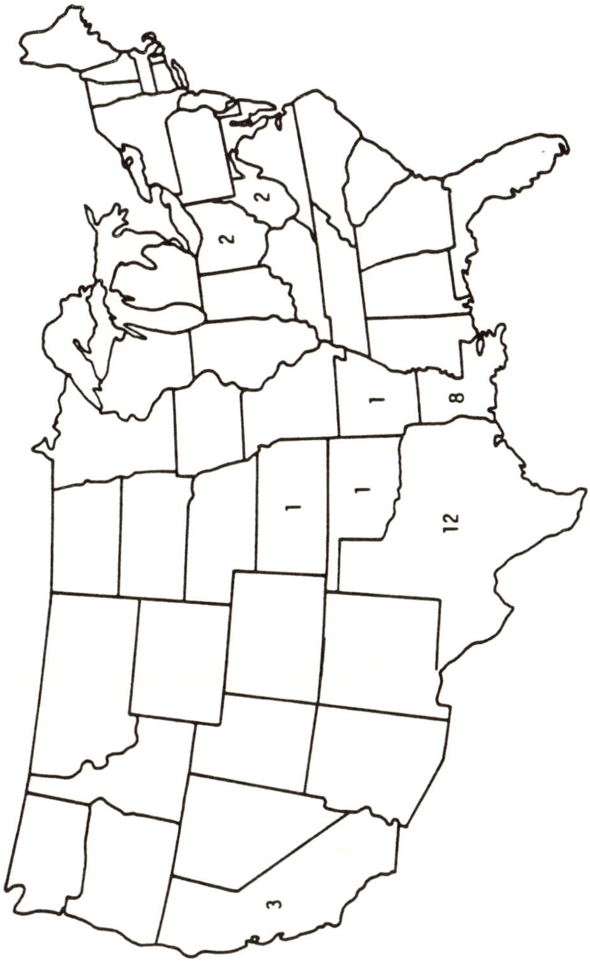

Figure 4.2-3. Geographic distribution of U.S. carbon black manufacturing plants.[23]

destructible waste category (K015), however, contain chlorinated aromatics, and a wet scrubber may be required to remove HCl (and Cl_2) in the off-gases. Most of the chlorine is believed to evolve as HCl because of the atmosphere of hydrogen in the furnace. Quantities of HCl emitted will vary greatly, depending on the amount of chlorine in the waste incinerated. The design of a scrubber system to control these emissions must consider the highly corrosive nature of chlorides on the materials of construction. A water treatment system for the scrubber effluent must also be included.

Environmental impact--Atmospheric emissions from carbon black furnace processes consist of CO, hydrogen sulfide, gaseous hydrocarbons (primarily acetylene), and particulates. At a number of plants, off-gas combustion (afterburner or CO boiler) is used to reduce these emissions. Any additional impact on these gaseous emissions due to hazardous waste combustion should be minimized by such off-gas combustion. Sulfide emissions are proportional to the level of sulfur in the feedstock; hence, H_2S and other sulfide compound emissions (CS_2, COS, etc.) will be low if the sulfur content of the waste and oil is low.

No major liquid process waste is generated by the furnace process because it is basically a gas-phase process. If a wet scrubber is used, however, a liquid effluent is produced. This effluent could be used in the pelletizer (see Figure 4.2-1), which would preclude the discharge of carbon black into wastewater.

The impact of hazardous waste combustion on solid waste produced from carbon black furnaces is believed to be negligible because insignificant quantities of solid waste are produced in such processes. Limited quantities of refractory material, scrap, and worn fabric filter bags are discarded and landfilled.

Fugitive emissions from carbon black production processes may result from maintenance procedures, leaks in plant conveying systems, or loading and unloading operations.[24] This may be a problem if hazardous compounds are adsorbed on the carbon black product before such compounds are completely destroyed. Most fugitive emissions can be reduced by a plantwide vacuum system ducted to a fabric filter and good housekeeping procedures which assure that all spillage is picked up and combusted or recycled. If such preventive measures are taken or if tests show that no significant hazardous components are adsorbed on the carbon black product, the combustion of hazardous wastes in the furnace should pose no significant problems.

4.2.2.4 Institutional Factors--

Because there is no generally acceptable substitute for carbon black in major applications, such as automobile tires, a steady growth of the domestic industry is expected.[23]

The carbon black industry (SIC 2895) is already considering the possibility of using waste tires as substitute feedstock to help alleviate the solid waste problem in the United States,[23] and it may not object to the combustion of hazardous waste in carbon black furnaces. It is a favorable factor that most of the 33 carbon black plants in the United States are located in rural areas far removed from large population centers.

Advances in technology and improved process economics kept carbon black prices stable from 1950 to 1973, but in recent years the cost of feedstock has increased more than fourfold, which has caused about a twofold increase in the price of large-volume rubber grades. This price increase could be controlled somewhat if hazardous wastes could be substituted for a portion of the conventional feedstock.

4.3 GLASS INDUSTRY

The glass industry consists of four major segments: glass containers, flat glass, pressed and blown glass, and fiberglass. The fiberglass segment (SIC-3296) is considered separately because its manufacturing process is vastly different from that for the other glass industry segments. The glass container segment (SIC-3221), which is the largest in terms of shipments and employees, accounts for approximately 57 percent of all glass melted, and the remaining glass production (except for fiberglass) is roughly divided between the flat glass (SIC-3211) and pressed and blown glass (SIC-3229) segments, which account for 24 and 19 percent, respectively.[25]

The three general types of container glass are amber, clear, and green. Green glass and clear glass are sometimes grouped together in a single category designated as "flint." Flint glass represented 85 percent of glass container production in 1976, and amber glass, 15 percent.[25] Glass container plants are located near the markets they serve. Figure 4.3-1 presents the geographic distribution of the major domestic glass container plants. As the figure shows, a large number of the plants are concentrated in the East, North Central, and Middle-Atlantic regions, which also are areas in which significant quantities of hazardous wastes are generated. Table 4.3-1, which gives a breakdown of the companies who own the 119 domestic plants, shows that five of these companies operate roughly half of these glass container plants.

Flat glass manufactured in the United States is converted into float, sheet, rolled, or plate glass. Float glass accounts for more than 90 percent of the total flat glass production.[25] As shown in Figure 4.3-2, most flat glass manufacturing plants are located in Pennsylvania, Ohio, Tennessee, Michigan, Illinois, Texas, and California.

The pressed and blown glass segment consists of approximately 155 manufacturing establishments (Table 4.3-2). Forty of these produce mostly hand-pressed and blown glass. Figure 4.3-3, which shows the locations of the plants, indicates that the majority are in New York, Pennsylvania, West Virginia, Ohio, Indiana, and Illinois.

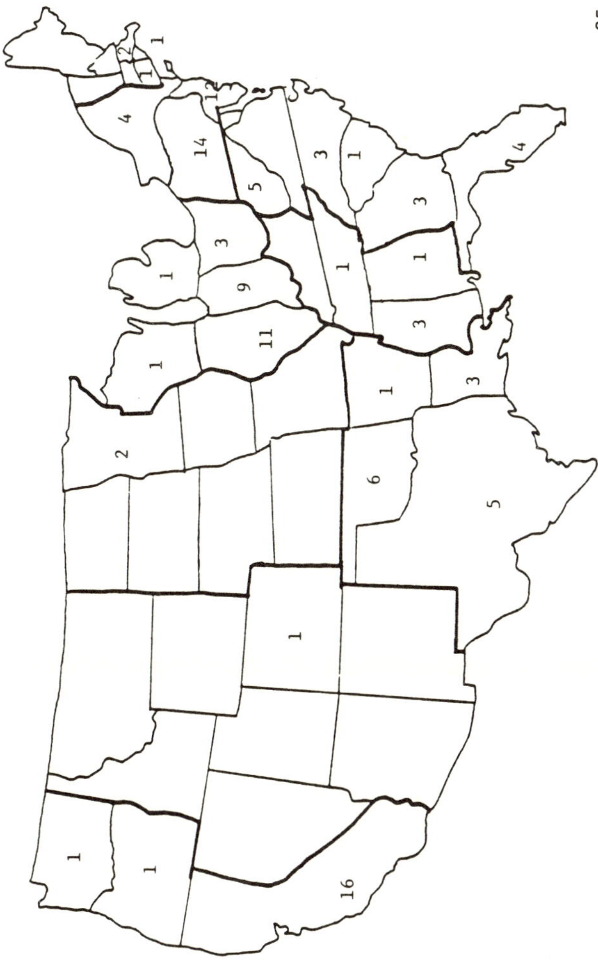

Figure 4.3-1. Geographic distribution of U.S. glass container manufacturers.[25]

TABLE 4.3-1. MAJOR GLASS CONTAINER MANUFACTURERS IN THE UNITED STATES, 1975[25]

Manufacturer	No. of plants	Manufacturer	No. of plants
Anchor Hocking	9	Kerr Glass	7
Arkansas Glass	1	Latchford Glass	1
Ball Corporation	4	Leone Industries	2
Bartlett Collins	1	Liberty Glass	1
Brockway Glass	14	Madera Glass Co.	1
Chattanooga Glass	7	Metro Glass	4
Columbia Gas	1	Midland Glass	4
Diamond Glass	1	National Bottle Corp.	4
Gallo Glass	1	National Can	4
Glass Containers Corp.	12	Owens-Illinois	20
Glenshaw Glass	2	Thatcher Glass	6
Hillsboro Glass	1	Underwood Glass	1
Indian Head	7	Wheaton Glass	2
Industrial Glass	1	Total	119

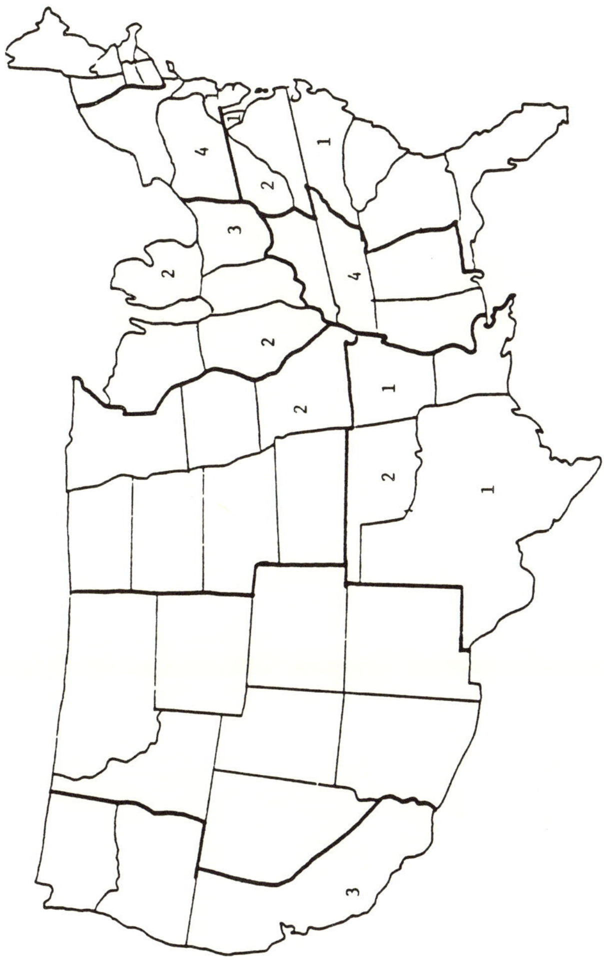

Figure 4.3-2. Geographic distribution of U.S. flat glass manufacturers.[25]

TABLE 4.3-2. MAJOR PRESSED AND BLOWN GLASSWARE MANUFACTURERS IN THE UNITED STATES

Manufacturer	No. of plants	Manufacturer	No. of plants
American Optical	3	Kessler, Inc.	1
Anchor Hocking	4	King Seeley Thermos Company	2
Arrowhead Puritas	1	Labino Glass Laboratories	1
Atlantic Optical Moulding	1	Lancaster Glass Corporation	1
B and J Optical	1	Lenox Crystal	1
Bartlett-Collins	1	Lewis County Glass	1
Bausch and Lomb	1	Louie Glass Company	1
Beaumont Company	1	Marble King	1
Big Pin Glass Company	1	Masden Ind.	1
Blenko Glass Company	1	Mayflower Glass	1
Brock Glass Company	1	Mid-Atlantic	1
Brockway Glass	2	J. H. Millstein	1
E. O. Brody Company	1	Minners	1
Cambridge Glass	1	Multicolor Glass	1
Canton Glass	1	Nuclear-Pacific	1
Cataphote	1	Overmyer-Perram	1
Champion Agate Company	1	Owens-Illinois	10
Chemglass, Inc.	1	Pacific Glass Works	1
Corning Glass Works	14	Peltier Glass Company	1
Crescent Glass Company	1	Penbarthy Electromelt Intl., Inc.	1
Crystal Art Glass	1	Pennsboro Glass Company	1
Davis-Lynch Glass Company	1	Phoenix Glass Company	1
EMC Glass Corporation	1	Pikes Peak Glass Company	1
Eastman Kodak Company	1	Pilgrim Glass Corp.	1
Emerson and Cuming	1	Pittsburgh Corning Corp.	2
Elite Company, Inc.	1	Pape Scientific, Inc.	1
Erie Glass Mfg. Company	1	Potter Industry, Inc.	4
Erskine Glass and Manufacturing	1	RCA Corp.	1
Federal Glass Company	1	Rainbow Art Glass	1
Fenton Art Glass	1	Ray-Lite Glass, Inc.	1
Fischer and Porter Company	2	Reha Glass Company	1
Fostoria Art Glass	1	Rocky Mountain	1
Friedrich and Dimmock, Inc.	1	Rodifer Gleason Glass Corp.	1
GTE Sylvania, Inc.	4	St. Clair Glass Works	1
General Electric	9	Scandia Glass Works	1
Gentile Glass Company	1	Schott Optical Glass Company	1
Gillinder Brothers	1	Scott Depot Glass Company	1
Gladding-Vitro-Agate Company	1	Sellstrom Mfg.	1
Glass Works, Inc.	1	Earl Shelly Glass Company	1
Guernsey Glass Company	1	Sinclair Glass	1
Haley Glass Company	1	L. E. Smith Glass Company	1
Hamon Handcrafted Glass	1	Super Glass Corp.	1
Harvey Ind.	1	Swift Glass Corp.	1
Holophane	1	Techniglass, Inc.	1
Houze Glass Corp.	1	Thermal American Fused Quartz	1
Imperial Glass Corp.	1	Thomas Ind., Inc.	1
Indiana Glass Corp.	1	Variety Glass, Inc.	1
Industrial Glass and Plastics	1	Venejian Art Glass	1
Interpace Corp.	1	Victory Glass Company	1
Jeanette Corp.	1	Viking Glass Company	1
Jeanette Shade and Novelty Co.	1	West Virginia Glass Specialty Co.	1
Johnson Glass	1	Westinghouse Electric Corp.	1
Kanawa Glass Company	1	Westmoreland Glass Company	1
Kapp Glass, Inc.	1	Wheaten, Ind.	1
Kaufman Glass Company	1	The Paul Wissmach Co., Inc.	1

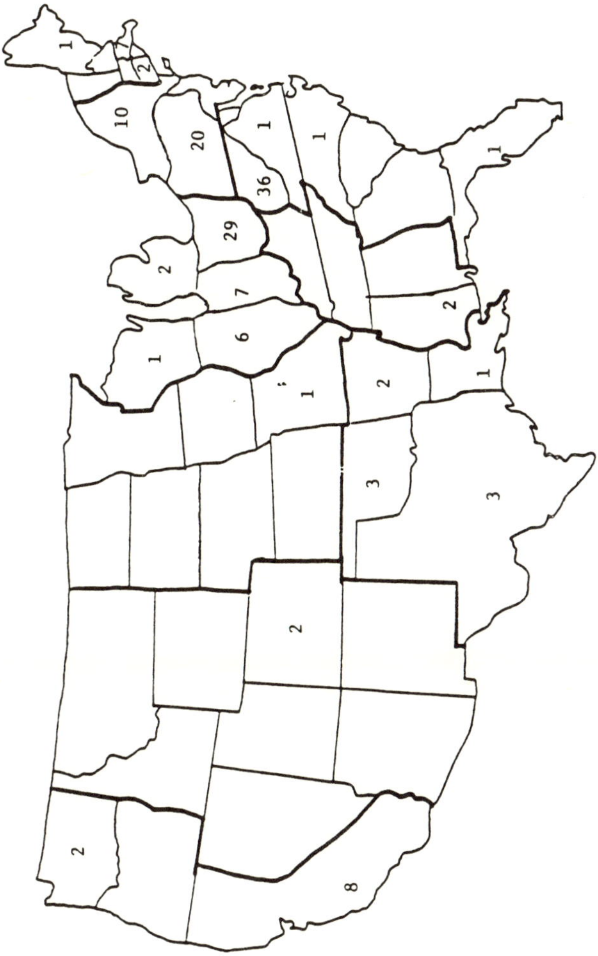

Figure 4.3-3. Geographic distribution of U.S. pressed and blown glassware manufacturers.[25]

The various locations of fiberglass plants include Parkersburg, West Virginia, Fairburn, Georgia, Santa Clara, California, and Kansas City, Missouri. A listing of all fiberglass manufacturers and plant locations was not readily available.

4.3.1 Process Descriptions

The major processes in glass manufacturing are as follows:[25]

- Batch weighing and mixing
- Melting
- Forming
- Postforming
- Product packaging

In the flat glass industry segment, forming and postforming are usually integrated into one continuous process.

Figure 4.3-4 presents a typical process flow diagram for glass manufacture. Although drawn specifically for common soda-lime glasses, this figure has general application to most commercial glasses.

Basically, the manufacture of glass is the high-temperature conversion of raw materials into a homogeneous melt that can be fabricated into useful articles.[26] The major raw materials are glass sand (SiO_2), soda ash, limestone, and cullet (crushed waste glass). Minor ingredients include salt cake (Na_2SO_4), niter ($NaNO_3$), feldspar (Al_2O_3), arsenic, barium oxide, and boron oxide.[27] Mixing systems range from hand-operated to fully automated.

4.3.1.1 Melting Process--

The bulk of the soda-lime glass production is melted in direct-fired continuous, regenerative furnaces, which range in capacity from 45 to 272 Mg/day (50 to 300 tons/day) of glass.[28] Two kinds of furnaces are used: a side-port melter and an end-port melter. Figure 4.3-5 depicts a side-port melting furnace. This furnace consists of a dual chamber filled with brick checkerwork. While the products of combustion from the melter pass through and heat one chamber, combustion air is preheated in the

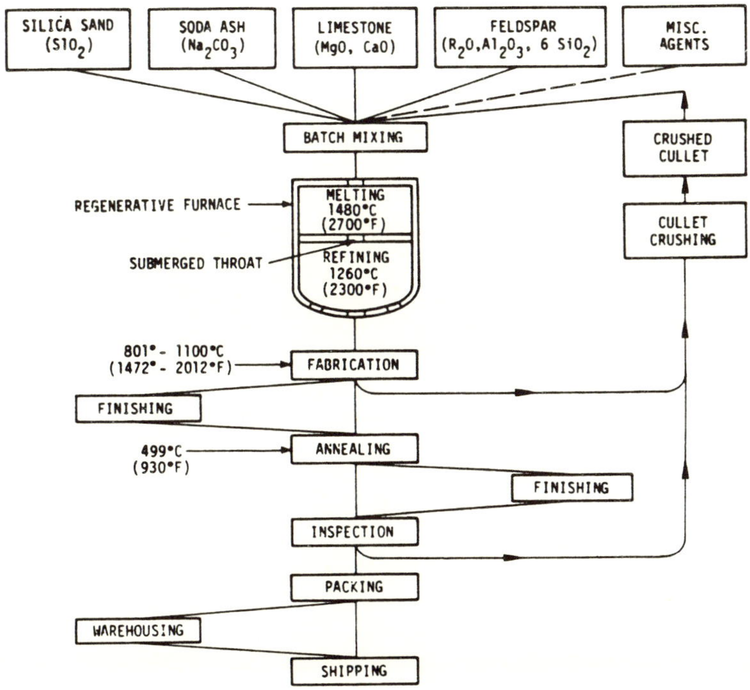

Figure 4.3-4. Flow diagram for soda-lime glass manufacture.[27]

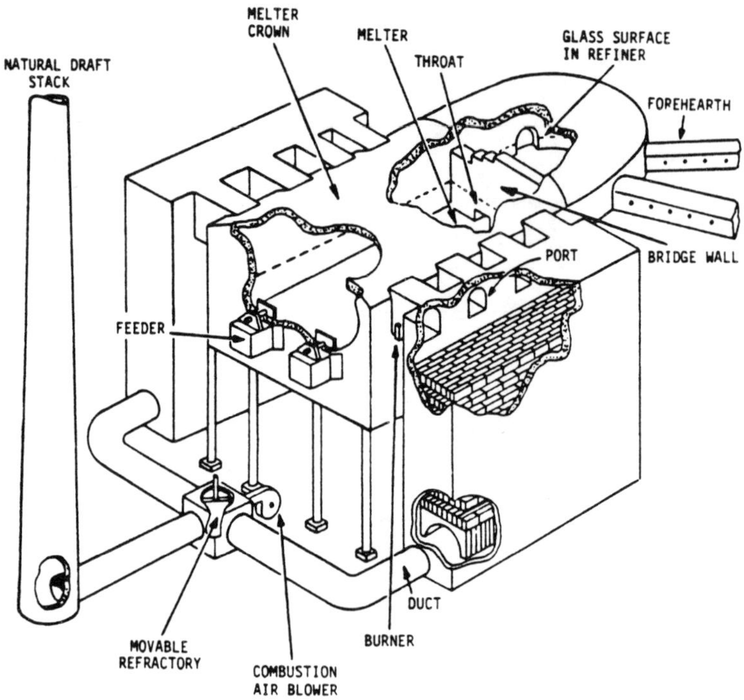

Figure 4.3-5. Regenerative side port glass-melting furnace.[27]

other chamber. The furnace is fired by gas or oil. The function of each chamber is interchanged by reversing the flow of air and combustion products every 15 to 20 minutes.[28] The batch-blended material is introduced into the furnace by screw- or reciprocating-type feeders. Dry material floats on top of the molten glass until it melts. As the carbonates decompose, they release carbon dioxide. Particles entrained in gas bubbles and other volatilized components of the feedstock are swept from the furnace by the hot gases passing across the molten surface. These materials are either deposited in the back checkerwork and refractory-lined passages or discharged into the atmosphere. In the refining section, the glass cools from 1480° (2700°) to about 1260°C (2300°F), and becomes viscous enough to form. It is then drawn from the furnace and sent to forming machines.

Table 4.3-3 shows typical emissions from glass melting furnaces. Generally, container glass melting furnaces have no pollution control equipment. Manufacturers usually rely on appropriate ingredient selection and good operating techniques to control emissions; however, many plants are now experimenting with such high-efficiency control devices as fabric filters and ESP's.[29] Pressed and blown glass furnaces, however, often use particulate control devices, predominantly ESP's.

TABLE 4.3-3. TYPICAL EMISSIONS FROM THE MANUFACTURE OF CONTAINER GLASS[27]

Emissions	Flint glass furnace	Amber glass furnace
Solids, g/Nm^3 (gr/scf)	0.071 (0.029)	0.101 (0.041)
Sulfur trioxide, ppm	17	15
Sulfur dioxide, ppm	250	315
Fluorine, ppm	2.2	1.9
Chlorine, ppm	4.9	4.1
Nitrogen oxides, ppm	340	640
Carbon monoxide, ppm	375	40

4.3.1.2 Forming Process--

Both glass containers and pressed and blown glass are formed in automated machines at temperatures of approximately 820° to

1090°C (1500° to 2000°F). This process uses various organic sprays and swabbing compounds to facilitate the release of the formed product from the hot metal mold.[25] Volatilization of these compounds results in emissions that exit through ventilators located in the roof of the building.

In flat glass manufacture, the forming process consists of allowing the molten glass to float onto a bath of molten tin and subsequently drawing it into a very large flat sheet.

4.3.1.3 Postforming Process--

The primary function in postforming is to remove detrimental, residual stresses by annealing. Natural gas is generally used as the fuel in the annealing step, and a temperature of about 930°F is attained. For glass containers and pressed and blown glass, decorative sprays and lubricity coatings are applied to the annealed product. Generally, a common ventilator is used for emissions from both forming and postforming processes.[25]

4.3.1.4 Fiberglass Manufacturing Process--

Both melting and forming operations in fiberglass manufacture differ from those in the other segments of the glass industry. The use of regenerative furnaces for production of fiberglass is declining, whereas the use of recuperative and electric furnaces is on the upswing. Melting temperatures in recuperators are in the range of 870° to 1090°C (1600° to 2000°F).[30] Figure 4.3-6 illustrates the operations involved in the production of fiberglass.

There are several types of fiberglass. Type E, the most common, has a composition similar to that shown for alumina-borosilicate glass. Because higher percentages of more volatile components are used, emissions from recuperative and regenerative fiberglass furnaces are much higher than those from container glass furnaces. Table 4.3-4, which lists typical emissions, shows that particulate loadings and fluoride levels are very high. The potential for hazardous pollutant emissions, such as

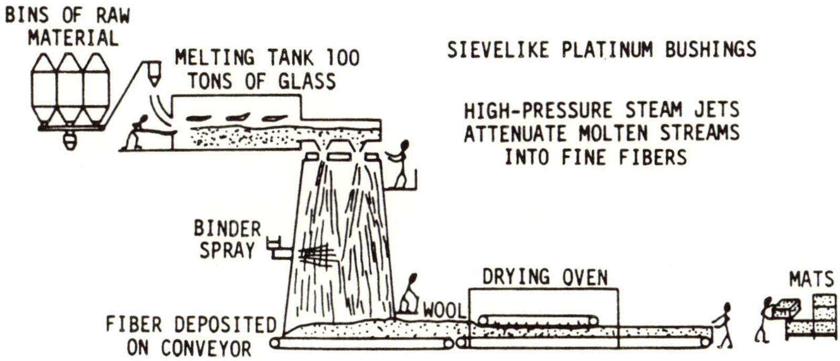

Figure 4.3-6. The operations in fiberglass production.[26]

compounds of beryllium, is greater in fiberglass manufacture than it is in container glass manufacture. Although fabric filters have been used to control emissions from fiberglass furnaces, they have met with only limited success. Electrostatic precipitators and wet scrubbers are also being used; however, the use of control measures is not an industrywide practice. The increased use of electric furnaces for fiberglass production is expected to reduce the generation of emissions considerably and thus preclude the need for control devices.

TABLE 4.3-4. TYPICAL EMISSIONS FROM RECUPERATIVE AND REGENERATIVE FIBERGLASS FURNACES[32]

Particulate loading	0.62 g/Nm^3 (0.25 gr/scf)
Gaseous fluoride	20-100 ppm
SO_x	40-80 ppm
B_2O_3	0.062 g/Nm^3 (0.025 gr/scf)
Opacity	30-80%

4.3.2 Process Assessment for Hazardous Waste Destruction

The large number of glass manufacturing plants and the high temperatures and moderate residence times achieved in the glass melting process make this industry attractive for thermal disposal of hazardous wastes.

The glass container segment, the largest in the industry, offers the greatest capacity for waste destruction. The pressed and blown glassware segment, which melts lower tonnages, also represents a viable alternative. In the flat glass industry, however, glass quality is a continuous problem,[31] and the introduction of extraneous materials, especially hazardous wastes, does not appear to be feasible. Although product quality is not a great concern in the fiberglass industry,[33] lower recuperator melting temperatures, high potential for hazardous emissions, and limited melting capacity present deterrents to consideration of fiberglass manufacturing processes for hazardous waste disposal.

In summary, the glass container and pressed and blown glassware segments of the glass industry appear to be good candidates for hazardous waste destruction. The preferred site of waste destruction would be in the melting furnaces, where the highest temperatures are attained. The following subsections address the melting processes of these two industry segments.

4.3.2.1 Technical Factors--

Compatibility of hazardous wastes with process fuel--Natural gas and oil are the primary fuels used. Producer gas is also used, but to a lesser extent. Propane and butane are also good for this application, but they are too expensive to be used much.[26] Natural gas is used more than oil because it is cleaner and because both the oil and its associated fuel systems are higher in cost. The Glass Packaging Institute indicated that fuel oil accounted for only 4 percent of fuel use at glass container plants in 1979 and 1980.[34] One major manufacturer of container and pressed and blown glass reported that fuel oil accounted for only 2 to 3 percent of their total fuel consumption in 1980.[35]

Liquid hazardous wastes would be more compatible with fuel oil than with natural gas because they could be blended with the oil and introduced into the melting furnace through the atomizing burners provided for firing oil. Because most furnaces are equipped with dual fuel capabilities, however, no compatibility problems should arise during the firing of natural gas.

Because both glass quality (primarily color) and emissions from glass furnaces are highly dependent on the composition of the metallic oxides that are charged to the furnace, it is important that the hazardous waste not contain any trace metals (e.g., nickel, iron, chromium, cobalt). One of the waste categories included in this study (K049) contains chromium (500 ppm) and lead (15 ppm); hence it would be eliminated from consideration.

Graphic determination of suitable waste categories--The dimensions of melting furnaces for the manufacture of container glass are similar to those of furnaces used for pressed and blown glass production. Outputs are as high as 270 Mg/day (300 tons/day). The maximum holding capacity of these furnaces is approximately 227 to 272 Mg (250 to 300 tons), corresponding with dimensions of 7.6 m x 15 m x 1.2 m (25 ft x 50 ft x 4 ft).[26] Typical flue gas volumetric flow rates are approximately 442 Nm^3/min (16,800 scfm).[27]

Pertinent process conditions for determining the time-temperature relationship in glass melting furnaces are shown below:

Volume of glass melting furnace:	140 m^3 (5000 ft^3)
Gas volumetric flow rate:	442 Nm^3/m (16,800 scfm)
Maximum reaction zone temperature:	1480°C (2700°F)
Exit gas temperature:[27]	616°C (1140°F)

Mean residence times (t) and fast-path residence times ($\bar{t}$) were computed for several temperatures between the maximum reaction zone temperature and the exit gas temperature. In Figure 4.3-7, the resulting process curves (plots of Δt and $\Delta \bar{t}$ versus T) are overlaid on the hazardous waste plots. Predictions regarding the potential of the melting process to destroy 99.99 percent of the hazardous wastes are shown in Table 4.3-5. It appears that wastes in 50 percent of the categories can be destroyed at the prevailing furnace temperatures.[30]

Effects on product quality and product application--The glass manufacturing industry has expressed a strong belief that hazardous wastes would be adequately destroyed at the temperatures and residence times attained in the melting furnaces.[33,34] The products of waste combustion would evolve in the flue gas, and no contamination of the glass product is expected as long as metallic constituents are not introduced in the charged wastes. Of the container glass types, product quality specifications are generally least demanding for amber glass.

Table 4.3-6 presents the major applications of pressed and blown glass. The nature of the end uses indicates that these

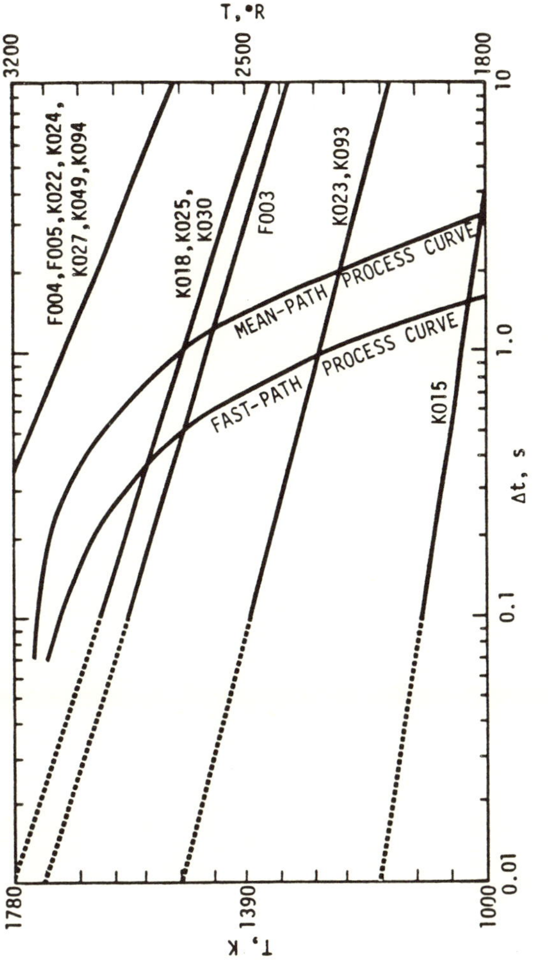

Figure 4.3-7. Hazardous waste destructibility predictions for melting furnaces in container glass and pressed and blown glass manufacture.

TABLE 4.3-5. FEASIBILITY OF GLASS MELTING PROCESS FOR THERMAL DESTRUCTION OF HAZARDOUS WASTES

Hazardous waste type													
F003	F004	F005	K015	K018	K022	K023	K024	K025	K027	K030	K049	K093	K094
Yes	No	No	Yes	Yes	No	Yes	No	Yes	No	Yes	No	Yes	No

TABLE 4.3-6. PRODUCT APPLICATIONS FOR PRESSED AND BLOWN GLASS, SIC-3229[25]

Process and major products	SIC	Percent of industry shipment (value)	
		1972	1976
Table, kitchen, and art ware			
Machine-made	32291	33	39
Handmade		8	6
Lighting and electronic	32292	35	32
Scientific and industrial	32294	24	23
Total industry (percent)		100	100
(actual)		$1.1 billion	$1.6 billion

products must be absolutely free of hazardous waste contamination; the same is true for container glass.

Impact on process hardware--One of the primary problems in glass manufacture is that of obtaining a material suitable for the construction of glass-melting equipment. All materials are soluble in molten glass. When refractory corrosion could affect properties important to end use of the ware, platinum is used because it is almost insoluble in most glasses. No single refractory material, however, is suitable for all purposes in a glass tank. An ideal refractory should have (1) the ability to resist the action of heat without deformation, (2) a low rate of solution in the particular glass, (3) no source of chemical discoloration, (4) low thermal conductivity, and (5) high resistance to thermal stresses and spalling. Major types of refractories used in the glass industry include zircon, silica, fireclay, and high alumina.[26]

Glass industry personnel have expressed concern about the possibility of hot spots developing in the furnace if wastes are introduced in the melting furnaces.[31,33-35] The occurrence of this phenomenon could significantly reduce the life expectancy of the furnaces. Hot spots would also result in a drop in regenerator efficiency and a rise in fuel consumption.

Acceptable amounts of hazardous waste feed--For a melting furnace that processes 272 Mg/day (300 tons/day) of glass and has dimensions of 7.6 m x 15 m x 1.2 m (25 ft x 50 ft x 4 ft), typical natural gas (fuel) requirements are about 200 m^3/Mg (6500 ft^3/ton) of glass.[28]

Assuming a caloric value of 3.7 MJ/m^3 (1000 Btu/ft^3) for natural gas,[26] heat input to the furnace through the fuel is 7.6 MJ/kg (6.5 x 10^6 Btu/ton) of glass produced.

The average heating value of the seven waste categories previously shown to be destructible is 19.8 MJ/kg (8500 Btu/lb). Based on the postulation that the hazardous wastes would account for 10 percent of the heat input to the furnace, the amount of hazardous waste feed would be:

$$\frac{0.1 \times 7.6 \text{ MJ}}{\text{kg of glass}} \times \frac{1 \text{ kg waste}}{19.8 \text{ MJ input}} = 38.2 \frac{\text{kg waste}}{\text{Mg glass}}$$

For a glass production rate of 272 Mg/day (300 tons/day), and an assumed process operation of 260 days/year, the hazardous waste disposal rate would be:

$$\frac{38.2 \text{ kg}}{\text{Mg glass}} \times \frac{272 \text{ Mg glass}}{\text{day}} \times \frac{260 \text{ days}}{\text{yr}} \times \frac{1 \text{ Mg}}{1000 \text{ kg}}$$

$$= 2700 \text{ Mg/yr}$$

Provisions for waste introduction--In those furnaces where fuel oil is the primary fuel, the hazardous wastes could be blended with the oil and the mixture could be preheated before being fed into atomizing burners. Blending of fuel and waste, however, is not feasible for natural gas-fired furnaces; therefore, additional burner ports might have to be installed to introduce the wastes. The locations of these burners should be such that the flame characteristics are not significantly affected.

Projected trends in process fuel--Table 4.3-7 presents a breakdown of the consumption of various types of fuel in glass container production. It should be noted that the reported consumption figures include operations other than glass melting, (e.g., space heating, glazing, and annealing); however, these other operations do not use fuel oil. Although natural gas is clearly the predominant fuel, the use of fuel oil appears to be increasing in recent years.[25]

4.3.2.2 Geographic Matching of Industry Locations and Waste Generators--

Table 4.3-8 presents the geographic distribution of the waste categories that conceivably could be destroyed in a glass melting furnace. Figures 4.3-1 and 4.3-3 (presented earlier) show that the majority of the glass container and pressed and blown glass plants are concentrated in the eastern United States, particularly New York, New Jersey, and Pennsylvania. Because the

TABLE 4.3-7. BREAKDOWN OF FUEL CONSUMPTION FOR CONTAINER GLASS MANUFACTURE, 1978[36]

Oil	590 Ml (155 x 10^6 gal)
Residual	320 Ml (84 x 10^6 gal)
Distillate	270 Ml (71 x 10^6 gal)
Natural gas	2.9 Gm^3 (103 x 10^9 ft^3)
LPG	15 Gg (33 x 10^6 lb)
Energy consumption (all fuels)	1.4 x 10^8 GJ (131 x 10^{12} Btu)
Electrical energy consumption	4872 x 10^6 kWh
Total energy consumption (fuels and electrical energy)	1.6 x 10^8 GJ (148 x 10^{12} Btu)

TABLE 4.3-8. SIC CODES AND GEOGRAPHIC DISTRIBUTION OF INDUSTRIES GENERATING HAZARDOUS WASTE DESTRUCTIBLE IN GLASS MANUFACTURING FURNACES[19]

Hazardous waste type	SIC code for waste generating industry	Geographic distribution of waste
F003	-	Widespread
K015	2865	Two plants (locations not known)
K018	2869	Houston and Freeport, Texas; Lake Charles and Baton Rouge, Louisiana; Deepwater, New Jersey; Long Beach, California
K023	2869	Richmond and El Sequendo, California; Kearney and Bridgeport, New Jersey; Baton Rouge, Louisiana; Bridgeville and Neville Island, Pennsylvania; Cicero and Millsdale, Illinois; Texas City, Texas
K025	2865	Bound Brook, New Jersey; Willow Island, West Virginia; Beaumont, Texas; Gibbstown, New Jersey; Geismar, Louisiana; New Martinville, West Virginia
K030	2869	Freeport, Deer Park, and Corpus Christi, Texas; Baton Rouge, Lake Charles, and Plaquemine, Louisiana; Pittsburg, California; Louisville, Kentucky; Wichita, Kansas
K093	2869	Texas, California, New Jersey

State of New Jersey is also a generator of four of the destructible waste categories (K018, K023, K025, K093), there is very good matching between waste generator and glass industry locations for these wastes. Waste K023 is generated in Pennsylvania and K025, in West Virginia, where many glass plants (especially pressed and blown glass) are located. Most of the destructible wastes are also generated in California, where a total of 24 container glass and pressed and blown glass plants are situated. Significant quantities of the destructible wastes are also generated in Texas and Louisiana, but few glass plants are located in these states.

4.3.2.3 Environmental Factors--

Impact on air pollution control devices--Control technology for glass furnaces is in a developmental state. The ESP appears to be emerging as the preferred control device.[25] Only a few control device installations are in continuous operation. Two tableware glass plants operated by Owens-Illinois, one in California and the other in Illinois, use a fabric filter and an ESP, respectively, for controlling particulates from their glass furnaces; an Anchor Hocking four-furnace tableware plant in Ohio has been using ESP's for this purpose since 1977. Combustion of hazardous wastes in the furnace is not expected to have any harmful effect on such air pollution control devices.

High carbon monoxide emissions (375 ppm) have been measured in flint glass furnaces.[27] The potential for increased CO levels and hydrocarbon emissions exists during the destruction of hazardous wastes, and an afterburner may be necessary to control such emissions. The only known afterburner installation is at an Owens-Corning fiberglass plant in Fairburn, Georgia. Adequate combustion control can also be exercised to control CO and HC levels.

Chlorine and fluorine levels in the exhaust from container glass furnaces are about 2 to 4 ppm (see Table 4.3-3).[27] Destructible waste types K015, K018, and K030 contain halogenated compounds, which could cause an increase in halogen emissions.

Test burns definitely should be conducted to determine the nature of emissions resulting from the combustion of hazardous wastes in glass melting furnaces. The results of such tests would dictate whether additional air pollution control equipment is required.

Environmental impact--In glass manufacture, the melting process is the major source of emissions. The volume of emissions varies from furnace to furnace and depends on the composition of the glass being produced. To a lesser degree, the volume also depends on the design and operating characteristics of the furnace and the type of fuel used. Flint glass has the lowest emission rate. Container glass furnaces can usually meet both process weight regulations and 40 percent opacity limitations without controls, but they have difficulty staying consistently below the 10 percent opacity limitations imposed by some agencies. The atmospheric impact will have to be closely monitored when charging hazardous wastes into the melting furnace.

Minor quantities of liquid/solid wastes are generated in the glass industry. These include cullet quench water, mold degreasing waste, and waste lubricants. Hazardous waste combustion should have no impact on the generation of these wastes.

4.3.2.4 Institutional Factors--

Historically, the growth of the glass container industry has fluctuated considerably, but shipments have steadily increased at an average annual rate of 6 percent since 1967. A large proportion of this growth is attributable to the popularity of the nonreturnable beverage bottle. The growth in shipments has slowed down in recent years, and future growth may be tied to legislation restricting the use of nonreturnable containers.[25] Production levels in 1980 were about 20 percent higher than 1974 levels.

Future growth trends are somewhat difficult to estimate for the pressed and blown glass segment. The diversity of this sector tends to exempt it from the fluctuations experienced by other glass industry segments. An annual growth rate of 5 to 8 percent appears reasonable for this segment.[25]

The concept of burning wastes in glass melting furnaces is not entirely new to the glass industry. The Glass Packaging Institute indicated that a glass container plant operated by Glass Containers Corporation has used the glass melting process to destroy hazardous components in waste lubricating oils generated at the plant.[34] These waste oils, which contained about 1 percent trichloroethylene, were initially centrifuged to remove sludges and metallic components before being charged to the furnace. No problems were encountered with the burning, nor was product (glass) quality affected. In spite of this positive experience, glass manufacturers still appear to be somewhat reluctant to accept the idea of charging hazardous wastes in their melting furnaces.[31,35]

Customers also may not readily welcome the idea of introducing hazardous wastes in glass melting furnaces, especially when the glass products are used for storing or processing food materials. The presence of most glass plants in highly populated areas is another unfavorable factor.

4.4 IRON AND STEEL INDUSTRY

Figure 4.4-1 presents a general material flow diagram for the iron and steel industry. The following high-temperature processes were considered for the feasibility of thermal destruction of hazardous wastes at high temperatures.

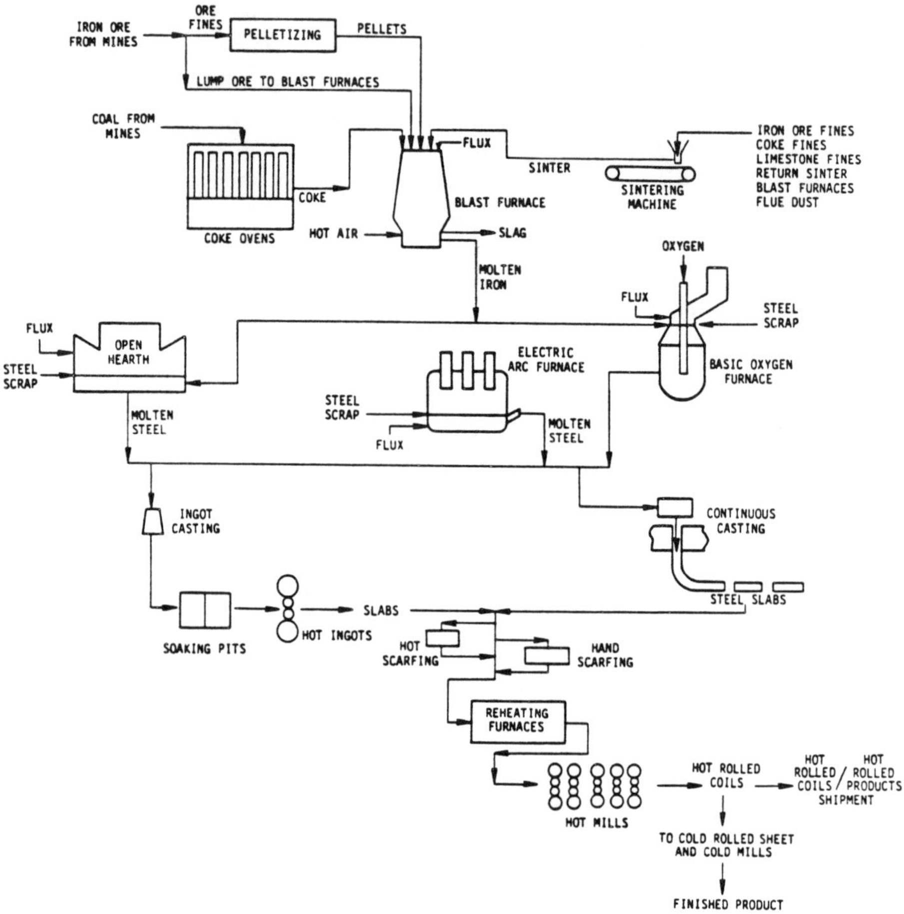

Figure 4.4-1. General material flow diagram for iron and steel industry.

Process	Temperature range °C	°F
Coke ovens	1260-1330	(2300-2400)
Sinter plants	1290-1450	(2350-2600)
Pelletizing plants	1220-1370	(2200-2500)
Basic oxygen furnaces (BOF)	1650-1820	(3000-3300)
Electric arc furnaces (EAF)	1570-1650	(2850-3000)
Soaking pits	1180-1340	(2150-2450)
Blast furnaces (BF)	1110-1870	(2000-3400)
Open hearths (OH)	1610-1650	(2900-3000)

Each process was examined with respect to the following issues:

- Suitability of time/temperature profiles to achieve 99.99 percent destruction of certain waste categories.
- Hazardous waste handling, charging limitations, and safety considerations.
- Handling of the incinerated products and other possible environmental impacts.
- Impact on the refractory lining of the furnaces.
- Effects of hazardous waste incineration on product quality.

4.4.1 Process Description

Each of the processes not considered feasible is briefly discussed first. The feasible processes are then discussed in more detail.

4.4.1.1 Coke Ovens--

Nearly all metallurgical coke, which is the major fuel used in blast furnaces, is made by the byproduct oven process. In this process, coal is heated in the absence of air, which drives off most of the volatile matter, part of the sulfur, and the contained moisture. The outgoing gases from this process are collected as byproduct gas. The residue is about 90 percent carbon, and the remainder is ash and sulfur. After it is screened, this metallurgical coke is sent to the blast furnace.

In the coke-making process, finely ground metallurgical coal is charged into a coke rectangular oven. Older ovens may be from 9 to 12 meters (30 to 40 feet) long, 1.8 to 4.3 meters (6 to 14 feet) high, and 0.28 to 0.56 meter (11 to 22 inches) wide. More recent ovens are 17 meters (55 feet) long, 4 to 6 meters (13 to 20 feet) high, and about 0.46 meter (18 inches) wide. Groups of these ovens are called batteries. As many as 70 to 100 ovens may be grouped together in a battery to facilitate the charging of the coal into the ovens and the pushing out of the coke produced when the coking process is complete. The coal is charged through three or four charging holes situated in a row on top of each oven. Specially designed hopper cars (known as larry cars) discharge coal from the bottom into the charging holes. The coal is carbonized in a reducing atmosphere for a period of 15 to 18 hours. The ovens are heated indirectly by burning cleaned coke oven gas in the combustion chambers on both sides of the oven. Combustion air is preheated by passing it through a regenerative brick matrix (known as checkerwork) which is located under the battery.

The gases released during carbonization are ducted to the byproduct plant, where useful byproducts such as tar, ammonia, and light oil are recovered from the gas. The removal process in the byproducts plant also cleans the gas, which still has heating value. This cleaned gas is used for combustion in the coke ovens and in other fuel burning units in the steel plant.

The temperature attained in the oven during carbonization of furnace coke is about 1300°C (2400°F). The carbonization temperatures in manufacturing foundry coke are lower, and the coking time varies from 24 to 32 hours. When the coke is ready, the doors on both sides of the oven are removed, and a ram mounted on a pusher machine pushes the hot coke into a receiving quench car. The quench car takes the coke to a quench tower where it is cooled with water sprays.

Considerable fugitive emissions can be generated in the coke-making process. Fugitive emissions occur during charging (from the charging holes and the larry car hoppers); during carbonization (from the doors, charging hole lids, and offtake pipes); and during the pushing of coke. These fugitive emissions are difficult to control. This emission potential and the lack of any feasible method for introducing hazardous wastes make it impossible to use today's coke oven process for waste pyrolysis.

Although cofiring of liquid hazardous wastes in the coke oven combustion chamber is conceptually possible, it is not considered technically feasible. The majority of coke ovens fire recycled coke oven gas in a complex heating system that uses multiple burners in each flue; temperature control is critical. Cofiring would require major retrofitting. Because any unincinerated products would be vented directly to the atmosphere through the stack, which generally does not have any air pollution control equipment, coke ovens were eliminated from further consideration.

4.4.1.2 Sinter Plant--

Sintering plants convert iron-ore fines, limestone fines, recycled metallurgical dusts, and mill scale (oily metal wastes from finishing operations) into an agglomerated product for charging into the blast furnace. The iron-bearing materials are mixed with coke breeze (coke not suitable for blast furnace feed) and then deposited in a layer on a slow moving grate. The combustion is initiated by igniting the surface of the bed by use of a gas or oil-fired ignition box at the head end of the machine. An induced draft applied at the bottom of the bed moves a narrow, high-temperature burning zone through the bed. Within this narrow fusion zone, the surfaces of adjacent particles reach fusion temperature and the gangue constituents form a semiliquid slag. Bonding is effected by a combination of fusion, grain growth, and liquid formation. Air pulled through the bed from the top quenches the rear edge of the advancing fusion zone as it moves downwards through the bed. The final product consists of a cellular mass of ore bonded in a slag matrix. The sintered material is cooled, crushed, and screened to remove the undersize, which is recycled.

Sintering machines process a wide variety of feed materials, and the raw material handling systems generally consist of open bins and conveyors. Also, fusion is incomplete in this process and does not necessarily encompass the entire mass. Because retention times are short, any hazardous waste added to the sinter feed might not be completely incinerated.

The maximum temperatures attained in the sinter bed are 1290° to 1430°C (2350° to 2600°F), but these temperatures are achieved only in the narrow fusion zone; thus, the residence time at these high temperatures is inherently short. By the time the draft air passes through the 25 to 30 cm (10 to 12 inches) of bed, the temperature falls below 650°C (1200°F). This means unincinerated or partly incinerated hazardous wastes could escape through the gas collection system and into the atmosphere. Cofiring wastes in the ignition hood is not feasible because of the short retention time, the relatively small fuel consumption 109 MJ/Mg sinter (about 160,000 Btu/ton), and the open structure of the hood, which could allow combustion products to escape into the workplace. These factors make sinter plants unsuitable for hazardous waste destruction.

4.4.1.3 Pelletizing--

Low-grade iron ores are subjected to various ore-treatment processes to increase their iron content. The ore fines obtained after crushing and froth flotation are agglomerated into pellets. The main advantages of pelletizing are achievement of controlled chemical analysis by mixing desired materials in the pellet mix, resistance to weathering and freezing, relative size uniformity, physical strength, and good reducibility. The concentrates are usually formed into balls (pellets) ranging in size from 1.0 to

1.3 cm (3/8 to 1/2 inch) in a balling mill, and the green pellets are then fired in kilns. In rotary kiln and grate kiln furnaces, essentially all the heat to the system is supplied countercurrently to the feed by a burner at the discharge end of the kiln. The hot gases from the kiln pass over and through the grate and supply the heat necessary to dry and pre-fire the pellets. Auxiliary burners are sometimes used near the grate to develop adequate strength in the pellets to withstand the fall into the kiln. The maximum temperature attained in the kiln is 1400°C (2500°F). The fuel requirements for induration (heat treatment of pellets) is estimated to be about 730 MJ/Mg (625,000 Btu per ton) for magnetite and about 860 MJ/Mg (893,000 Btu per ton) for hematite.[37]

The potential exists for incinerating selected hazardous wastes in the rotary kiln by cofiring the liquid wastes with the fuel when time and temperature conditions are suitable. Almost all the pelletizing plants that have kilns, however, are in the State of Minnesota, close to the iron ore mines (four plants). These plants generally are in inaccessible areas, far from the hazardous waste generators. Even though it is possible to incinerate hazardous wastes in these kilns, the logistics of transportation and the limited number of kilns preclude pelletizing kilns as a viable option for hazardous waste incineration.

4.4.1.4 Basic Oxygen Furnace--

In this process, a water-cooled lance is used to blow pure oxygen into the molten iron in the BOF vessel. The oxygen reacts with the carbon contained in the iron to supply the process heat. A temperature of 1670°C (3000°F) is developed in the body of the vessel, and temperatures at the lance tip reach 2220° to 2500°C (4000° to 4500°F). Ordinarily, no external fuel is used. The reaction product gases exit through a hood to gas-cleaning equipment. During the oxygen blow, the gases contain about 90 percent CO and 10 percent CO_2. In the case of open-hood designs, excess air enters the annulus between the hood and the mouth of the vessel and the CO is burned to CO_2, which generates temperatures in excess of 2800°C (5000°F). The gases are cooled by direct water sprays and cleaned by electrostatic precipitators. In the case of closed-hood systems, the CO combustion is suppressed by the close fit between the hood and the vessel, which minimizes air infiltration. The CO-rich exhaust gases pass through a quencher for cooling and then to venturi scrubbers. These exhaust gases can be stored for use as a fuel or flared and exhausted to the atmosphere.

Steel scrap is charged into the BOF vessel by tilting the vessel forward and charging the scrap directly into the vessel. After the scrap is charged, the hot metal is charged by tilting the vessel forward while metal is poured into the vessel from a transfer ladle by means of an overhead crane. After the hot metal is charged, the vessel is brought into an upright position, and the oxygen lance is introduced from the top. Fugitive emissions occur during the charging operations and, generally, any time the vessel is tilted away from the primary collection hood. Newer shops have secondary capture systems for control of these emissions. Given the high temperatures, relatively long gas retention time, and the pollution control devices used, this process has potential for hazardous wastes incineration, in spite of the potential for fugitive emissions from the open-hood systems.

Charging hazardous waste into the BOF vessel, however, could be potentially dangerous and time-consuming. Wastes would have to be segregated and charged separately, which would involve additional vessel tilting and an additional materials-handling step. The segregation and drying of wastes would also require additional space, which may not be available in an operating BOF shop. Also, the charging of uncombined water and organic wastes into the BOF vessel could pose an explosion hazard. The productivity of the BOF process is high because the average tap-to-tap time (time from charging the iron until removal of the steel) is typically less than one hour. A regular supply of wastes to meet the charging schedule could impose delays, particularly in light of the strict precautions necessary to handle these wastes. Such constraints would slow down the steelmaking process on these high-volume production units, which would probably not be acceptable to the steel industry.

In some BOF shops, scrap is preheated to reduce the hot metal requirement. This is accomplished by charging extra scrap and then heating the scrap by means of natural gas or oil and oxygen introduced through an auxiliary lance. Depending on the amount of scrap charged, the preheating lasts 10 to 20 minutes. This is one potential practice where hazardous wastes might be used. The use of wastes in this procedure should not unduly affect the quality of steel, as the incineration of wastes would be completed during the scrap preheating period. Unfortunately, the number of BOF shops with open-hood configuration, where scrap preheating coupled with waste incineration could be practiced, is limited. Only 9 of the 36 BOF shops in the United States have an open-hood configuration, and not all of them practice scrap-preheating. Those that do, only practice it periodically. Because of these constraints, this method of waste disposal is not considered a viable option.

4.4.1.5 Electric Arc Furnaces--

Electric arc furnaces are refractory-lined cylindrical basins. Electrical energy is supplied through large graphite electrodes extending down through the roof of the furnace. The charge consists of select steel scrap, cast iron scrap, pig iron, alloying elements, and flux to achieve the desired composition. Melting is accomplished by the heat of the arc between the graphite electrodes and the charge. Scrap is charged by raising the electrodes and swinging the roof and electrodes to one side. This method of charging generates heavy emissions, which generally are controlled by direct-shell evacuation, canopy hoods, or enclosed buildings evacuated to a fabric filter. Complete capture of the emissions is not achieved.

Because of the high degree of furnace atmosphere control (oxidizing or reducing) that can be achieved, low-carbon, high-carbon, and alloy steels are all made in electric furnaces. This is a batch operation with tap-to-tap (time from materials charge to steel removal) times of 2 to 8 hours, depending on the quality of steel and the use of oxygen. Temperatures range from 1570° to 1650°C (2850° to 3000°F). The capacities of electric arc furnaces range widely, from 4.5 to 320 Mg (5 to 350 tons) per heat (one production cycle), depending on the quality of the steels made and the desired finished products.

These high-efficiency furnaces could not accommodate the charging of large quantities of hazardous wastes without disruption to production. The electric furnace is not considered a viable option for hazardous waste incineration for several reasons: the fugitive emissions that escape during charging, emissions that escape through electrode holes during meltdown, fugitive emissions generated by oxygen lancing, emissions generated by slagging and tapping, and finally, the logistic and safety problems involved in the handling and charging of hazardous wastes.

4.4.1.6 Soaking Pits--

Steel ingots are cast in molds at the melting shop and allowed to solidify until a sufficiently thick skin is formed to contain the molten interior. The molds are then stripped from the ingots by means of a specially designed crane. The hot ingots are transferred to the soaking pits before they are rolled into slabs. The purpose of the soaking pits is to heat the ingots to the desired rolling temperature and to achieve a uniform temperature of 1150° to 1330°C (2100° to 2400°F) throughout. The soaking pits are rectangular boxes with removable covers.

The pit cover is opened, and 6 to 10 ingots (depending on the size of the pit) are charged by crane. After the cover is replaced, the pit is fired by burners in the end walls. The fuels are natural gas, coke oven gas, and oil. The exit gases pass through openings in the lower part of the pit, and proceed through recuperators to a stack. Each batch is heated for 4 to 12 hours, depending on the elapsed time between stripping and charging. Theoretically, it is possible to incinerate selected liquid wastes as an admixture with oil firing; however, such a practice is not considered feasible for the following reasons:

1. Natural gas and coke oven gas are the preferred fuels in soaking pits.
2. No soaking pits have any type of air pollution control other than good combustion practices.
3. Leaks can occur at the cover of the soaking pit.
4. Fuel switching is practiced in some pits, which requires that fuel/air ratios be adjusted.

4.4.2 Detailed Evaluation of Feasible Processes

Because of their greater potential for hazardous wastes disposal, the following processes are examined in more detail:

- Blast furnaces using oil/tar injection
- Open hearth furnaces

4.4.2.1 Blast Furnaces--

Process description--The blast furnace is a tall, cylindrical, countercurrent flow reactor. Figure 4.4-2 shows a cross section of a typical blast furnace. Iron ore, pellets, sinter, limestone, and coke are charged into the blast furnace from the top. Hot air [780° to 1220°C (1400° to 2200°F)] is introduced at the bottom of the furnace through tuyeres. The coke burns at the tuyere level, and the hot gases rise through the descending mass and transfer the sensible heat to the descending raw materials. A temperature of around 1930°C (3500°F) is attained at the tuyere

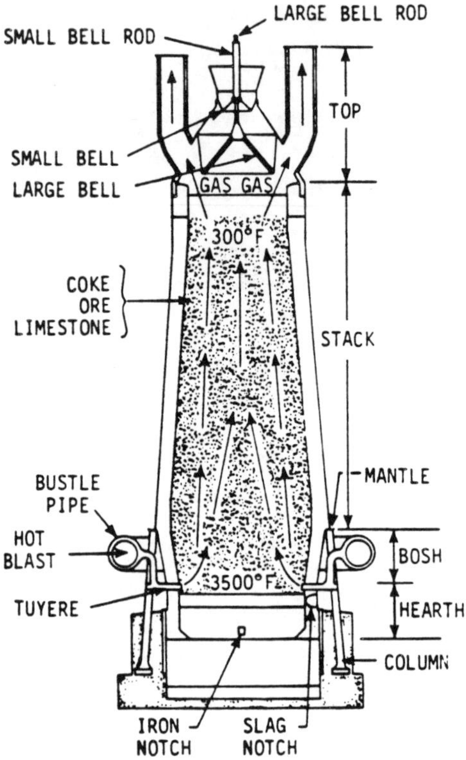

Figure 4.4-2. Cross section of a blast furnace.

level, and the top gases leave the furnace at around 150°C (300°F). The exit gases pass through a dust catcher (cyclone) and then through a scrubber system. Part of this thoroughly cleaned gas is fired in the blast furnace stoves for preheating the combustion air. The molten iron accumulates in the hearth, and the lighter slag forms a layer on top of the molten iron. Iron and slag are periodically drained from the furnace.

Uniform raw material size and chemical consistency play a major role in achieving high productivity and trouble-free operations. The raw material requires extensive preparation to achieve proper consistency of size and chemical composition. A systematic pattern of charging different raw materials is practiced to achieve uniform distribution inside the furnace. It takes approximately 8 hours for the charged raw materials to descend to the hearth zone.

The reactions that take place as the raw materials descend through the furnace are essentially endothermic. Iron ore is progressively reduced to iron oxide (FeO) as the mass descends to the bosh area, where the final reduction to iron (Fe) takes place. A reducing atmosphere is prevalent throughout the descent of the materials in the furnace. Requirements such as uniformly sized burden, reducing atmosphere, and temperature conditions make this process unsuitable for the destruction of solid hazardous wastes by charging such wastes at the top of the furnace. The vapors produced at low temperature would leave with the exit gases without complete incineration.

Technical factors--

Compatibility of hazardous wastes with process fuel--In most blast furnaces in the United States some kind of fuel is injected through the tuyeres to reduce coke consumption and increase productivity. The fuels used for this purpose include oil, tar, pitch, natural gas, coke oven gas, and pulverized coal or combinations thereof. Selection is governed by availability and cost.

Table 4.4-1 shows the pattern of injected fuel usage in U.S. and Canadian blast furnaces in 1976. Table 4.4-2 indicates the location, capacity, and fuel injection data for blast furnaces in the United States. Figure 4.4-3 shows where these blast furnaces are located. It is technically feasible to cofire selected wastes with the fuel oil or tar injected at the bottom of the blast furnace.

Graphical determination of suitable waste categories--Smooth and trouble-free blast furnace operation with fuel injection requires that the adiabatic flame temperature and blast volume be kept as constant as possible. In a typical operation the flame is maintained at a temperature of 1870° to 2040°C (3400° to 3700°F)[41] and a blast volume of 1720 Nm^3/min per ton (60,000 scfm per ton) of hot metal. Pertinent process conditions for determining the time-temperature relationship in blast furnaces are based only on the lower portion of the furnace; it is assumed that about one-third of the furnace volume lies below an imaginary line where the gas temperature drops below 1090°C (2000°F). For purposes of constructing a time-temperature profile, it is conservatively assumed that the upper portion of the furnace will be inactive in destroying hazardous compounds. The process conditions for the lower zone are:

Volume of furnace	589 m^3 (20,800 ft^3) (1/3 of total)
Gas volumetric flow rates	5150 Nm^3/min (196,000 scfm) (1.4 x blowing rate)
Maximum reaction zone temperature	1870°C (3400°F)
Exit gas temperature	1090°C (2000°F)

The resulting process curves (plots of Δt and $\Delta\bar{t}$ versus T) are overlaid on the hazardous waste plots in Figure 4.4-4.

Table 4.4-3 indicates the categories of wastes that can be dissociated in the blast furnace. The high flame temperature maintained at the tuyeres and the relatively long residence time

TABLE 4.4-1. USE OF VARIOUS FUELS INJECTED INTO THE BLAST FURNACE IN THE UNITED STATES AND CANADA DURING 1976[38]

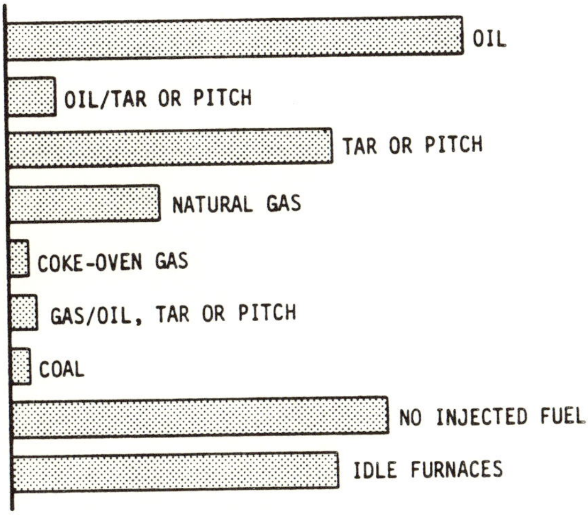

	%
OIL	26.5
OIL/TAR OR PITCH	2.7
TAR OR PITCH	18.5
NATURAL GAS	8.6
COKE-OVEN GAS	1.1
GAS/OIL, TAR OR PITCH	1.5
COAL	1.1
NO INJECTED FUEL	21.4
IDLE FURNACES	18.6
	100.0

TABLE 4.4-2. BLAST FURNACES IN THE UNITED STATES[39,40]

Company	Location	No. of blast furnaces	Capacity, 10^6 Mg (10^6 tons)	Fuel injection
Armco, Inc.	Ashland, Kentucky	2	1.8 (2.0)	Coal/O_2[a]
	Houston, Texas	1	0.7 (0.8)	Natural gas
	Middletown, Ohio	1	1.8 (2.0)	Natural gas
	Hamilton, Ohio	2	1.8 (2.0)	Tar in one furnace
Bethlehem Steel	Bethlehem, Pennsylvania	4	3.2 (3.5)	-[b]
	Burns Harbour, Indiana	2	3.2 (3.5)	-
	Lackawanna, New York	6	3.8 (4.2)	-
	Sparrows Point, Maryland	5	5.7 (6.3)	-
CF&I Steel Corp.	Pueblo, Colorado	4	0.9 (1.0)	Oil/tar
Crucible Steel	Midland, Pennsylvania	2	0.9 (1.0)	-
Cyclops Corp. Empire-Detroit Division	Portsmouth, Ohio	1	0.6 (0.7)	Oil
Ford Motor Co.	Dearborn, Michigan	3	2.0 (2.2)	Natural gas/O_2
Inland Steel Co.	Indiana Harbour Works East Chicago, Illinois	9	5.9 (6.5)	Oil/tar
Interlake, Inc.	S. Chicago, Illinois	2	1.1 (1.2)	-
Jim Walter Resources, Inc.	Birmingham, Alabama	1	0.4 (0.4)	-
J&L Steel	Aliquippa, Pennsylvania	5	3.3 (3.6)	Oil/O_2
	Cleveland, Ohio	2	1.6 (1.8)	Oil
	E. Chicago, Indiana	4	3.2 (3.5)	-
Kaiser Steel	Fontana, California	4	2.5 (2.7)	-
Lone Star Steel	Lone Star, Texas	1	0.4 (0.4)	-
McLouth Steel Corp.	Trenton, Michigan	2	1.5 (1.7)	-

(continued)

TABLE 4.4-2 (continued)

Company	Location	No. of blast furnaces	Capacity, 10^6 Mg (10^6 tons)	Fuel injection
National Steel	Granite City, Illinois	2	1.6 (1.8)	Oil/O_2
	Great Lakes Steel Div., River Rouge, Michigan	4	3.4 (3.7)	Gas/oil/O_2
	Hanna Furnaces, Buffalo, New York	3	0.6 (0.7)	Oil in one furnace
	Weirton Steel Division, Weirton, West Virginia	4	2.7 (3.0)	Oil/tar/pitch
Republic Steel Corp.	Buffalo, New York	2	0.7 (0.8)	-
	Cleveland, Ohio	5	2.7 (3.0)	-
	Gadsden, Alabama	2	0.3 (0.8)	-
	S. Chicago, Illinois	1	1.13 (1.25)	-
	Warren, Ohio	1	1.13 (1.25)	-
	Youngstown, Ohio	3	2.2 (2.4)	-
Sharon Steel Corp.	Sharon, Pennsylvania	2	0.9 (1.0)	-
Shenango, Inc.	Neville Island, Pennsylvania	2	0.4 (0.4)	-
U.S. Steel Corp.	Braddock, Pennsylvania	4	2.3 (2.5)	-
	Duquesne, Pennsylvania	2	1.4 (1.5)	-
	Fairfield, Alabama	6	2.4 (2.7)	-
	Fairless Hills, Pennsylvania	3	2.4 (2.7)	-
	Gary, Indiana	11	5.9 (6.5)	-
	Geneva, Utah	3	1.8 (2.0)	-
	Lorain, Ohio	5	2.4 (2.7)	-
	McKeesport, Pennsylvania	2	0.3 (0.3)	-
	Rankin, Pennsylvania	2	1.8 (2.0)	-
	S. Chicago, Illinois	5	3.7 (4.1)	-
	Youngstown, Ohio	4	1.4 (1.6)	-
Wheeling - Pittsburgh Steel	Monessen, Pennsylvania	3	1.4 (1.5)	-
	Steubenville, Ohio	5	2.2 (2.4)	-

[a] O_2 = Oxygen.

[b] Dash means no information available.

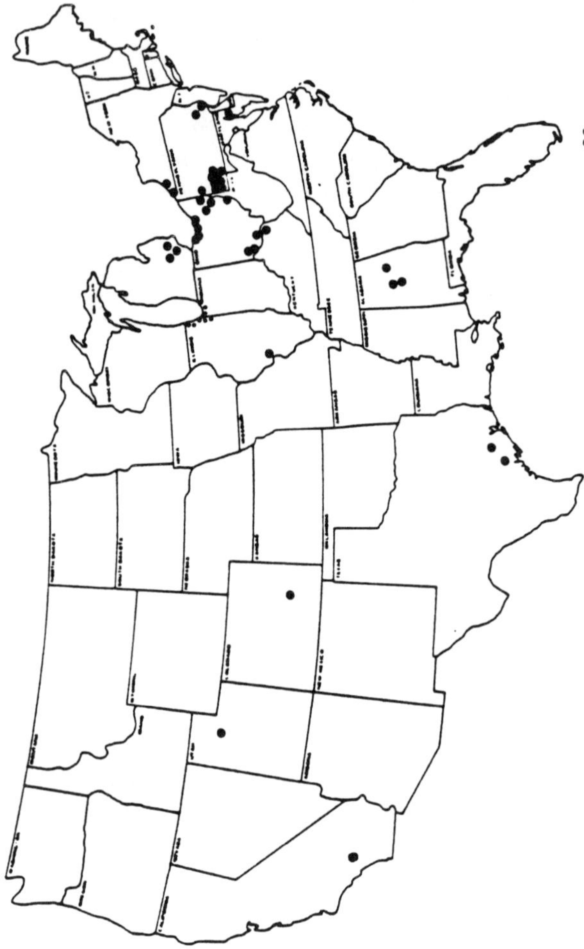

Figure 4.4-3. Location of blast furnaces in the United States.[39]

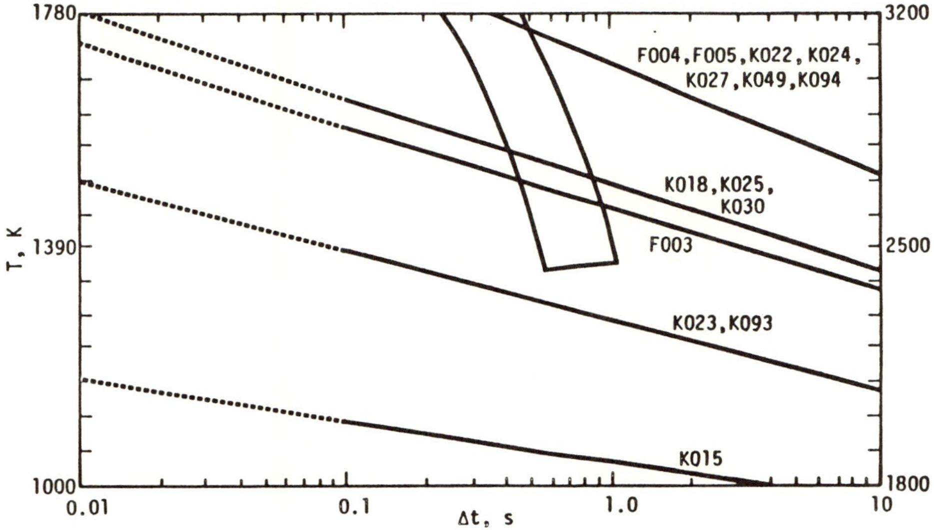

Figure 4.4-4. Hazardous waste destructibility predictions for blast furnaces in iron and steel manufacture.

provide conditions suitable for the thermal destruction of most K and F Series wastes. The reason for excluding some of the wastes is given in the remarks column.

TABLE 4.4-3. APPLICABILITY OF BLAST FURNACES FOR THERMAL DESTRUCTION OF HAZARDOUS WASTES

Type of wastes	Applicability	Remarks
K015	No	Highly chlorinated compounds
K023	Yes	
K093	Yes	
F003	Yes	
K018	No	Highly chlorinated compounds
K025	No	Organic sludge
K030	No	Chlorinated
F004	Yes	
F005	Yes	
K022	No	Sludge
K024	No	Sludge
K027	No	Chlorinated sludge
K049	No	Aqueous sludge
K094	No	Sludge

Effects on product quality and product application--The only impact on iron or slag composition would be from metals or inorganics in the waste. Most of the elements that form stable oxides (aluminum, calcium, titanium) will be taken up by the slag. For the most part, metals such as iron, manganese, and chromium will be dissolved in the molten iron. In either case, the contribution of metals from the waste categories under consideration is expected to be insignificant and to have no measurable effect on the iron or slag. Sulfur-containing wastes should be avoided, however, because sulfur is an undesirable element in the iron.

Impact on process hardware--The dissociation products in the reaction zone at the tuyeres are carbon oxide, hydrogen, nitrogen, and a variety of volatile metal oxides, halides, and sulfides, depending on the composition of the hazardous wastes. Volatile metallic species can form at intermediate to high temperatures in the blast furnace. The most troublesome are zinc, lead, and alkali metal oxides. The volatile species formed at the lower level in the furnace ascend to the cooler portions of the furnace where condensation and readsorption can occur. In this case, a recirculating load is established with increasing concentration of volatile species, which eventually leads to discharge of these species in the off-gases or slag-metal system. The condensation of large amounts of alkali metals and halides in the cooler part of the furnace can lead to sticking and bridging and thus impede the smooth flow of materials in the blast furnace. In addition, the alkali and halides attack the refractory lining of the furnace, which leads to physical deterioration of the lining and reduced lining life. Lead and zinc deposits also can cause refractory damage; great care is taken to avoid these metals in the feed to blast furnaces.

Most furnaces have scrubbers with recirculating water systems. Inorganic residuals originating from hazardous wastes, particularly alkali metal salts and volatile metal salts, might increase the load on the scrubber water in these systems and make it necessary to increase blowdown rates. The blast furnace is normally the keystone of steel mill operation. Because upsets in operation can be disastrous and can require days or weeks for recovery, injection of wastes would have to be done very cautiously.

Acceptable amounts of hazardous waste feed--The oil injection rates achieved in the industry vary widely according to the individual practices followed. The highest reported consumption is around 115 kg/Mg (230 lb/net ton) of hot metal (maximum) and 90 kg/Mg (180 lb/net ton) on the average.[42] Assuming that 25 percent of the furnaces using oil or tar injection could burn 9 to 14 kg (20 to 30 lb) of liquid wastes per ton of hot metal (or 10 to 15 percent of it), it is estimated that 180,000 to 270,000 Mg/yr (200,000 to 300,000 tons/yr) of liquid wastes could be disposed of in blast furnaces. This is based on 79 million Mg (87 million tons) of iron produced in 1979 in the U.S. steel industry.

Provisions for waste introduction--Oil is generally pumped through a separate lance introduced through the tuyere. Steam or air may be used to atomize the oil. A typical lance design consists of two concentric pipes. Oil flows through the central pipe and steam goes through the outer pipe.

Oil is generally stored in a storage tank near the furnace, which is equipped with a heating and pumping system for delivery to the individual injection lances. Wastes could be mixed with the oil prior to injection. Use of a separate lance is possible, but this would require new blast pipes at each tuyere and a separate control system.

Projected trends in process fuels--Coke consumption in blast furnaces has steadily declined over the past 20 years because of improved burdens, higher blast air temperatures, and the use of fuel injection. Although fuel injection is expected to continue as a common practice, the trend is towards the replacement of several smaller furnaces with a few larger furnaces. Consequently, the total number of blast furnaces is declining. With iron production assigned to fewer furnaces, upsets or delays in the operation become more critical. This will tend to discourage such auxiliary operations as hazardous waste injection.

<u>Geographic matching of industry locations and waste generators</u>--The generation of wastes in categories F003, F004, and F005 is widespread in industrial areas. As Figure 4.4-3 shows, most of the blast furnaces also are located in the industrialized

states. Category K023 wastes are generated in Illinois, Pennsylvania, and New Jersey, and K093 wastes are generated in Illinois, Texas, New Jersey, and California. Thus, the geographical matching between waste sources and blast furnaces is reasonably good.

Environmental factors--The effluent streams from a blast furnace include slag, off-gases, scrubber water blowdown, and fugitive emissions during casting. The slag is normally sold as a byproduct for use as aggregate. Although ash components and certain metallic oxides in the waste could find their way into the slag, the amounts would probably be small and have no significant effect on the slag leaching characteristics.

Blast furnaces are equipped with high-efficiency (ΔP = 15 to 17 kPa) scrubbers. The off-gases are cleaned to 0.023 g/m^3 (0.01 gr/ft^3) or better for use as a byproduct fuel in the stoves and boilers and for other uses. The gas has a heating value of about 3.4 MJ/m^3 (90 Btu/ft^3); therefore, gaseous compounds formed from the destruction of hazardous wastes would be further exposed to combustion conditions. Because excess gas is sometimes bled to the atmosphere, the off-gases would have to be monitored during test burns to determine what compounds could be expected in them.

Most furnaces are equipped with recirculating scrubber water systems. These systems must meet stringent blowdown rates to comply with guideline discharge limitations for effluents. Any increase in chlorides resulting from waste destruction would necessitate an increase in the blowdown rates.

Fugitive emissions can occur at the top of the furnace as a result of leaking seals in the charging system or sudden increases in furnace pressure. Normally these releases involve relatively small amounts of off-gas. The casting and flushing of slag are more common sources of fugitive emissions. Because gases escape from the tap hole, injection would have to be stopped during these operations. If wastes are used, close attention would have to be given to the injection system. This would require instrumentation for monitoring the flow at each tuyere and automatic shutoff valves. Auxiliary lances and related equipment require frequent inspection and maintenance to avoid plugging, leaks, and blowouts. No fugitive emissions are associated with oil injection.

Economic factors--Fuel injection is done for economic reasons; it increases productivity and reduces coke consumption. Any credits for the use of hazardous wastes would be an added incentive. Hazardous wastes with high Btu values could replace part of the injected fuel and thus provide an economic incentive.

Special considerations--Halogenated wastes and wastes containing alkalies should be avoided completely because they would have an adverse effect on the blast furnace lining. Similarly, wastes with a high-sulfur content should be avoided because sulfur is an undesirable element in the metal product. The safety precautions necessary in the receiving, storage, handling, and atomizing of liquid hazardous wastes would have to be meticulously followed.

Institutional factors--In general, steelmaking facilities are not likely to accept hazardous wastes for incineration. Labor unions may object on the grounds of worker exposure to the hazardous wastes. Inasmuch as the industry has invested heavily in air and water pollution controls to treat its own waste streams, it is not likely to consider the introduction of new potential environmental problems. In addition, the decreases in productivity associated with the handling, storage, and destruction of hazardous wastes would not be consistent with the industry's goal to increase productivity. Moreover, most steel plant facilities are close to population centers, which means potential adverse community impacts would have to be considered. One steel plant considered experimenting with incineration of PCB's by injection through the tuyeres in a blast furnace and later evaluated the use of a kiln, but both plans were eventually rejected.[43]

4.4.2.2 Open Hearth Furnaces--

Process description--Open hearth (OH) furnaces are used to refine scrap and hot metal into steel. Up to 100 percent steel scrap or a combination of scrap and molten iron can be charged in this versatile process. Limestone and fluorspar are used as fluxes to absorb impurities. Fuels (coke oven gas, natural gas, oil, tar, or a combination of gas and liquid fuel) are alternately fired from both ends of the furnace. As the combustion gases pass through brick checkerwork, they impart sensible heat to the brickwork. The incoming combustion air is preheated by passing it through the checkerwork. Approximately every 30 minutes, the firing is reversed to recover the heat through the brickwork. Many OH furnaces are equipped with waste heat boilers to conserve energy. The exit gases are first cleaned in either electrostatic precipitators or scrubbers and then exit through the stack. Figure 4.4-5 shows a cross section of an open hearth furnace.

The scrap and flux are charged into the furnace through the doors of specially designed charging boxes. The hot metal is poured into the furnace through the hydraulically operated doors by means of a hot metal chute. Steel and slag are tapped from the tap hole at the rear of the furnace opposite the charging doors. Tap-to-tap time, which varies from 6 to 12 hours, depends on the ratio of hot metal and scrap and the use of oxygen. Productivity has increased with the advent of oxygen lancing through the roof, which hastens the refining process. The operating OH furnaces in integrated steel plants have capacities ranging from 250 to 420 tons per heat. Table 4.4-4 lists currently operating furnaces in the United States. Figure 4.4-6 shows the location of these furnaces. No new OH furnaces are being constructed, and many are being phased out because of the high cost of production per ton of steel and lower productivity compared with basic oxygen furnaces. Once the major method for steel production in the United States, OH furnaces accounted for only 1.7 percent of total production in 1980.

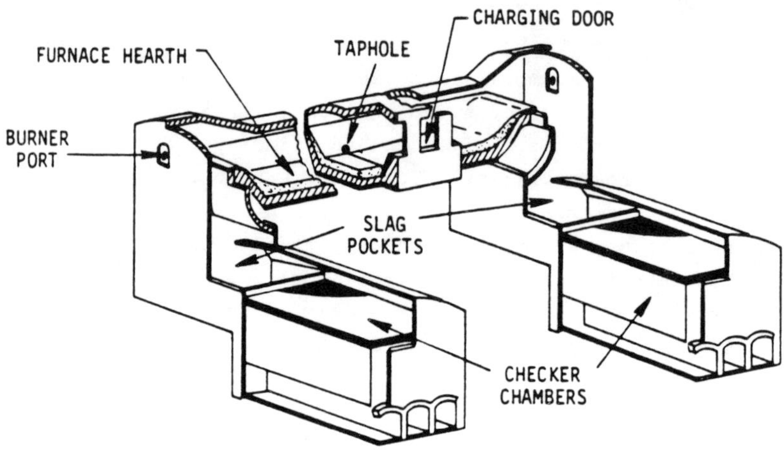

Figure 4.4-5. Cross section of an open hearth furnace.

TABLE 4.4-4. LIST OF OPEN HEARTH FURNACES IN THE UNITED STATES[39,42]

Company	Location	Number of furnaces	Annual capacity, 10^6 Mg (10^6 tons)
Armco, Inc.	Middletown, Ohio	6	1.11 (1.225)
Bethlehem Steel Corp.	Sparrows Point, Maryland	7	3.2 (3.50)
Inland Steel Co.	East Chicago, Illinois	7	2.2 (2.40)
Lone Star Steel Co.	Lone Star, Texas	5	1.4 (1.50)
Pacific States Steel	Union City, California	4	0.45 (0.50)
United States Steel Corp.	Fairless Hills, Pennsylvania	9	3.3 (3.60)
	Homestead, Pennsylvania	11	4.2 (4.60)
	Geneva, Utah	10	2.5 (2.75)

Two methods of hazardous waste destruction can be considered in OH furnaces:

1. Liquid wastes having high Btu value could be cofired with oil or tar.

2. Highly viscous, semisolid wastes could be charged into the furnace through the doors.

Technical considerations--

Graphical determination of suitable waste categories-- Pertinent process conditions for determining the time-temperature relationship in open hearth furnaces are:

Volume of furnace	200 m^3 (7000 ft^3)
Gas volumetric flow rate	790 Nm^3/min (30,000 scfm)
Maximum reaction zone temperature	1790°C (3250°F)
Exit gas temperature	1480°C (2700°F)

Figure 4.4-7 presents a temperature profile for a typical open hearth furnace. Figure 4.4-8 shows a time-temperature curve for a typical OH furnace overlaid on the hazardous waste plots. A flame temperature of around 1930°C (3500°F) is achieved in OH furnaces. Coupled with a retention time of 2 to 3 seconds, this temperature is suitable for destroying several hazardous wastes.

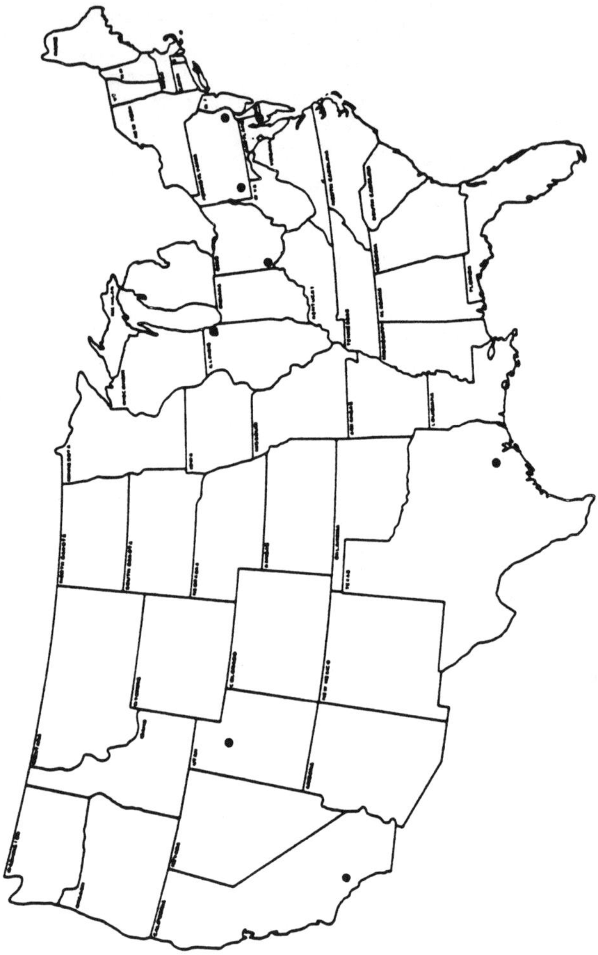

Figure 4.4-6. Location of open hearth furnaces in the United States.[39]

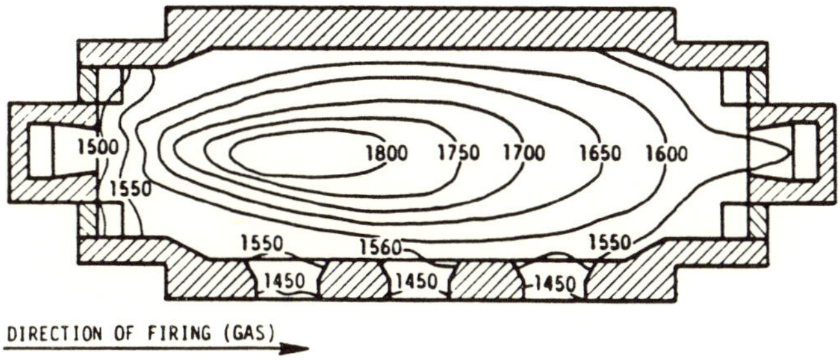

Figure 4.4-7. Temperature profiles within a typical open-hearth furnace.[44]

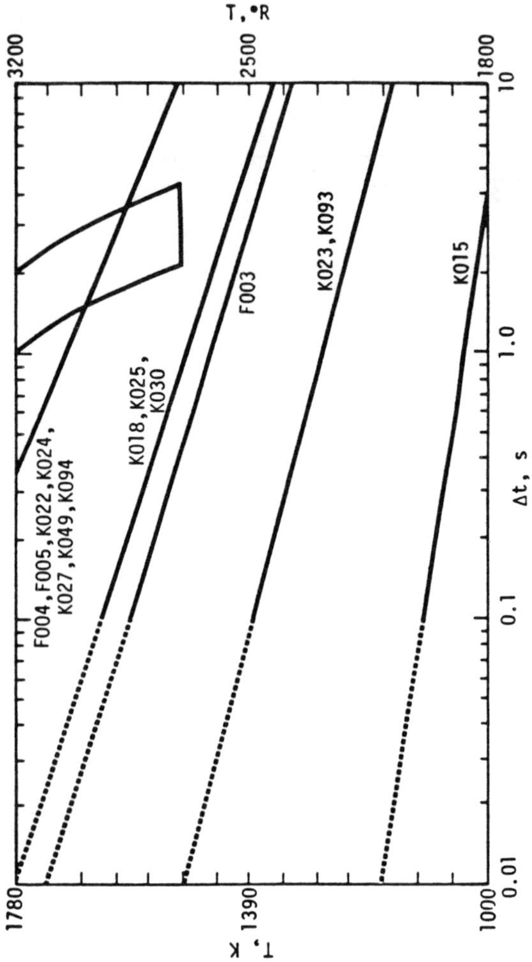

Figure 4.4-8. Predictions of hazardous waste destructibility in open-hearth furnaces in steelmaking.

Gases generated farthest from the burner will travel through the checkerwork, where the temperature is still sufficiently high [1150° to 1210°C (2100° to 2200°F)][43] to complete destruction.

Effects on product quality and product application--Table 4.4-5 shows the distribution of elements between slag and metal in the OH furnace process, based on oxygen potential. When present in slag, the elements are in their oxide or sulfide form. Relatively small amounts of elements such as Cn, nickel, or antimony significantly affect steel properties. Any waste would therefore have to be monitored for the presence of such tramp elements before being accepted.

TABLE 4.4-5. DISTRIBUTION OF CONTAMINANT ELEMENTS BETWEEN SLAG AND METAL[44]

Group 1 - Elements almost completely taken up by slag: Si, Al, Ti, Zr, B, and V; Ca, Mg
Group 2 - Elements distributed between slag and metal: Mn, P, S and Cr
Group 3 - Elements almost completely in solution in metal: Cu, Ni, Sn, Mo, Co, W, and probably As and Sb
Group 4 - Elements out of slag and metal: Zn, Cd, Pb, and Ag and Hg

The small amounts of contaminants that would normally combine in the slag are not expected to have an adverse effect on slag properties. On occasions when slag is recycled to the blast furnace, lead and zinc would have to be monitored. As in the case of the blast furnace, sulfur-bearing wastes would be unacceptable.

Impact on process hardware--Most open hearth furnaces have a basic refractory lining. The basic slag contains a high percentage of calcium oxide, which would promote retention of heavy metal oxides that may be generated during disposal of some wastes.

The operating OH furnace shops in the United States are equipped with either high-efficiency electrostatic precipitators or venturi scrubbers. Gaseous removal efficiencies in ESP's are negligible; however, venturi scrubbers can collect a portion of the gaseous pollutants, particularly halogens. As discussed under blast furnace fuel injection, both halogenated wastes and wastes containing alkali metals should be avoided because they could reduce the life of the refractory lining. Lead and tin residuals also should be avoided because they, too, adversely affect the lining life.

Acceptable amounts of hazardous waste feed--Fuel consumption in OH furnaces varies according to the proportions of scrap and hot metal charged. A typical value for a 50/50 charge is 2.9 GJ/Mg (2.5 x 10^6 Btu/ton). In a 180-Mg (200-ton) oil-fired furnace, this is equivalent to about 13,200 liters (3500 gallons) of oil per heat. Assuming a substitution rate of 10 percent for hazardous waste fuel, about 6.4 liters (1.7 gallons) of waste could be consumed per ton of steel. Given an annual steel production rate in OH furnaces of 12 x 10^6 Mg (13 x 10^6 tons), the maximum potential waste disposal rate would be 83 x 10^6 liters (22 x 10^6 gallons per year). The actual potential is probably closer to half this rate, however, because wastes could not be fired during charging operations or periods of bath turbulence. Furthermore, not all furnaces could accommodate this practice.

Provision for waste introduction--Fuel oil and tar are stored near the furnace building in storage tanks equipped with a heating and pumping system to individual furnaces. Wastes could be stored in a separate tank and metered into the fuel before it reaches the burner. If the waste is cofired with coke oven gas or natural gas, burner modifications would have to be made to the furnace and separate waste fuel lines would have to be provided from the storage tank to the furnace(s).

Projected trends in process fuel--No major shifts in process fuel use are expected in the OH furnace process. As discussed previously, however, the use of the OH furnaces is declining. Existing shops will continue to operate as long as they are economical, but they will be retired as they near the end of their useful life and will not be replaced.

Geographical matching of industry locations and waste generators--Table 4.4-6 presents the categories of liquid wastes that probably can be destroyed in open hearth furnaces, based on the predictions from Figure 4.4-8. Categories K023 and K024 are generated in the States of Pennsylvania, Illinois, and New Jersey; K093 and K094 are generated in Illinois, Texas, New Jersey, and California; generation of F003, F004, F005, K022, and K049 is widespread. Waste categories K015, K018, K027, and K030 are shown as not applicable because the chlorinated compounds will deteriorate the furnace refractory. Based on the distribution shown in Figure 4.4-6, there is some degree of matching between waste generators and open hearth furnace locations. The limited number of furnaces overall, however, rules out widespread national application.

TABLE 4.4-6. APPLICABILITY OPEN HEARTH FURNACES FOR THERMAL DESTRUCTION OF HAZARDOUS WASTES

Waste category	Applicability	Remarks
K015	No	Highly chlorinated
K023	Yes	
K093	Yes	
F003	Yes	
K018	No	Heavily chlorinated organics
K025	Yes	
K030	No	Chlorinated
F004	Yes	
F005	Yes	
F022	Yes	
F024	Yes	
K027	No	Chlorinated sludge
K049	Yes	
K094	Yes	

Environmental factors--Fugitive emissions during charging of raw materials and emissions through the doors during flame reversal are potential problems. Waste injection would have to be cut off during charging and periods of bath turbulence. Furnace pressure would have to be controlled at near zero or slightly negative to avoid puffs of fugitive emissions, which is counter to the normal practice of maintaining furnace pressure at zero to slightly positive. Any residues left after destruction of the hazardous waste would enter the metal or the slag. The slag could be handled in the normal manner, but heavy metal concentrations in slag, scrubber sludge, and ESP dust would have to be monitored closely.

Economic factors--Fuel costs are not an overwhelming factor in the total cost structure of open hearth steelmaking. Although fuel credits from high-Btu wastes could be used, it is doubtful they could affect total costs significantly. There is a potential for converting retired shops into waste facilities operated independently of the steelmaking operation by other entrepreneurs. Analysis of such a scenario, however, is beyond the scope of this study.

Institutional factors--Considerable resistance can be expected to the acceptance of hazardous wastes into the open hearth furnaces. The operating open hearth furnaces are competing to remain in business, and productivity and quality are of prime consideration. This factor may preclude any consideration of hazardous waste destruction in these furnaces.

4.5 LIGHTWEIGHT AGGREGATE INDUSTRY

Lightweight aggregates encompass a variety of raw materials that provide the bulk in concrete products. They are distinguished from other mineral aggregate materials by their ability to deliver strength with less dead load.[45,46] The various lightweight aggregate materials can be classified as follows:[45]

- Natural lightweight aggregates - Prepared by crushing and sizing pumice, volcanic cinders, tuff, scoria, and breccia.
- Byproduct lightweight aggregates - Prepared by crushing and sizing foamed and granulated slag, coke breeze, and cinders.
- Manufactured structural lightweight aggregates - Prepared by pyroprocessing clay, shale, or slate in rotary kilns or on traveling grate sintering machines.

This section focuses on the pyroprocessing of expandable clay, shale, or slate to produce manufactured structural lightweight aggregates. Estimates indicate that the rotary kiln method is used in the production of 80 percent of all structural lightweight aggregates.[47]

4.5.1 Process Descriptions

4.5.1.1 The Rotary Kiln Method--

In most cases, the lightweight aggregate plant is composed of a quarry, a raw material preparation area, a kiln, a cooler, and a product storage area. These facilities are usually located on a tract of land that contains a quarry and encompasses several hundred acres.[48] Figure 4.5-1 provides a flow diagram of a lightweight aggregate plant.

The first step in the manufacture of lightweight aggregate is to establish a plant on a tract of land that has suitable raw material. Only some deposits of clay, shale, or slate have the desired characteristic of bloating. The raw material must be expandable and contain small quantities of gas producing compounds.[49] The clay, shale, or slate is mined in the quarry and then transferred to the raw material preparation area. These areas vary from plant to plant, but they generally include drying, crushing, stock piling, final reduction, and sizing.[50]

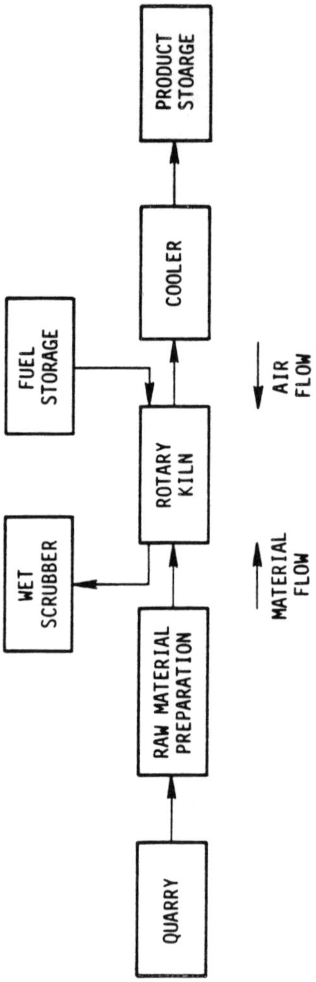

Figure 4.5-1. Flow diagram of a modern lightweight aggregate plant.

After it has been prepared, the sized raw material is sent by conveyor to the kiln for firing. The rotary kiln is the heart of a lightweight aggregate plant. Figure 4.5-2 is a diagram of a modern lightweight aggregate rotary kiln. As the figure shows, the kiln is a refractory-lined, cylindrical, steel furnace equipped with an electrical drive that rotates it at 2 to 3 rpm.[51] Typically, its dimensions are 3.7 to 4.6 m (12 to 15 feet) in diameter by 46 to 61 m (150 to 200 feet) in length.[48,51] The kiln is slightly inclined to the horizontal so that material fed onto the upper end travels slowly by gravity until it discharges into the cooler at the lower end. The burners, located at the discharge end, can accommodate natural gas, fuel oil, pulverized coal, or a combination thereof.[51,52]

The dry raw material fed into the kiln is initially preheated by hot combustion. When it leaves the preheating zone, the raw material enters a second zone, where it melts to a semiplastic state and begins to generate gases which serve as the bloating agent. In this zone, specific compounds (hematite, calcite, dolomite, pyrites, and lemonite) begin to decompose and form gases such as SO_2, CO_2, SO_3, and O_2 that eventually trigger the desired bloating action.[49] As temperatures reach their maximum [~1150°C (~2100°F) for clay and ~1210°C (~2200°F) for shale or slate] in the firing or bloating zone, the semiplastic raw material becomes sufficiently viscous to contain the expanding gases. This bloating action produces a mass of small, unconnected air cells, which the material retains after it cools and solidifies.[47]

Such kiln operating parameters as flame temperature, excess air, feed size, material flow, and speed of rotation vary from plant to plant and are determined by the general characteristics

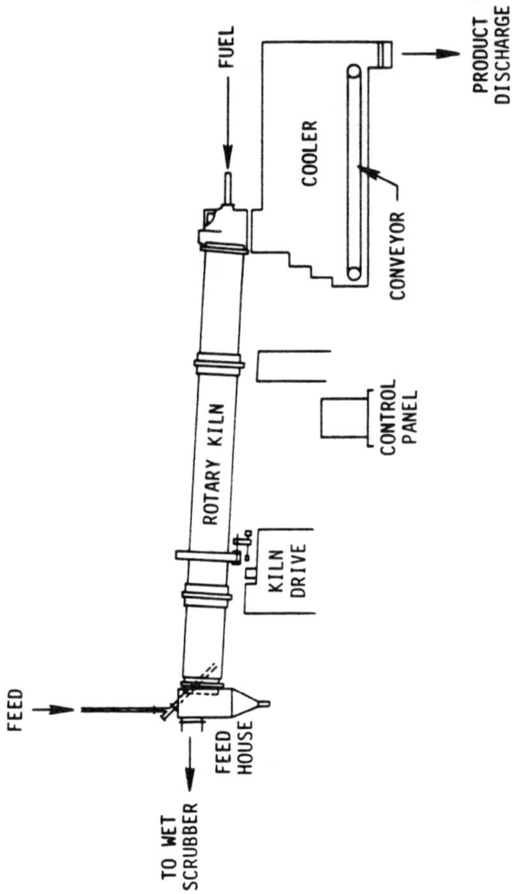

Figure 4.5-2. Lightweight aggregate rotary kiln and cooler.[53]

of the raw material. Maximum temperatures in the rotary kiln vary from 1120° to 1260°C (2050° to 2300°F), depending on the type of raw material being processed and its moisture content.[48,51] Exit temperatures may range from 150° to 650°C (300° to 1200°F), again depending on the raw material and on the kiln's internal design.[48,51] Approximately 80 to 100 percent excess air is forced into the kiln to aid in expanding the raw material.[51]

Originally, most lightweight aggregate plants were small, loosely run operations that used secondhand equipment from other industries.[51] Some of these marginal plants have ceased operations because of low productivity and rising energy and pollution control costs, whereas others have increased their productivity.[51] Approximately 50 to 80 percent of the remaining lightweight aggregate plants have modernized.[51] Some kilns (approximately 50 percent) have incorporated such internal design features as lifters, dams, and heat exchangers to modify the retention time of raw materials, to control the heat transfer over the length of the kiln, and to reduce energy costs.[52,53] Coolers also have been added to reduce energy costs and simultaneously increase productivity. The number of kilns at a lightweight aggregate plant varies from one to six. The industry appears to be moving toward large-volume, high-efficiency plants, which can be more profitable.[54]

The design of an appropriate burner system is perhaps the most critical feature in the lightweight aggregate rotary kiln.[52] Maintaining the correct relationship between the retention time of the feed and the temperature along the axis of the kiln is essential to achieving the desired product quality. A proper time-temperature relationship ensures that the raw material does not disintegrate or fuse to either adjoining particles or to the kiln's refractory lining. The flame shape and directional control of the flame must provide sufficient heat release to meet the time-temperature requirements while maintaining a material temperature within relatively narrow limits in the bloating zone.[52] A thermocouple assembly is normally installed and monitored in the control room to ensure that appropriate time-temperature requirements are maintained.

4.5.1.2 Sintering Process--

In the sintering process, clay or shale is mixed with pulverized coal and burned.[47] During burning, the raw material on the moving grate expands under controlled conditions. Because of the lower temperatures used in this method [870°C (~1600°F)], sometimes 20 to 30 percent of the burnt material must be remixed with the raw material and reprocessed on the traveling grate.[45,47]

4.5.2 Process Assessment for Hazardous Waste Destruction

The sintering process is not considered further because of its lower temperatures, the brief residence time in the combustion zone, and the difficulty of introducing hazardous waste into the furnace. Only the rotary kiln method for pyroprocessing expandable clay, shale, or slate is investigated. This latter process appears to meet the time, temperature, and turbulence requirements needed for thermal destruction of hazardous wastes.

4.5.2.1 Graphical Determination of Suitable Waste Categories--

Process data for the lightweight aggregate rotary kiln--The residence time (4 seconds, based on maximum temperature) and temperature range [370° to 1150°C (700° to 2100°F)] of a typical lightweight aggregate rotary kiln indicate that destruction of some hazardous waste categories may be feasible.[48] A typical rotary kiln processes approximately 450 Mg (500 tons) of lightweight aggregate per day.[51] This kiln has an average furnace volume of 1000 m^3 (35,343 ft^3) and an average volumetric flow rate of 1920 Nm^3/min (73,170 scfm).[48]

Figure 4.5-3 presents mean and fast-path process curves for the lightweight aggregate rotary kiln. These plots are overlaid on the hazardous waste curves to determine which hazardous waste categories are destructible given the kiln's time-temperature

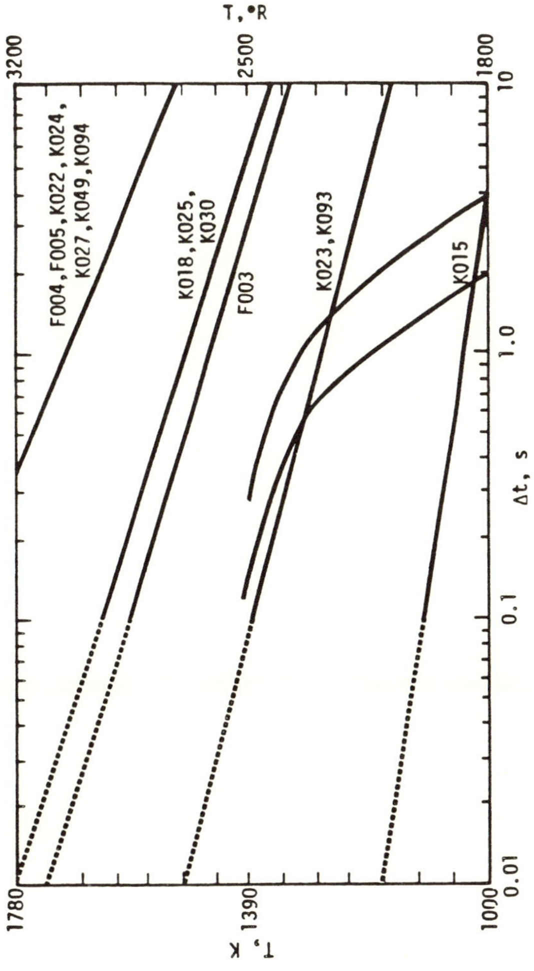

Figure 4.5-3. Hazardous waste destructibility predictions for the lightweight aggregate rotary kiln.

relationship. Based on data from this graph, Table 4.5-1 provides predictions regarding this kiln's ability to destroy completely (99.99%) the selected hazardous waste categories. Because the exit gas temperature is less than the lowest temperature on the graph (Figure 4.5-3), the conservative process line is not shown. Based on the extended case, it appears that waste categories K015, K023, and K093 can be destroyed at the conditions prevailing in this kiln.

TABLE 4.5-1. APPLICABILITY OF THE LIGHTWEIGHT AGGREGATE ROTARY KILN FOR THERMAL DESTRUCTION OF HAZARDOUS WASTES

K015	K023 K093	F003	K018, K025 K030	F004, F005, K022, K024 K027, K049, K095
Yes	Yes	No	No	No

4.5.2.2 Effects on Product Quality and Product Application--

Lightweight aggregate is consumed almost entirely in the construction industry in lightweight concrete or masonry block. It is used in the construction of high-rise complexes and other buildings, ships and barges, bridges and bridge decking, road pavement, and sports tracks.[47] This product has found wide application in the construction industry because of its many advantages over other aggregate materials. These advantages include:[47]

- Lighter weight and greater strength
- Fire resistance
- Thermal insulation and moisture resistance
- Sound insulation
- Wall strength
- Textural finish and workability

The introduction of organic hazardous wastes will have little or no impact on the product quality. The consumption of lightweight aggregate materials is closely linked to construction codes and other specifications dealing with unit weight, size graduations, concrete-making properties, and deleterious substances.[45] The quantities of organic impurities and iron oxide are limited to avoid staining and discoloration. These limits, set by the American Society for Testing and Materials (ASTM), most likely will not be negatively affected by the thermal destruction of hazardous wastes. All of the organic material in the waste should decompose when subjected to the high kiln temperatures.

The introduction of chlorinated wastes could increase dust loading. Some of this dust may precipitate within the kiln. Although chlorinated impurities in the lightweight aggregate will not have an adverse effect on the product, they may have a negative impact on the application of the product. Chlorinated impurities in lightweight aggregate used in cement may cause the setting time of the cement to accelerate.[54] Calcium chloride has been known to have this impact on concrete.

4.5.2.3 Impact on Process Hardware--

A protective coating of material such as that found in cement or lime manufacture is not normally formed on the lightweight aggregate kiln refractory.[52] This lack of a protective coating allows the refractory to be subject to continuous abrasion and to considerable thermal shock during periods of temperature variation. Highly corrosive waste fuels could increase the abrasive damage to the refractory lining.

In addition, the presence of trace quantities of metals and the formation of acids [e.g., hydrogen chloride (HCl)] could cause corrosion of ducting and possibly interfere with the operation of the air pollution control equipment. One company currently incinerating hazardous wastes in these kilns strictly limits both the chlorine and solids content because they believe highly corrosive wastes will have a negative impact on both their wet scrubber and fan.[48]

4.5.2.4 Acceptable Amounts of Hazardous Waste Feed--

According to an industry expert, 3.0 GJ are required to produce 1 Mg of lightweight aggregate in a typical rotary kiln.[51] Based on this estimate and assuming that 10 percent of the required heat input into the kiln can be substituted with hazardous waste fuels, the heat input of hazardous waste can be calculated as:

$$0.10 \times 3.0 \text{ GJ/Mg} = 0.3 \text{ GJ/Mg or } 300 \text{ MJ/Mg of aggregate}$$

It was previously shown that the lightweight aggregate rotary kiln is capable of destroying wastes in categories K015, K023, and K093. By using an average heating value of 20 MJ/kg of waste or 20,000 MJ/Mg (obtained from Table 2-3), the rate of hazardous waste feed can be estimated as:

$$300 \text{ MJ/Mg of aggregate} \times \left(\frac{1}{10{,}000 \text{ MJ/Mg}}\right)$$

$$= 0.015 \frac{\text{Mg waste}}{\text{Mg aggregate}}$$

For a typical lightweight aggregate kiln process rate of 450 Mg per day, assuming 260 days per year of operation, the hazardous waste disposal rate is estimated to be 1750 Mg/yr.

4.5.2.5 Trends in Process Fuel Consumption--

The rising cost of natural gas and oil over the last several years has caused the lightweight aggregate industry to turn to coal as the main fuel source.[51] The industry has been looking for cheap alternative fuel sources, as evidenced by the one firm that is already thermally destroying hazardous wastes.[48]

4.5.2.6 Geographical Matching of Industry Locations and Waste Generators--

Figure 4.5-4 shows the distribution of domestic lightweight aggregate plants by state. Texas, North Carolina, Virginia, and California account for 33 percent of these plants.

Table 4.5-2 presents the SIC codes and geographic distribution of the industries generating each category of destructible

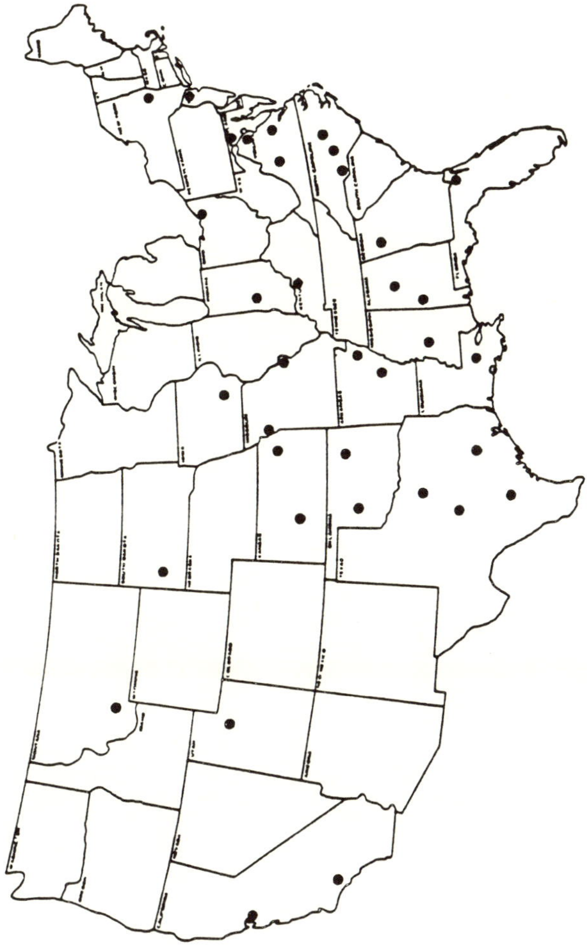

Figure 4.5-4. Lightweight aggregate plant locations in the United States.[55]

wastes. The areas specifically identified as generating wastes in categories K015, K023, and K093 (California, New Jersey, Louisiana, Pennsylvania, Illinois, Texas, and Puerto Rico) contain 8 of the 36 operating lightweight aggregate plants. Although Pennsylvania, Illinois, and Pureto Rico have no lightweight aggregate plants, Pennsylvania and Illinois are surrounded by states having one or more plants.

TABLE 4.5-2. SIC CODES AND GEOGRAPHIC DISTRIBUTION OF INDUSTRIES GENERATING HAZARDOUS WASTES DESTRUCTIBLE IN THE LIGHTWEIGHT AGGREGATE MANUFACTURING ROTARY KILN[1,19]

Waste category	SIC code of waste generating industry	Geographic location
K015	2865	Two plants (locations not known)
K023	2869	Richmond and El Sequendo, California; Kearney and Bridgeport, New Jersey; Baton Rouge, Louisiana; Bridgeville and Neville Island, Pennsylvania; Cicero and Millsdal, Illinois; Texas City, Texas
K093	2869	Illinois, Texas, New Jersey, Puerto Rico, and California

4.5.2.7 Environmental Factors--

Impact on air pollution control devices--Atmospheric emissions from a typical lightweight aggregate rotary kiln consist of particulates, SO_2, NO_x, and CO. The quantity of these emissions varies with the type of fuel in use. Unfortunately, emission factors for this process are not available. Generally either a wet scrubber or a fabric filter is used to control these emissions; the wet scrubber is more popular.[48,51]

Depending on the characteristics of the pollutant and the increase in pollutant loadings, the addition of hazardous wastes may necessitate some adjustment in the control equipment. Highly corrosive wastes such as HCl may have a detrimental effect on pollution control devices. When a corrosive waste is being thermally destroyed, a neutralizing step may be required before the exhaust gases reach the wet scrubber and glass filters may be required in the fabric filter.[51]

Chlorinated wastes can also increase particulate loadings. The amount of increase would depend upon the mineralogical constituency in the raw material. Although raw materials vary, most will contain some sodium, calcium, potassium, and silicon compounds.[55] These compounds may react with the chlorine to form volatile chlorides that will exit into the atmosphere with the exhaust gases.

A potential exists for trace quantities of incomplete combustion products to escape into the stack gases. In addition, any heavy metals in the waste fuel will volatilize and exit through the stack. One firm is now successfully destroying easily combustible waste fuels (such as solvents) in many of their lightweight aggregate kilns.[48] This firm has established strict limits on the percent chloride, solids, and ash content to avoid operational and environmental problems.

4.5.2.8 Institutional Factors--

Because of rising energy costs, the lightweight aggregate industry seems more than willing to accept and adapt to the use of waste fuels. The largest firm in the industry has already become active in this area.[48] The three most inhibiting factors toward using waste fuels are:

- Public concern
- Impact on process and environmental control equipment
- Environmental impact

The industry is very sensitive to maintaining good public relations and avoiding public concern over incinerating hazardous wastes. Many older plants that were once located well beyond the city limits have found residential suburban neighborhoods growing up around the perimeter of their facilities. This has increased their sensitivity to the use of waste fuels and the potential impact of an accidental spill.

4.6 LIME INDUSTRY

The production of lime involves quarrying the raw limestone, preparing it by crushing and sizing, and calcining it to produce quicklime (CaO), which may be processed further by hydration. During calcination, the limestone is subjected to high temperatures that cause it to decompose. This process produces quicklime and releases CO_2. This reaction can be represented as:

$$CaCO_3 + heat \rightarrow CaO + CO_2$$

Quicklime can be manufactured in a variety of kilns: the long rotary kiln, the short rotary kiln with an external stone preheater, the vertical or shaft rotary kiln, the Calcinatic or rotary hearth kiln, and the fluidized-bed kiln. Rotary kilns, however, account for approximately 90 percent of the domestic lime production.[56]

According to the Bureau of Mines, U.S. Department of the Interior, lime producers at 153 plants in the United States and Puerto Rico sold or used 16,400 Gg (18.1 million tons) of lime in 1980.[57] The leading lime-producing states were Ohio, Pennsylvania, Missouri, Texas, and Alabama; together these states accounted for 48 percent of the total domestic output.

4.6.1 Process Descriptions

4.6.1.1 Rotary Kilns--

This manufacturing process involves the calcination of limestone at temperatures of approximately 1260°C (2300°F) in either a long rotary kiln or a short rotary kiln with an external stone preheater. Figure 4.6-1 presents a process flow diagram of a modern rotary kiln and hydration plant.

The major difference between the long kiln and the short rotary (other than dimensions) is that the short kiln is preceded by an external stone preheater, which increases productivity and decreases energy requirements. Hot exhaust air exiting the kiln exchanges heat with the limestone in an external preheater.

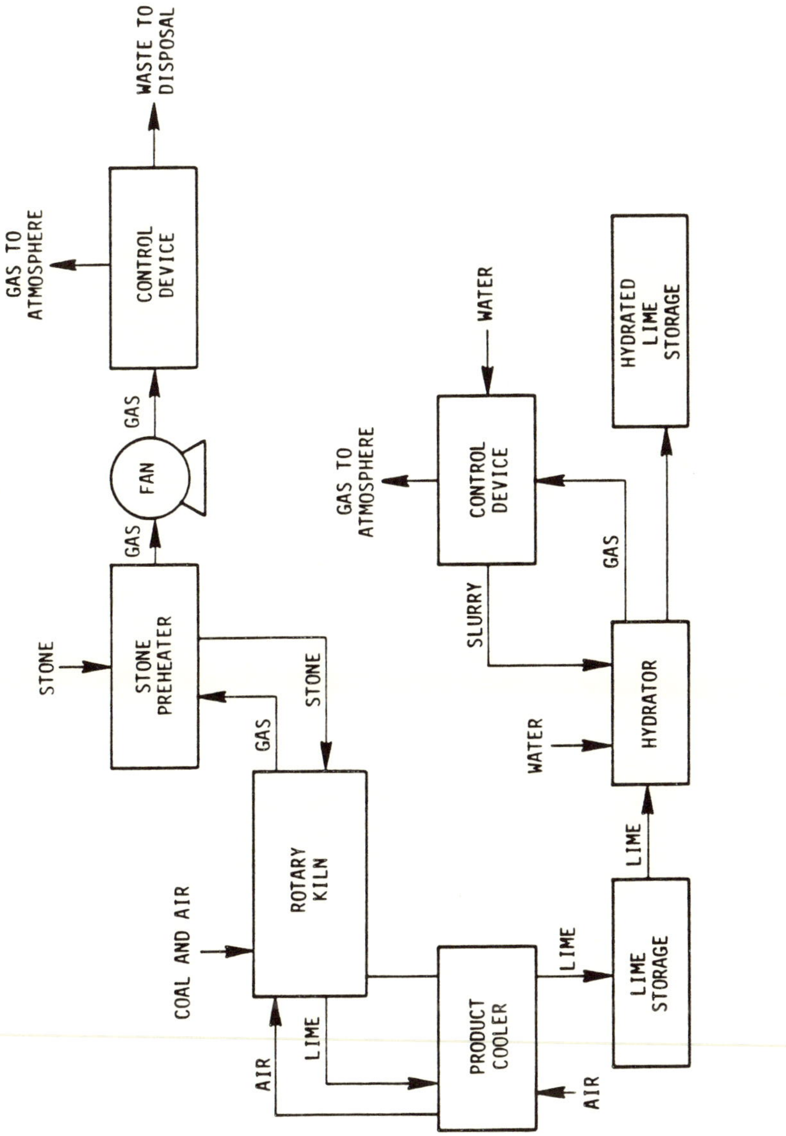

Figure 4.6-1. Flowsheet for modern lime calcination and hydration plant.[56]

Both the long and short rotary kilns operate in similar fashion. The kilns are made of heavy steel plate lined with approximately 15 to 24 cm of refractory brick.[58] Although most rotary kilns are fired with pulverized coal, they can be adapted to firing natural gas or oil. The kilns vary in size from about 3 to 6 meters (10 to 20 feet) in diameter and 46 to 150 meters (150 to 500 feet) in length.[59] The kilns are installed at a 3- to 5-degree inclination on four or six foundation piers and revolve on trunnions at 1 to 2 rpm.[58] As shown in Figure 4.6-2, limestone is fed into the elevated end of the kiln and is discharged as quicklime into the cooler at the lower end. Approximately 10 percent of the kiln is filled with limestone or lime as the feed tumbles slowly and gently through the long cylindrical furnace. Cool air is usually introduced into the discharge end of the product cooler and subsequently into the kiln to serve as secondary combustion air.[58]

4.6.1.2 Vertical or Shaft Kilns--

Once the most widely used kiln in the United States, the total capacity of the vertical kiln has fallen well behind that of the rotary kiln. Basically, this kiln is an upright heavy steel cylinder lined with refractory material. Dimensions of the kiln may vary from 0.3 to 0.7 meter (0.9 to 2.4 feet) in diameter and 1.0 to 2.9 meters (3.4 to 9.4 feet) in height.[56] Vertical kilns of various design are in operation.

Most modern vertical kilns are divided into the following four generalized zones, often indistinguishable from each other:

1) Storage
2) Preheating
3) Calcining
4) Cooling

In the preheating zone, recirculated exhaust gases preheat the stone; in the calcining zone, 95 percent of lime burning occurs; and in the cooling zone, the incoming cool air, which flows upward countercurrent to the descending red-hot lime, cools the product and provides secondary combustion air for the calcining zone.[58]

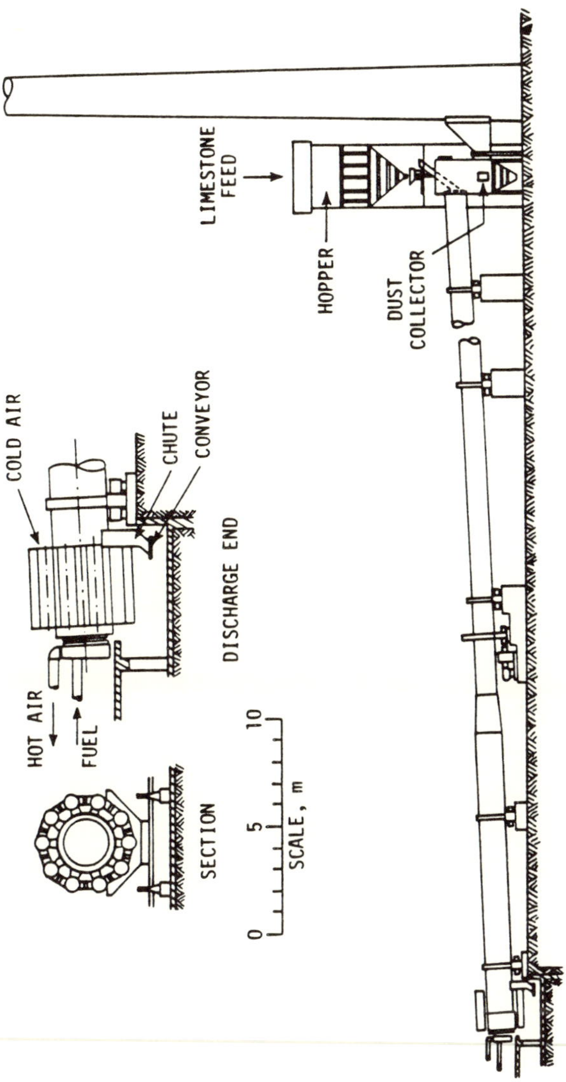

Figure 4.6-2. Rotary kiln for processing lime.[58]

A higher average fuel efficiency is the primary advantage of the vertical kiln over rotary kilns. The primary disadvantages are its relatively low production rate and inability to burn solid fuel (coal) without degrading the quality of the lime.[56]

4.6.1.3 Calcimatic Kilns--

The Calcimatic (or rotary hearth) kiln is circular-shaped and has a slowly revolving doughnut-shaped hearth. The hearth is divided into zones, and precise temperatures are maintained in each of the different zones.[56] The quicklime product is discharged after it makes one revolution around the kiln. The heated gases from the calcination zone of the kiln are used to preheat the feed. The major disadvantage of this kiln is that it cannot be operated with solid fuel.

4.6.1.4 Fluidized-Bed Kilns--

Because of their specialized feed size requirements, the application of fluidized-bed systems has been limited in the domestic lime industry.[56] In this kiln, a very fine feed is fluidized by controlled air and combustion gases as it descends through the preheating and calcining zones. The finely divided limestone is physically agitated by turbulent air as it comes into direct contact with the combustion gases.

4.6.2 <u>Process Assessment for Hazardous Waste Destruction</u>

In this report only rotary kilns are considered as candidates for hazardous waste destruction. Because these kilns represent more than 90 percent of the domestic lime production, they appear to be a likely alternative for disposing of significant quantities of hazardous wastes. The other types of kilns were eliminated from further consideration for two reasons; the limited availability of process data and the relatively small number in operation.

4.6.2.1 Technical Factors--

<u>Compatibility of hazardous wastes with process fuel</u>--It appears feasible to inject hazardous wastes into rotary kilns along with the fuel (in most cases coal). Some retrofitting would be required to allow for the injection and subsequent atomization of

liquid waste. This would entail the installation of a liquid burner/atomizer system, complete with the appropriate pump and piping. Appendix A provides a more detailed description of the appropriate equipment and procedures required for safe feeding and firing of hazardous waste.

Graphical determination of suitable waste categories--

Process data for the long rotary lime kiln--The temperature range [690° to 1890°C (1265° to 3440°F)] of a typical long rotary kiln indicates that destruction of hazardous wastes may be feasible. A typical long rotary kiln processes approximately 540 Mg (600 tons) of quicklime per day, has an average furnace volume of 720 m^3 (25,518 ft^3), and has an average volumetric flow rate of 910 Nm^3/min (34,560 scfm).[60]

Figure 4.6-3 plots the residence time for the mean- (Δt), and fast-path ($\Delta\bar{t}$) process curves for the long rotary lime kiln. These plots are overlaid on the hazardous waste curves to determine which hazardous waste categories are destructible in the kiln, given the time-temperature relationship. Based on data from this graph, Table 4.6-1 predicts the ability of a typical long rotary kiln to completely (99.99%) destroy the selected hazardous wastes. Since the average exit gas temperature of about 690°C (1265°F) falls below the lowest temperature in the graph (Figure 4.6-3), a conservative process line is not shown. Based on the extended case, it appears that the wastes in all the selected categories can be destroyed at the prevailing conditions in this kiln.

TABLE 4.6-1. APPLICABILITY OF THE SHORT AND LONG ROTARY LIME KILNS FOR THERMAL DESTRUCTION OF HAZARDOUS WASTES

K015	K023 K093	F003	K018, K025 K030	F004, F005, K022, K024 K027, K049, K094
Yes	Yes	Yes	Yes	Yes

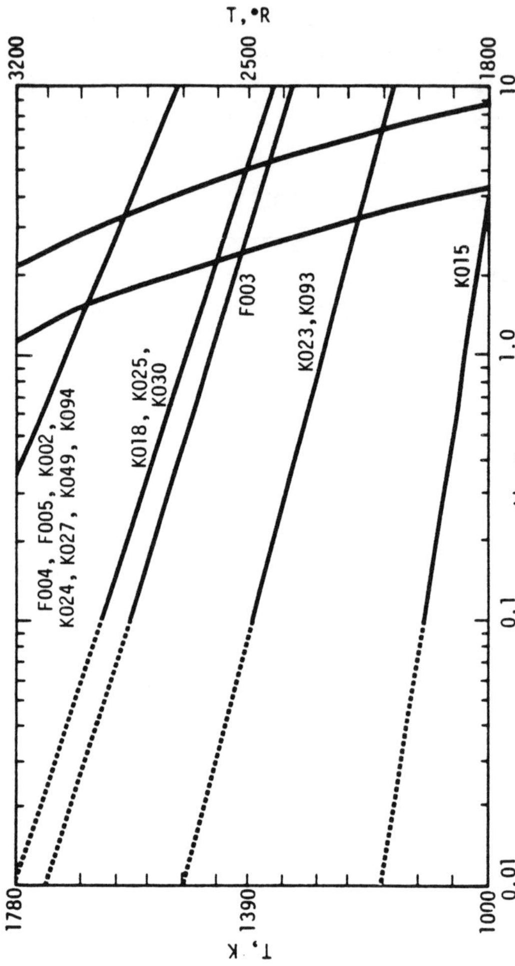

Figure 4.6-3. Hazardous waste destructibility in the long rotary lime kiln.

Process data for the short rotary kiln with an external stone preheater--The temperature range [1160° to 1840°C (2115° to 3340°F)] of a typical short rotary kiln with an external stone preheater suggests that thermal destruction of hazardous wastes in this kiln may be feasible. A typical short rotary kiln processes 540 Mg (600 tons) of quicklime per day, and has an average furnace volume and volumetric flow rate of 470 m^3 (19,792 ft^3), and 700 Nm^3/min (26,460 scfm), respectively.[60]

The Figure 4.6-4 process curves for the short kiln are overlaid on the hazardous waste curves to determine which hazardous waste categories are destructible given the kiln's time-temperature relationship. Table 4.6-1 presents predictions regarding the ability of this kiln to destroy completely (99.99%) the selected hazardous wastes. In the most conservative case, only those waste categories falling below the conservative process line are shown as being applicable. In the extended case it appears that the wastes in all the selected categories are destructible given the kiln's time-temperature relationship.

Effects on product quality and product application--Limestone and lime are used either directly or indirectly in more industries than any other natural substance. Among the leading individual end uses of lime are in steel manufacturing, water purification, alkali production, paper and pulp manufacturing, and sewage treatment; together these uses account for approximately 63 percent of the total 1979 consumption.[61] Lime is also used in medicine, insecticides, sugar refining, plant and animal food, gas absorption, and as a raw material in the production of calcium salts.

Many end uses of lime, such as in steel making, require a high product quality. Therefore, precautions are taken to select limestone and other raw materials that are low in impurities.

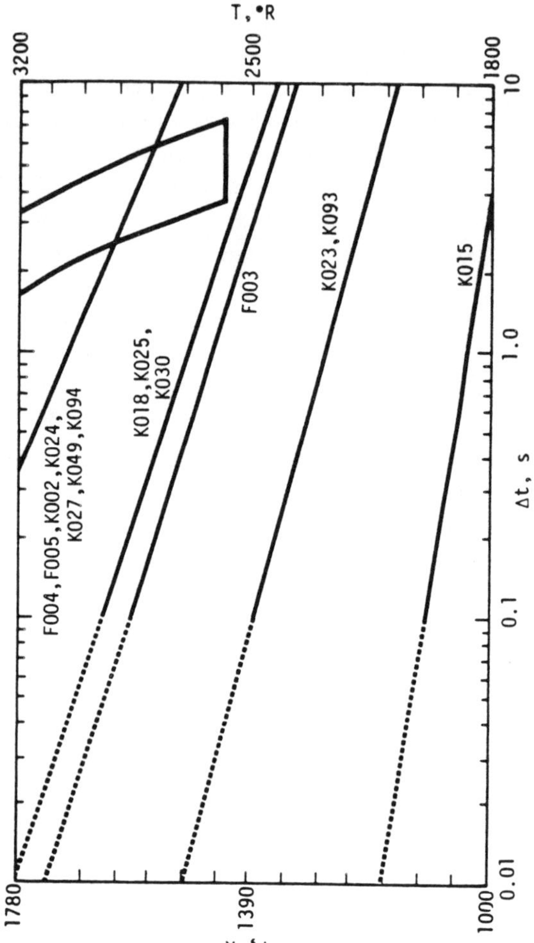

Figure 4.6-4. Hazardous waste destructibility in the short rotary lime kiln.

Impurities such as alumina, iron, and silica tend to combine chemically with lime and form aluminates, ferrites, and silicates.[58] The level of these impurities in the waste fuel must be minimal to ensure a high purity product. Chlorinated wastes are also likely to react with the calcium and form calcium chloride.[59] The boiling point of calcium chloride is 760°C (1400°F), whereas the exit gas temperature in a lime kiln ranges from 260° to 760°C (500° to 1400°F). With this temperature range, some calcium chloride probably would precipitate in the kiln and be retained in the lime product as an impurity.

The coal is most likely to contain sulfur, and to some extent so will the waste fuels. Most of this sulfur will be converted to SO_2 during fuel combustion and calcination. The extent to which the SO_2 could interact with the lime product will depend on the chemical composition of the stone, the temperature in the kiln, and the amount of excess oxygen in the kiln.[56] Studies have shown that most of the sulfur impurities in the lime product originate in the limestone raw material.[56]

The actual behavior of waste fuels in a lime kiln has not been reported in literature, and the actual impact of hazardous wastes on lime quality must be known to determine the feasibility of thermally destroying hazardous wastes in these kilns.

Impact on process hardware--The formation of silicates from the use of waste fuels could have some adverse effects on the process hardware. Silicates could be incorporated into the lime kiln coatings and lead to the formation of rings in the kiln, which could eventually disrupt kiln operations.[59] Ring formation could also be caused by fuels with a high ash content.[62] Because hazardous wastes generally have a lower ash content than coal, the ash ring problem could be minimized by destroying hzardous wastes. Highly chlorinated wastes could cause a buildup and baking problem in the short rotary kiln's stone preheater.[60] The preheater may need to be retrofitted with a bypass duct to minimize this problem.

In addition, the presence of trace quantities of metals and the formation of acids (such as HCl) could cause corrosion in the ducting and possibly interfere with the operation of air pollution control equipment. Highly corrosive wastes could also have a deleterious effect on the refractory lining of the kilns,[63] and excessive wear and tear of the refractories could increase kiln downtime for repairs.

Acceptable amounts of hazardous waste feed--According to industry experts, one-third megagram of coal with a heating value of 30 MJ/kg is required to produce one megagram of lime.[64] Based on this estimate and the assumption that 10 percent of the required heat input into the kiln can be provided by hazardous waste fuels, heat input through hazardous wastes is:

$$0.1 \times 0.33 \text{ Mg coal/Mg lime} \times 30{,}000 \text{ MJ/Mg coal} = 990 \text{ MJ/Mg lime}$$

For the most conservative case, the short rotary kiln is capable of thermally destroying only a few hazardous wastes.

By using an average heating value of 23,000 MJ/Mg, the quantity of hazardous waste feed can be estimated as 43 kg hazardous waste per Mg lime (86 lb/ton).

The typical lime kiln processes 540 Mg (600 tons) per day; therefore, up to 23.2 Mg (51,000 pounds) of waste fuel could be disposed of per day (based on the most conservative case for the short kilns). Assuming 260 days per year of operation, the annual disposal rate is estimated to be 6000 Mg/yr (6600 tons/yr).

In the extended case, both the long and short rotary kilns are shown to be capable of thermally destroying all of the selected hazardous wastes. Based on an average waste fuel heating value of 22 MJ/kg (9,400 Btu/lb), the quantity of waste feed for the extended case is estimated to be 45 kg waste/Mg lime (90 lb/ton).

For a typical lime kiln process rate of 600 tons per day (assuming 260 days of operation per year), the annual hazardous waste disposal rate is estimated to be 7000 Mg/yr (7700 tons/yr).

Trends in process fuel consumption--The rising cost of oil and natural gas over the last several decades has caused the domestic lime industry to use coal as their main fuel source, which necessitated the choice of the rotary kiln. Approximately 69.6 percent of all domestic lime kilns now burn coal and coke, an increase of 147 percent over 1972 usage.[65]

4.6.2.2 Geographical Matching of Industry Locations and Waste Generators--

Figure 4.6-5 shows the distribution of operating domestic lime plants by state. As mentioned earlier, Ohio, Pennsylvania, Missouri, Texas, and Alabama account for 48 percent of the total domestic lime production.[57]

Table 4.6-2 presents the SIC code for each waste category and the waste's geographic distribution. Appendix B presents data concerning the annual generation rate in each waste category. In the short kiln's most conservative case, only the wastes in categories K015, K023, K093, and F003 are destructible. The areas where these wastes are generated (California, Illinois, Louisiana, New Jersey, Pennsylvania, Puerto Rico, and Texas) contain 41 of the 155 operating lime plants.

In the extended case, both the long and short rotary kilns are able to destroy all selected hazardous wastes. The areas specifically identified as generating these wastes (California, Illinois, Kansas, Kentucky, Louisiana, New Jersey, Ohio, Oklahoma, Pennsylvania, Puerto Rico, Texas, and West Virginia) contain 63 of the 155 operating lime plants.

The table shows that the sources of some wastes are widespread, although specific states are not mentioned. Such widespread geographical distribution greatly enhances the possibility of destroying these hazardous wastes in lime kilns.

4.6.2.3 Environmental Factors--

Impact on air pollution control devices--Atmospheric emissions from a typical rotary kiln consist of particulates, SO_2,

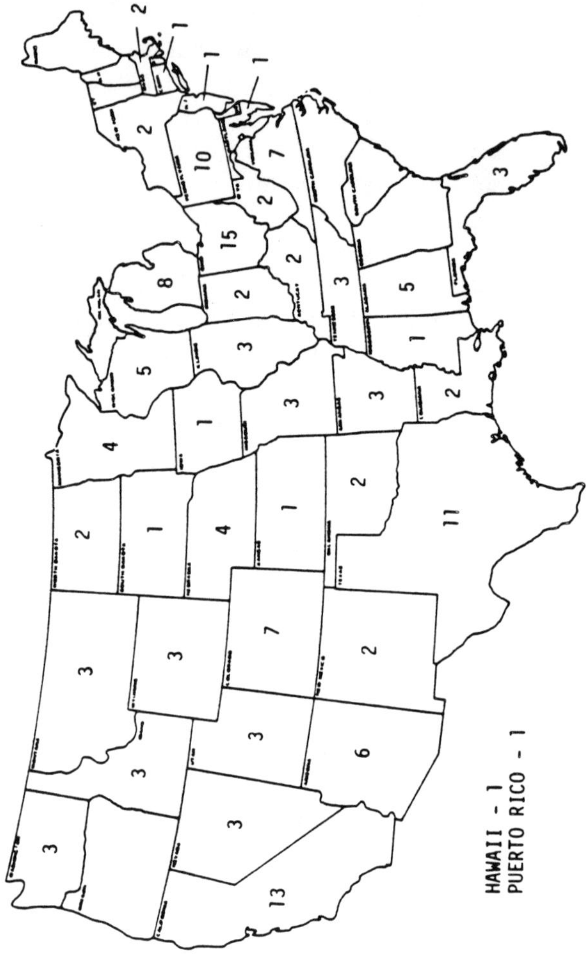

Figure 4.6-5. Distribution of domestic lime plants, by state.[61]

TABLE 4.6-2. SIC CODES AND GEOGRAPHIC DISTRIBUTION OF INDUSTRIES THAT GENERATE HAZARDOUS WASTES DESTRUCTIBLE IN SHORT AND LONG LIME KILNS[1,19]

Waste category	SIC code	Geographic location
F003		Widespread
F004		Widespread
F005		Widespread
K015		Two plants (locations not known)
K018	2869	Houston and Freeport, Texas; Lake Charles and Baton Rouge, Louisiana; Deepwater, New Jersey; Long Beach, California
K022	2869	Widespread
K023	2869	Richmond and El Sequendo, California; Kearney and Bridgeport, New Jersey; Baton Rouge, Louisiana; Bridgeville and Neville Island, Pennsylvania; Cicero and Millsdale, Illinois; Texas City, Texas
K024	2869	Richmond and El Sequendo, California; Kearney and Bridgeport, New Jersey; Baton Rouge, Louisiana; Bridgeville and Neville Island, Pennsylvania; Cicero and Millsdale, Illinois; Texas City, Texas
K025	2865	Bound Brook, New Jersey; Pas and Willow Islands, West Virginia; Beaumont, Texas; Gibbstown, New Jersey; Geismar, Louisiana; New Martinville, West Virginia
K027	2869	Moundsville and New Martinsville, West Virgnina; Geismar and Lake Charles, Louisiana; Freeport and Baytown, Texas; Astabula, Ohio
K030	2869	Freeport, Deer Park, and Corpus Christi, Texas; Baton Rouge, Lake Charles, and Plaquemine, Louisiana; Pittsburg, California; Louisville, Kentucky; Wichita, Kansas
K049	2911	Widespread
K093	2869	Illinois, Texas, New Jersey, Puerto Rico, California
K094	2869	Illinois, Texas, New Jersey, Puerto Rico, California

NO_x, and CO. The industries use a wide variety of air pollution control equipment. Generally, primary particulate control is accomplished by multiple cyclones, which entrap 85 to 93 percent of the dust loading from rotary kiln systems.[58] A secondary system is required to abate most of the remaining dust. Among the secondary systems used, 45 percent are fabric filters, 40 percent are high-energy wet scrubbers, 10 percent are electrostatic precipitators, and 5 percent are gravel-bed filters and miscellaneous control equipment.[58] Table 4.6-3 shows the emission factors for a hypothetical 450-Mg (500-ton) per day rotary kiln plant that uses coal as its fuel and applies various control equipment. Solid waste quantities emitted from control devices are estimated to be 3200 kg/h at this hypothetical rotary kiln plant.[56]

If chlorinated wastes are burned, a corrosion-resistant scrubbing system would be required to capture the gaseous pollutants. Some plants now equipped with scrubbing systems may opt to burn only those wastes having a low chlorine content rather than retrofitting the plant to handle highly corrosive wastes.

A potential exists for trace quantities of incomplete combustion products such as organic chlorides to escape into the stack gases. The formation of such chlorinated acids should be avoided because of the possibility of the formation of calcium chloride. The actual behavior of the hazardous wastes and products of incomplete combustion will need to be investigated further by conducting a test burn to determine details of their environmental impact.

Fugitive emissions from the production of lime and the thermal destruction of waste fuels will occur primarily from transfer points, screening operations, hazardous waste pumps and piping, and loading and unloading operations.[56,59] Proper plant design and layout (e.g., by minimizing the height and number of drop points, enclosing elevators and screens, and installing packed seals on pumps) will greatly reduce these emissions. The installation of hoods over exit sources and the establishment of

TABLE 4.6-3. ROTARY KILN EMISSION FACTORS[56]

	Particulate, kg/Mg (lb/ton)	SO_2, kg/Mg (lb/ton)		Electrical energy factors, kWh/ton (Mg/ton) stone
		Low-sulfur coal	High-sulfur coal	
Uncontrolled (cyclone only)	17 (34.1)	-	-	18 (16[a])
Fabric filter	0.15 (0.3)	0.5 (1.0)	1.0 (2.0)	2.7 (2.42)
High-energy ESP	0.15 (0.3)	-	-	3.5 (3.19)
Low-energy ESP	0.25 (0.5)	-	-	2.4 (2.21)
Water scrubber (22 in. H_2O)	0.15 (0.3)	0.005 (0.1)	0.005 (0.1)	16.8 (15.2)
Water scrubber (15 in. H_2O)	0.25 (0.5)	0.005 (0.1)	0.005 (0.1)	11.9 (10.8)
Water scrubber (9 in. H_2O)	0.5 (1.0)	0.005 (0.1)	0.005 (0.1)	7.1 (6.45)
Gravel-bed filter	0.5 (1.0)	-	-	-

[a] Electrical energy required for kiln and associated equipment.

good housekeeping procedures could reduce fugitive emissions even further. These preventive measures need to be taken to ensure the safe incineration of hazardous wastes.

4.6.2.4 Institutional Factors--

Immediate industrywide acceptance of the idea of incinerating hazardous wastes in lime kilns is not likely. Such acceptance will be based on the costs associated with retrofitting a plant to accept hazardous waste, on public attitude, and on the savings associated with burning a relatively inexpensive fuel. The potential fuel savings may provide the incentive for industry to adapt to using hazardous wastes. The incinerability of the wastes and the impact on both the plant operation and the environment will have to be determined. The three major problems that may discourage the industry's use of waste fuels are the impact on product quality, the possible formation of kiln rings, and the impact on the external stone preheater (applies only to short kilns).

4.7 PRIMARY COPPER INDUSTRY

The high-temperature processes in the copper smelting industry are as follows:

Process	Process temperature, °C (°F)
Multiple-hearth roasting	760 (1400) at the bottom to 200 (400) at the top
Fluidized-bed roasting	650 to 760 (1200 to 1400)
Reverberatory smelting	980 to 1320 (1800 to 2400)
Flash smelting	1090 (2000)
Noranda furnace	1210 to 1320 (2200 to 2400)
Peirce-Smith converters	1180 to 1210 (2150 to 2200)
Hoboken converter	1200 (2200)
Electric smelting	980 to 1320 (1800 to 2400)

Figure 4.7-1 presents a generalized process flow diagram of the pyrometallurgical copper industry.

The following subsections describe each process and present an assessment of its potential for hazardous wastes incineration. Also presented is the rationale for eliminating most of the processes from further consideration.

4.7.1 Process Descriptions

4.7.1.1 Multiple Hearth Roasting--

The copper content of as-mined copper ores ranges from 0.5 to 2 percent, but usually runs less than 1 percent. Most copper ores are sulfide ores, which are concentrated by froth flotation. The concentrates contain copper (20 to 50%), sulfur (30 to 38%), iron (20 to 30%), and smaller amounts of various other metals. The purpose of roasting is to decrease the sulfur content by partial oxidation of the copper sulfide concentrates with air. Roasting also removes the water from the concentrates; volatilizes much of the arsenic (As), some of the antimony (Sb), and other elements in the feed; and preheats the concentrate before it is charged as calcined feed to the reverberatory furnaces.

In the multiple-hearth roaster, the concentrates are introduced into the top of a cylindrical vessel fitted with a series of round horizontal trays or hearths. As the concentrates are raked across each hearth they fall onto the next lower hearth and are subsequently discharged from the bottom of the roaster. The long retention time of the solids in the roaster results in an intimate contact between hot gases and solids. The bottom hearth temperature is around 760°C (1400°F). A limited amount of air is admitted into the roaster, along with sufficient fuel to maintain an adequate temperature. The varying fuel requirement depends on the sulfide content of the concentrates. The burning of sulfur to SO_2 is exothermic and releases heat. If the heat is sufficient to sustain the required temperature, the process is known as autogenous roasting.

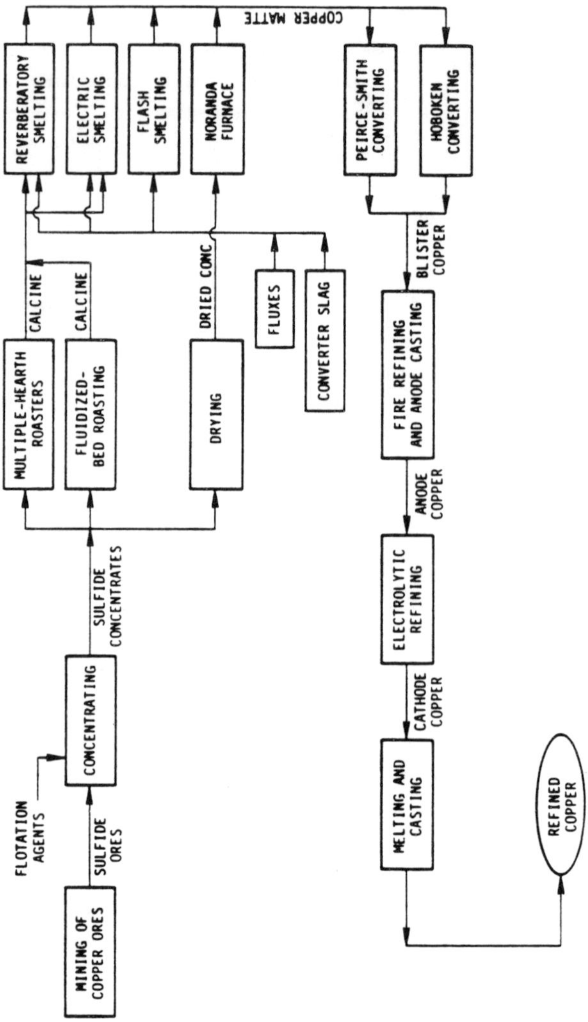

Figure 4.7-1. Simplified flow diagram of the primary pyrometallurgical copper process.[69]

Only four plants in the U.S. primary copper industry use multiple-hearth roasters. Three of these plants are in Arizona (at Hayden, Douglas, and Tacoma), and one is in El Paso, Texas. The highest temperature achieved in roasting is 760°C (1400°F). No new multiple-hearth roasters have been installed in recent years; all new roasters since 1960 have been fluidized-bed units.[68] Those multiple-hearth furnaces still in operation usually generate large amounts of fugitive emissions and are poorly controlled. This process is not considered further for destroying hazardous wastes because its relatively low operating temperature, fugitive emission potential, and the small number of operating units make it a poor candidate.

4.7.1.2 Fluidized-Bed Roasting--

The function of fluidized-bed roasting is the same as that of multiple-hearth roasting. In this process, the gas is a recycled stream of flue gas, into which regulated streams of air (no excess air) and fuel gas are introduced. The copper concentrate solids are continuously fed and the smaller roasted concentrate particles are collected with cyclones after they are carried out of the fluidized-bed area with the exhaust gases, and the larger concentrate particles are collected as they overflow the fluid bed portion of the vessel. The temperature of the bed is around 650° to 760°C (1200° to 1400°F). Intimate contact between solid particles and gases is achieved through the constant agitation of the fluidized-bed. Any chemical reaction that takes place in the fluidized-bed roaster between the solids and the gas happens very quickly, with no cold pockets or hot spots. Both the fuel and the oxygen are completely consumed, and because there is no excess air, the SO_2 content of the flue gas stream is high (12 to 16 percent of the exhaust). This high SO_2 content is beneficial for the production of sulfuric acid, elemental sulfur, or liquid SO_2.

Only four plants in the United States have fluidized-bed roasters. These plants are located at Hayden, Arizona; Morenci,

Arizona; Anaconda, Montana (currently shut down); and Copperhill, Tennessee. Although specific matching of certain waste suppliers and smelters might be arranged, the locations and limited number of these plants rule out widespread national application.

The fluidized-bed roaster process is eliminated from further consideration for hazardous waste disposal because of the limited number of facilities and the relatively low process temperature.

4.7.1.3 Reverberatory Furnace Smelting--

The reverberatory furnace is the most widely used unit for matte smelting. Matte smelting produces separable liquid phases of slag and a copper-rich sulfide phase (matte) by the melting of concentrates or roaster calcines. The smelting is done in a arch-roofed horizontal hearth furnace at 980° to 1320°C (1800° to 2400°F). The slag is discarded if the copper content is below 1 percent. Reverberatory furnaces use large quantities of hydrocarbon fuels. Reverberatory furnaces are used extensively because of their versatility; all types of material (lumpy, fine, wet, or dry) are readily smelted in a continuous operation.

The heat for smelting and for maintaining the temperatures of the liquid slag and matte is provided by combustion of oil, natural gas, or coal in burners situated at one end of the furnace. The flame and hot combustion gases travel the entire length of the furnace and achieve an efficient heat transfer with the solid charge and molten pool of slag and matte. Based on the time/temperature profiles inside the furnace, it is possible to consider the incineration of hazardous liquid wastes in this process.

4.7.1.4 Peirce-Smith Converters--

The molten matte is transported from smelting and charged directly into the Peirce-Smith converter along with flux. This converter is a horizontal cylindrical vessel that blows compressed air into the molten matte layer via tuyeres along the longitudinal length of the vessel. The purpose of converting is to remove iron, sulfur, and other impurities from matte; this produces a liquid metallic copper known as crude blister copper (98.5 to 99.5% Cu) and slag.[69] This is achieved by oxidizing the molten matte at a high temperature [1180° to 1210°C (2150° to 2200°F)]. The two products occur at different stages of the process. The slag is removed first, and then the blister copper. They are poured separately from the converter mouth by rotating the converter about its longitudinal axis. The converting process produces hot SO_2-laden gases, which are collected by means of a fixed hood above the converter. The gases, which contain 2 to 6 percent SO_2, usually pass through a waste heat boiler and an electrostatic precipitator for particulate control before being sent to an acid plant for recovery of SO_2 as sulfuric acid. The converting reactions are exothermic and the process is autogenous.

Peirce-Smith converters generate considerable quantities of fugitive dust, fumes, and SO_2. Whenever flux is added or copper and slag are to be poured out of the mouth of the converter, the converter has to be tilted away from the collection hood, which is when most of the fugitive emissions occur. These emissions, which occur between the hood and converter mouth during blowdown, are minimized by placing the hood as close as possible to the converter mouth. Because of the high fugitive emissions and the lack of a proper method for charging of hazardous wastes (i.e., matte is poured from ladles), the Peirce-Smith converter is considered unsuitable for hazardous waste incineration.

4.7.1.5 Hoboken Converter--

The Hoboken converter is another type of furnace used to produce crude blister copper from the matte formed in the smelting process. Its function is identical to that of the Peirce-Smith converter. The principal difference is that the flue that removes the gas from the converter is an integral part of the converter construction rather than a hood mounted above it. This design minimizes infiltration of air, minimizes local losses of

SO_2 from the converter mouth, and allows production of a gas with a higher and more uniform SO_2 content. The main operating advantages are minimal fugitive emissions during charging of molten matte, tapping, and slagging.

The furnace operates at around 1210°C (2200°F) and at atmospheric pressure. The SO_2 generation is intermittent, but a battery of several converters will produce a stream of SO_2 of sufficient concentration and flow to allow production of sulfuric acid. The converter gases pass through cyclones and an electrostatic precipitator before finally arriving at the acid plant. Only one plant (in Inspiration, Arizona) uses the Hoboken converter. This plant has five, only four of which operate at any given time.

Because there is only one U.S. plant that uses Hoboken converters and there is no proper method for charging hazardous wastes into these converters (i.e., matte is poured into the furnace from ladles and there are no burners), this process has been eliminated from further consideration for the destruction of hazardous wastes.

4.7.1.6 Noranda Furnace--

The Noranda furnace is a continuous smelter. In theory, this design is a single furnace that combines most of the functions of roasting, smelting, and converting. This process approaches a one-step method of producing copper metal from ore concentrates. Only one plant in the United States (in Garfield, Utah) operates Noranda furnaces. The Noranda furnace is a horizontal cylinder about 21 meters (70 feet) long, into which a mixture of concentrate and flux is continuously fed, along with fuel and oxygen. The furnace is fired from both end walls. The mixture reacts to form blister copper, matte, and slag, which separate into layers. Additional oxygen-enriched air is blown through side-mounted tuyeres into the matte layer to produce blister copper. Slag and blister copper are tapped intermittently from the furnace. The one U.S. facility produces a high-grade matte (70% copper), which is sent to a Peirce-Smith converter for the production of blister copper and the removal of most of the impurities that would be in the blister copper if it were made in the Noranda furnace directly. The operating temperature ranges from about 1210° to 1320°C (2200° to 2400°F). Particulate loads, which are dependent on the size distribution of the feed, are quite substantial. These particulates flow with the high-temperature exhaust gases through a connected waste heat boiler and are collected as the exhaust gas flows through hot cyclones and an ESP. The dust is either pelletized or directly recycled. The gas stream is cleaned and used for sulfuric acid production.

Because only one plant in the United States uses the Noranda furnace, this process is not considered further.

4.7.1.7 Flash Smelting--

In flash smelting, finely pulverized ore concentrates are injected into a combustion chamber along with ground flux and preheated air or oxygen. Part of the sulfur in the concentrates burns in a "flash" combustion as the particles travel through the chamber. The exothermic reactions taking place normally maintain the smelting temperature, but some concentrates may require additional fuel. External fuel is required during startup of the furnace. Matte and slag form into separate layers within the chamber, as in a reverberatory furnace. The matte is sent to the conventional converters for further processing into blister copper. The slag is further processed to recover any copper content.

Temperature in a flash chamber is maintained at approximately 1090°C (2000°F). Fuel consumption in the Outokumpu type is about two-thirds of that required by a reverberatory furnace of equal capacity. Except for startup and abnormal operations, fuel is required for preheating the combustion air. Flash smelting removes a large percentage of sulfur as SO_2 (50 to 80%).[69]

The gas stream leaving the flash smelter contains 10 to 20 percent SO_2 suitable for sulfuric acid manufacture.[69] Only one plant operating in the United States (in New Mexico) uses an Outokumpu flash smelting unit. This flash smelting plant has an acid manufacturing unit. Complete removal of particulates from the gas stream is achieved by the use of a waste heat boiler, cyclones, balloon flues, and electrostatic precipitators.

Because of the highly selective nature of the raw material feed and because there is only one U.S. flash smelting plant, this process was eliminated from further consideration for destroying hazardous wastes.

4.7.2 Detailed Evaluation of Reverberatory Furnace Smelting

4.7.2.1 Process Description--

Copper smelting is the process of removing iron and other impurities from a roasted or dried ore concentrate and producing a molten mixture that can be efficiently processed by a copper converter. This process is most often accomplished with a reverberatory furnace. Figure 4.7-2 shows a cross section of a typical reverberatory smelting unit. The term "reverberatory" refers to the configuration of the flame, which enters the chamber from one end, reverberates off the roof, and strikes the charge from above. As the charge melts, it separates into two layers: matte and slag. The matte consists of cuprous sulfide-ferrous sulfide (Cu_2S-FeS) (35 to 65% Cu).[68] The slag, which contains iron oxide, combines with silica flux to form silicates and other compounds of calcium, magnesium, and aluminum. Slag is drained periodically from a skimming bay at one end of the furnace; matte is withdrawn periodically through tap holes in the lower furnace wall at the other end of the furnace. Energy consumption in this process is about 3.5 to 7.6 GJ/Mg (3.0 to 6.5 million Btu/ton) of charge. Natural gas, pulverized coal, and oil are used as fuel. The temperature of the slag and matte during operation ranges from 1150° to 1250°C (2100° to 2300°F), and the gases leave the furnace 38° to 93°C (100 to 200°F) hotter than the slag.[68]

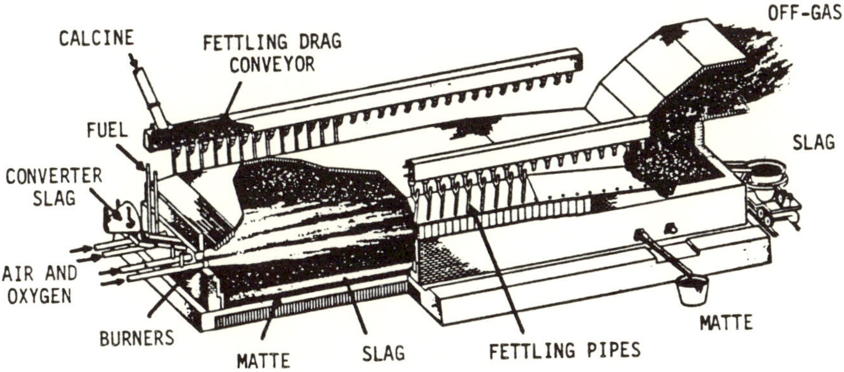

Figure 4.7-2. Cross section of a reverberatory furnace.

Waste heat boilers are typically used to recover some of the heat in the exhaust gases.

A total of 11 plants in the U.S. primary copper industry use reverberatory furnaces. Table 4.7-1 lists the plants and number of furnaces at each. Figure 4.7-3 shows the geographical location of these plants; six are located in Arizona and one each in the states of Washington, Nevada, New Mexico, Texas, and Michigan.

TABLE 4.7-1. LIST OF REVERBERATORY FURNACE LOCATIONS IN THE U.S. PRIMARY COPPER INDUSTRY

Company	Location	No. of reverberatory furnaces
Asarco, Inc.	El Paso, Texas	1
	Hayden, Arizona	2
	Tacoma, Washington	2
Copper Range Company	White Pines, Michigan	2
Kennecott Copper Corp.	Hurley, New Mexico	2
	Hayden, Arizona	1
	McGill, Nevada	2
Magma Copper Co.	San Manuel, Arizona	3
Phelps Dodge Corp.	Morenci, Arizona	5
	Douglas, Arizona	2
	Ajo, Arizona	1

Two methods of hazardous waste destruction in a copper smelting furnace can be evaluated: 1) liquid wastes could be co-fired with the fuel, and 2) solid and semisolid hazardous wastes could be charged into the furnace. Because of the high potential for fugitive emissions from handling and conveying solid and semisolid hazardous wastes, this section addresses only the co-firing of liquid wastes and fuel.

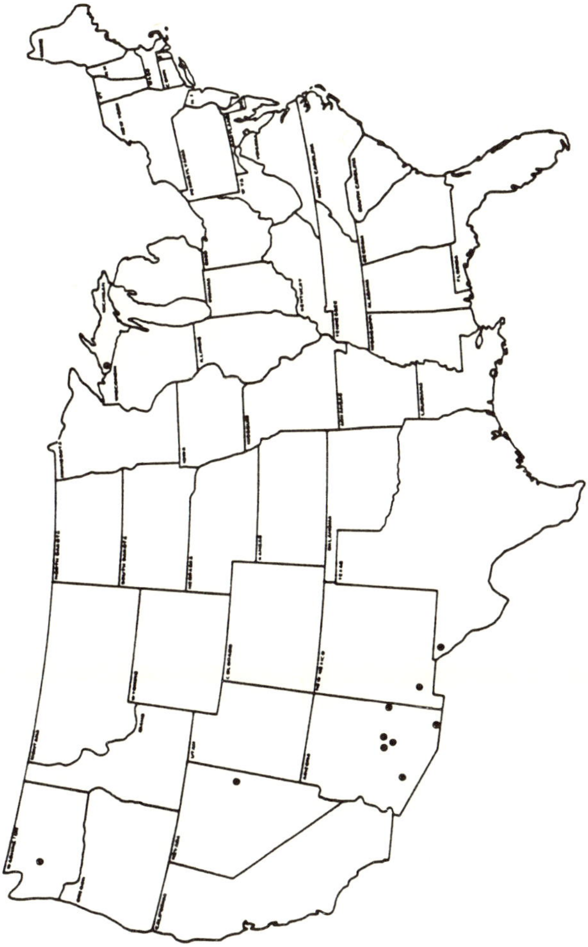

Figure 4.7-3. Distribution of reverberatory furnaces in the U.S. primary copper industry, by state.[69,70]

4.7.2.2 Technical Considerations--

Graphical determination of suitable waste categories--Pertinent process conditions for determining the time/temperature relationship in reverberatory copper smelting furnaces are:

Volume of furnace	566 m^3 (20,000 ft^3)
Gas volumetric flow rate	2630 Nm^3/min (100,000 scfm)
Maximum reaction zone temperature	1430°C (2600°F)
Exit gas temperature	1320°C (2400°F)

Figure 4.7-4 shows a gas temperature profile for a reverberatory furnace with and without oxygen enrichment of the air. Figure 4.7-5 presents a time/temperature curve for a typical reverberatory smelting furnace. The high furnace temperature and the gas retention time of about 2 seconds [as estimated by the process model (Section 3)] are suitable for the destruction of the hazardous wastes in several categories.

As Figure 4.7-5 shows, K-015, K-023, K-093, F-003, K-018, K-025, and K-030 wastes are projected to be 99.99 percent destroyed if used as part of the fuel in a reverberatory furnace.

Effects on product quality and product application--Table 4.7-2 indicates the distribution of elements between the matte and slag in a typical smelter. The matte contains most of the heavy metals present in the charge, practically all the gold and silver, and part of the arsenic and antimony. Any of these elements contained as impurities in the hazardous waste that might find their way into the slag or matte layers should not affect the quality of the matte or prevent the slag from being discarded in the normal manner. The products of combustion would be emitted with the flue gas, and no matte contamination is expected.

Major uses of copper are for electrical equipment and supplies, construction (plumbing, fixtures, and roofing), and industrial machinery. Any possible minor contamination is not expected to affect these applications.

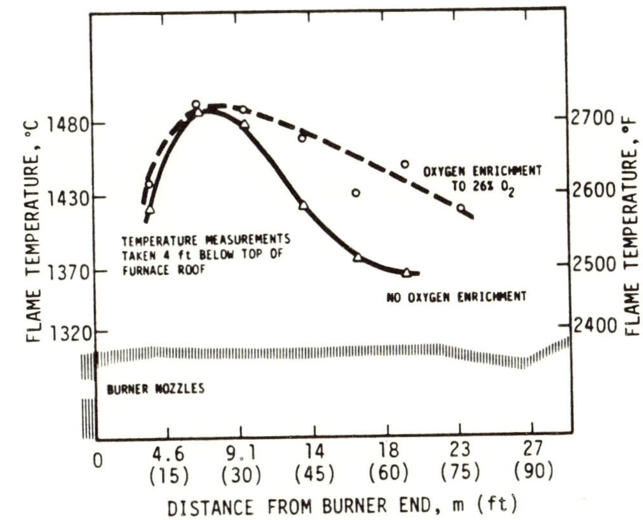

Figure 4.7-4. Temperature profile in a reverberatory furnace.

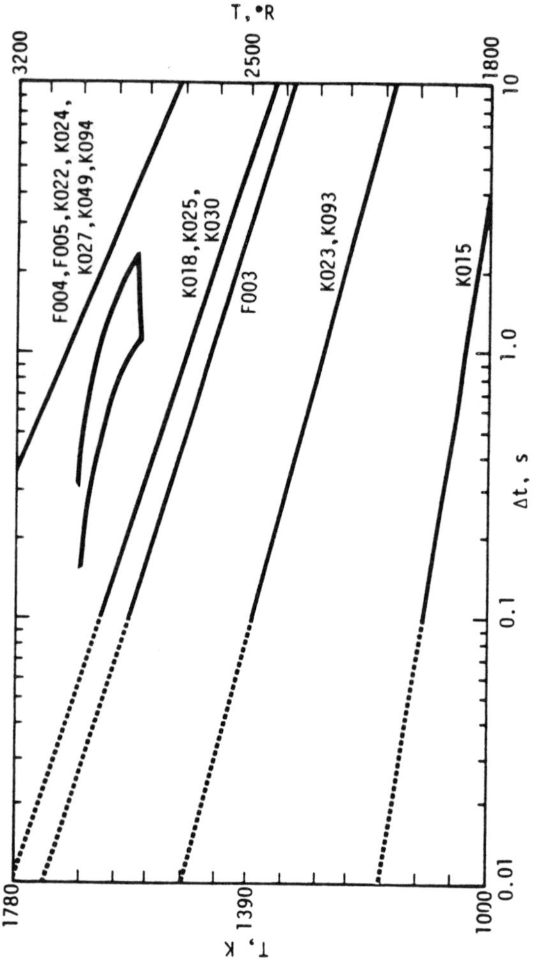

Figure 4.7-5. Hazardous waste destructibility predictions for reverberatory furnaces in primary copper smelting.

TABLE 4.7-2. DISTRIBUTION OF ELEMENTS DURING MATTE SMELTING[68]

Metal	Distribution		
	Matte	Slag	Volatilized[a]
Alkali and alkaline earth metals, aluminum, titanium		100	
Silver, gold, platinum metals	99	1	
Antimony	30	55	15
Arsenic	35	55	10
Bismuth	10	10	80
Cadmium	60	10	30
Cobalt	95	5	
Lead	30	10	60
Nickel	98	2	
Selenium	40		60
Tellurium	40		60
Tin	10	50	40
Zinc	40	50	10

[a] Not including solids blown from the furnace (dust losses).

Impact on process hardware--The refractory used in copper smelting reverberatory furnaces is typically either magnesite (MgO) or chrome magnesite ($Cr_2O_3 \cdot MgO$).[68] Halogenated wastes and wastes containing alkali metals should be avoided because they could drastically shorten the life of the refractory lining.

Acceptable amounts of hazardous waste feed--Fuel consumption in the reverberatory copper smelting furnace varies according to whether the charge is a wet or dry concentrate or calcine (from the roaster). A typical heating requirement is about 5.8 GJ/Mg (5 million Btu/ton) of charge. In a furnace processing 1360 Mg (1500 tons) of charge per day, this is equivalent to about 200 kl (52,000 gal) of oil per day. Assuming a substitution rate for hazardous waste fuel of 10 percent, about 18,900 liters/day (5000 gal/day) of waste could be disposed of per day or about 14 liters/Mg (3.3 gal/ton) of charge.

Provisions for waste introduction--Pulverized coal, oil, and natural gas are used as fuels for reverberatory furnaces. Wastes could be stored in a tank and metered into the pulverized coal or the oil before it reaches the burner. If cofired with natural gas, burner modifications at the furnace and separate lines from the waste storage tank to the furnace(s) would be necessary. The existing burners usually have low-pressure air and fuel delivery systems and are often equipped to handle many types of fuel. In these cases, burner modifications may not be necessary.

Projected trends in process fuel--Although pulverized coal was once the predominant fuel, oil and natural gas are now the most used fuels because they are easier to transport, store, and burn. Rapidly increasing prices of oil and natural gas, however, may cause a return to pulverized coal. Increased fuel costs are also causing a trend away from reverberatory furnace smelting toward other processes because this process does not make significant use of the energy available from oxidizing the sulfide charge.

4.7.2.3 Geographical Matching of Industry Locations and Waste Generators--

Figure 4.7-5 shows that hazardous wastes in categories K015, K023, K093, F003, K018, K025, and K030 should be 99.99 percent destroyed if used as part of the fuel in a reverberatory furnace. Categories K015, K018, and K030 wastes probably will not be compatible with furnace operation because they contain halogenated compounds. The K023, K093, and F003 wastes generated in Illinois and F003, K025, and K093 wastes generated in the Texas area are relatively close to smelter locations. The limited number of furnaces, however, rules out widespread national application.

Table 4.7-3 shows the liquid waste categories that can be destroyed in copper smelting reverberatory furnaces.

TABLE 4.7-3. APPLICABILITY OF COPPER SMELTING REVERBERATORY FURNACES FOR THERMAL DESTRUCTION OF HAZARDOUS WASTES

Types of wastes	Applicability	Remarks
K015	No	Highly chlorinated
K023	Yes	Except dichlorethane
K093	Yes	
F003	Yes	
K018	No	Highly chlorinated organics
K025	Yes	
K030	No	Chlorinated

4.7.2.4 Environmental Concerns--

Gases from the reverberatory furnace typically pass through a waste heat boiler and then through an electrostatic precipitator for particulate removal. The gases may pass through a spray chamber or balloon flues before entering the ESP units. The degree of particulate removal ranges from 50 to 99.99 percent.[69]

Fugitive emissions that occur during charging and tapping may be a problem when firing hazardous waste. Any residues remaining after hazardous waste destruction will pass into the control system or enter the metal or the slag. The ESP dust is usually recycled to the furnace, whereas the slag from the reverberatory furnace is usually landfilled on site. Heavy metal concentrations in the slag would have to be monitored closely.

4.7.2.5 Economic Factors--

Fuel costs are a large part of the reverberatory smelter operating costs. Fuel credits from high-Btu wastes may decrease the rate of copper smelter switchover from reverberatory smelters to more fuel-efficient smelting techniques.

4.7.2.6 Institutional Factors--

Reverberatory furnace smelters are not located in highly populated areas. Thus, the major obstacle for hazardous waste destruction will probably come from within the industry. The threat of reduced productivity and quality may preclude consideration of hazardous waste destruction in reverberatory copper smelters.

4.8 PRIMARY LEAD INDUSTRY

The United States leads the world in lead production, and accounted for about 15 percent of the total world mine production in 1978.[71] Figure 4.8-1 shows the locations of U.S. primary lead smelters and refineries. Three of the five U.S. lead smelters are located in southeastern Missouri's "New Lead Belt." More than 80 percent of the ore mined specifically for lead in the United States also comes from this region. Table 4.8-1 shows the capacities of the individual smelters/refineries.

4.8.1 Process Descriptions

Lead is usually found in nature as the sulfide ore, galena (PbS), which also contains small amounts of copper, iron, zinc, and other trace elements. It is generally processed at the mine,

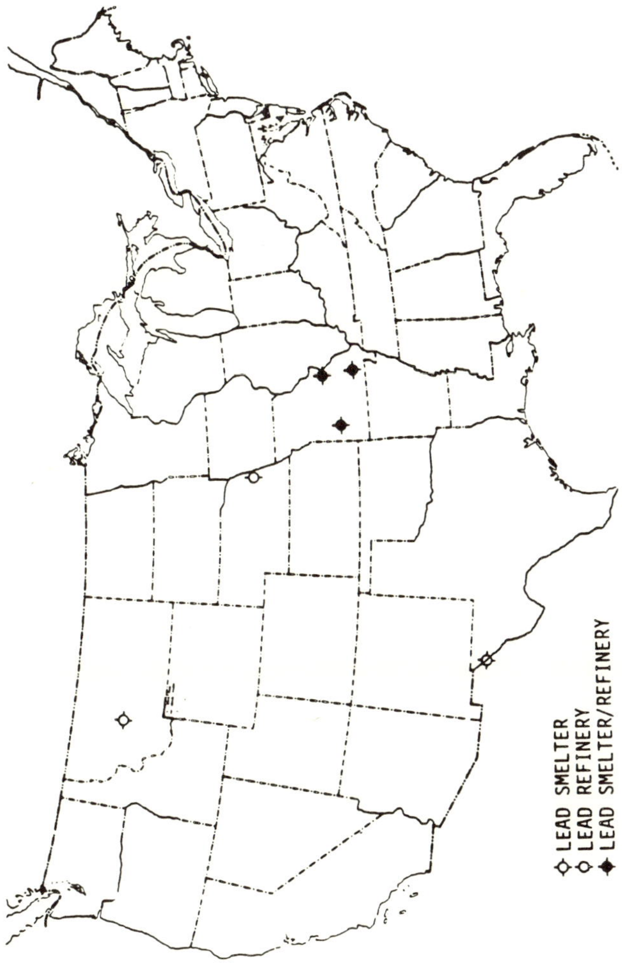

Figure 4.8-1. Distribution of primary U.S. lead smelting and refining facilities.[71]

TABLE 4.8-1. U.S. PRIMARY LEAD PRODUCERS[71]

Company	Location	Description	Design capacity, Mg/yr (tons/yr)
Amax-Homestake Lead Tollers	Boss, Missouri	Smelter/refinery[a]	127,000 (140,000)
Asarco, Inc.	East Helena, Montana	Smelter	109,000 (120,000)
	El Paso, Texas	Smelter	109,000 (120,000
	Glover, Missouri	Smelter/refinery[a]	99,300 (110,000)
	Omaha, Nebraska	Refinery	163,000 (180,000)
St. Joe Minerals Corporation	Herculaneum, Missouri	Smelter/refinery[a]	204,000 (225,000)

[a] The smelting and refining of Missouri lead ore takes place in the same plant.

from an ore of 3 to 8 percent lead to a concentrate of 55 to 70 percent lead. Lead content of some of the high-grade Missouri ores exceeds 70 percent even before the concentrating process.[71] Typically, 13 to 19 percent by weight of the concentrate is sulfur.[68]

The simplified flow sheet in Figure 4.8-2 shows the fundamental operations involved in the production of lead from lead sulfide concentrates. The primary lead industry uses various types of high-temperature furnaces (each of which has unique functions) in the process of recovering lead and other trace metals from the concentrates. A list of these furnaces, along with their primary functions, is presented below.[71]

1. Sintering furnace: Sintering (blending, pelletization and roasting) for subsequent operations.
2. Blast furnace: Reduction of sintered feed.
3. Slag fuming furnace: Recovery of metal values from blast furnace slag.
4. Drossing kettle: Initial step in lead refining.
5. Dross reverberatory furnace: Separation of components in dross.
6. Roasters: Recovery of cadmium from sinter dust.
7. Reverberatory softening furnace: Purification of drossed lead bullion.
8. Softening slag treatment furnace: Recovery of antimony from slag.
9. Faber-du-Faur type retorting furnace: Recovery of zinc.
10. Cupel furnace: Purification of retort metal.

The sintering machines, blast furnaces, and refining kettles are characteristic of all five domestic primary lead smelters, whereas the other types of furnaces are specific to certain individual smelters. Capacities of some of these other furnace types

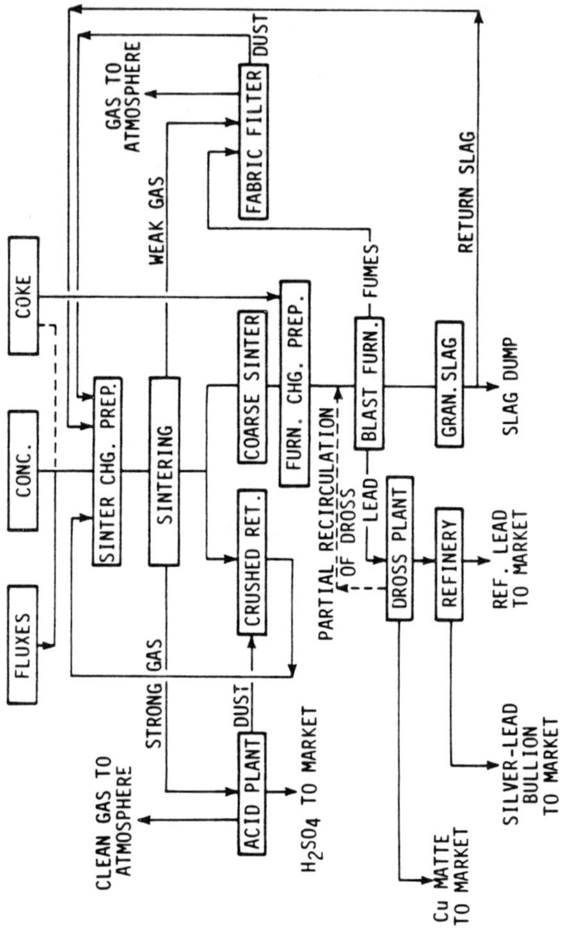

Figure 4.8-2. Primary lead smelting--general flow sheet.[72]

are fairly low: retorting furnaces have a capacity of about 0.6 Mg (0.7 ton), while cupels can handle a maximum of approximately 11 Mg (12 tons) per charge.[71]

A total of five sintering machines and seven blast furnaces are used in the five domestic primary lead smelters. The sintering and blast furnace reduction processes are discussed in further detail in the following subsections.

4.8.1.1 Sintering Process--

The basic purpose of sintering is to convert most of the lead sulfide concentrate (PbS) into the oxide form while simultaneously producing a hard, porous, clinker material suitable for the rigid requirements of the blast furnace. About 85 percent of the sulfur is removed from the concentrate during sintering.[71] Figure 4.8-3 presents a typical flow sheet of a sintering plant.

The sintering machine is essentially a continuous, steel pallet-type conveyor belt that is motor-driven with appropriate gearing and sprockets. Each pallet consists of slotted grates. Either beneath or above the moving pallets are windboxes, which are connected to fans that provide a draft on the moving sinter charge. Depending on the direction in which the draft is placed upon the charge, the sintering machines are termed "updraft" or "downdraft." Because of their greater capacities, their lower maintenance and fan power requirements, and their ability to produce sinter of higher lead content, the updraft systems are more popular.[68]

Input to the sintering machine consists of lead concentrate, collected flue dusts, recycled sinter, and smelter residues. Sulfide-free fluxes are also added in most processses to maintain a specified sulfur content (5 to 7 percent by weight) in the charge. Silica and limestone are used as needed. Coke fines are also mixed with the charge.[71] Gas- or oil-fired burners are used to ignite the charge in updraft sintering machines. Water may be added for pelletizing the charge if the moisture content falls below required limits. Air is injected through the charge for oxidizing in the sintering machine.

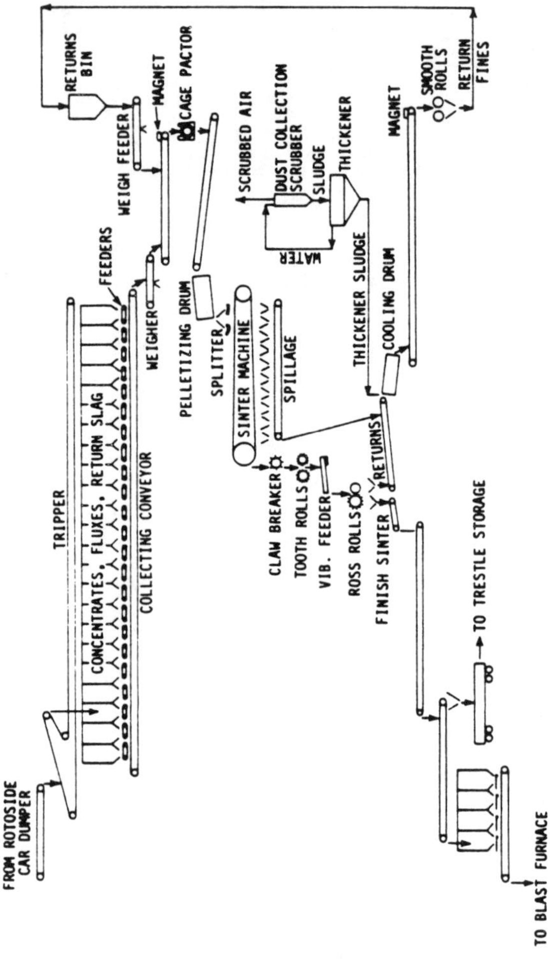

Figure 4.8-3. Flow sheet of a typical sinter plant.[73]

The sintering reaction is autogenous and creates temperatures of approximately 820°C (1500°F), the temperature basically controlled by the sulfur content of the charge. About 85 percent of the sulfur is removed from the concentrate during sintering. The exit gases may be removed in a single stream with an SO_2 concentration of about 2 percent, or split into two streams, one predominantly from the front and the other from the rear. The latter procedure produces both a weak (~0.5% SO_2) and a strong (~6.0% SO_2) SO_2 stream. The latter is amenable to sulfuric acid production.

An alternative solution to the elimination of the weak stream from the sintering machine is the use of weak gas recirculation, which can enable a single strong SO_2 gas stream (~6.0% SO_2) to be emitted.[68] This technique, however, is only used by foreign lead smelters.

Two of the existing domestic primary lead smelters use fabric filters for particulate control. Table 4.8-2 lists the controls used on lead sintering processes by U.S. smelters.

4.8.1.2 Blast Furnace Reduction Process--

The blast furnace is used in the primary lead industry to reduce sintered feed to a crude lead bullion. Figure 4.8-4 presents a flow sheet for the blast furnace reduction process.

Figure 4.8-5 allows a closer look at the blast furnace. The charge, which consists of sinter, metallurgical coke, and compounds such as limestone, silica, litharge, and other fluxing materials, is fed through a water-jacketed shaft at the top of the furnace. Coke comprises 8 to 15 weight percent of the furnace charge.[71] Combustion air is injected through tuyeres located near the bottom and evenly spaced on either side of the furnace. Oxygen is sometimes added to enrich the combustion air. Temperatures in the blast furnace range from 220°C (420°F) for the charge near the top of the furnace to 1210°C (2200°F) in the slag zone. Slag temperatures range from 980° to 1210°C (1800° to 2200°F), and bullion temperatures, from 890° to 960°C (1640° to 1750°F).

TABLE 4.8-2. ATMOSPHERIC CONTROL SYSTEMS ON DOMESTIC PRIMARY LEAD SINTERING MACHINES[71]

Plant and location	Control system
Amax, Boss, Missouri	Updraft sintering machine produces two gas streams: a strong gas stream that goes to the acid plant and a weak gas stream that is combined with blast furnace gas before it is discharged through the stack.
St. Joe, Herculaneum, Missouri	Updraft sintering machine produces two gas streams: a strong gas stream that goes to the acid plant and a weak gas stream that joins other gases and then passes through a fabric filter before exiting out the stack.
Asarco, Inc., E. Helena, Montana	Updraft sintering machine. Gases go to a water spray and an ESP; then dilution air is added and the gases are released out of the stack.
Asarco, Inc., Glover, Missouri	Updraft sintering machine. All gases go to a water spray and a fabric filter, and then out the stack.
Asarco, Inc., El Paso, Texas	Updraft sintering machine. A new control system is planned, which will make use of an acid plant for the strong stream.

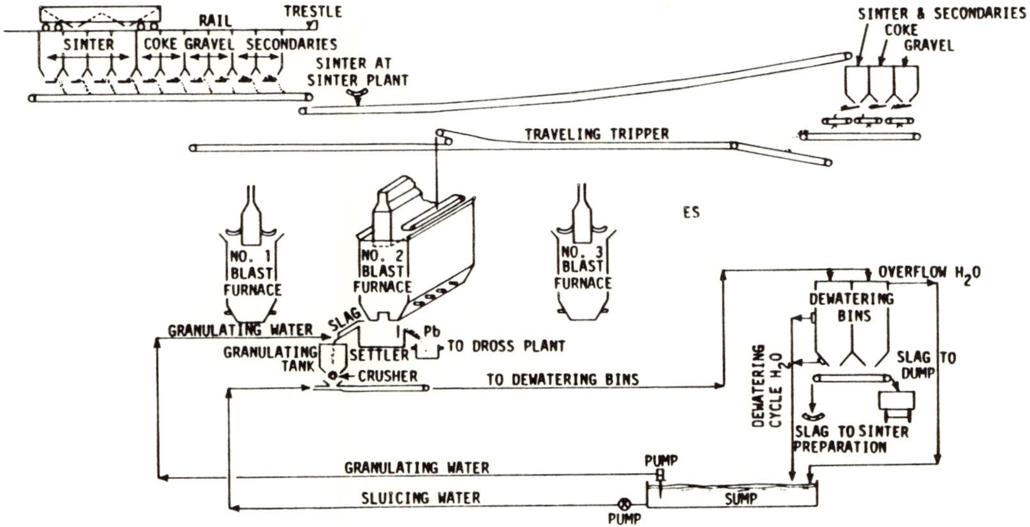

Figure 4.8-4. A flow sheet of a blast furnace reduction process.[73]

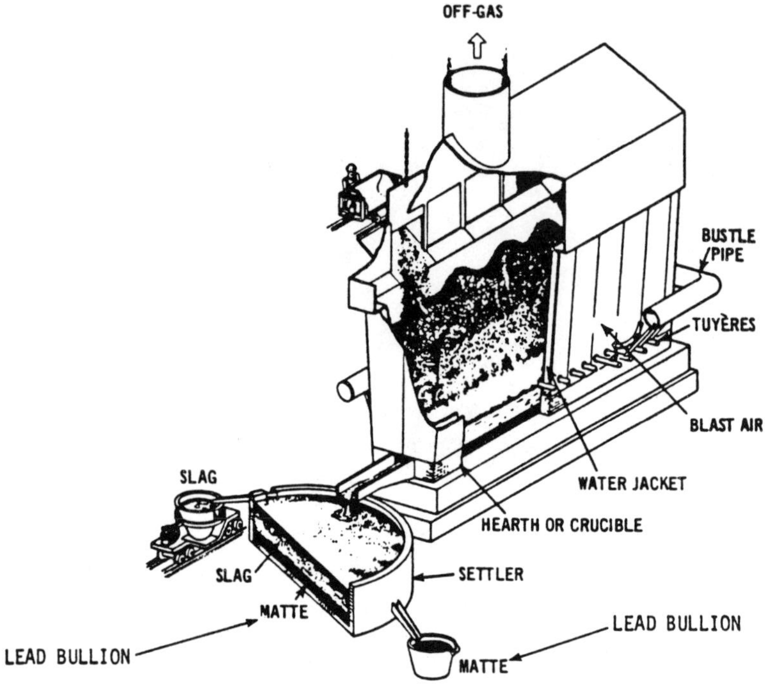

Figure 4.8-5. Diagram of a lead blast furnace.[68]

The principal reactions that take place in the blast furnace are:[68]

$$PbO + CO + heat \rightarrow Pb + CO_2$$

$$C + O_2 \rightarrow CO_2 + heat$$

$$C + CO_2 + heat \rightarrow 2CO$$

Because the sintering process is not 100 percent efficient in the conversion of lead sulfide (PbS) to lead oxide (PbO), some lead sulfate ($PbSO_4$) and small amounts of lead sulfide remain in the product sinter. These compounds lead to additional lead-forming reactions and the simultaneous generation of SO_2.

$$2PbO + PbS \rightarrow 3Pb + SO_2$$

$$PbSO_4 + PbS \rightarrow 2Pb + 2SO_2$$

Sulfur dioxide concentrations range from a few hundred ppm to as much as 2500 ppm.[68] Sulfur dioxide emissions are not currently controlled.

Particulate emissions from lead blast furnaces are controlled by fabric filters at all the domestic smelters. Control efficiency ranges from 95 to 99.9 percent.[71] Table 4.8-3 describes the controls used for blast furnace gases.

4.8.2 Process Assessment for Hazardous Waste Destruction

A feasibility assessment was made of the use of the main processes in primary lead smelting for thermally destroying hazardous wastes.

The sintering process achieves a maximum temperature of only about 920°C (1500°F), which is not sufficient to destroy most of the hazardous wastes being considered in this study. Charging hazardous wastes in a safe manner also would be a problem with

TABLE 4.8-3. ATMOSPHERIC CONTROLS SYSTEMS ON DOMESTIC PRIMARY LEAD BLAST FURNACES[71]

Plant location	Control system
Amax, Boss, Missouri	Blast furnace gases join sinter weak gases, pass through fabric filters, and then exit out to stack.
St. Joe, Herculaneum, Missouri	Blast furnace gases join sinter weak gases and other gases, pass through the fabric filters, and then exit out the stack.
Asarco, Inc., E. Helena, Montana	Blast furnace gases join reverb and ventilation gases and then pass through three fabric filters in parallel, each with its own stack.
Asarco, Inc., Glover, Missouri	Blast furnace gases pass through a water spray, a fabric filter, and three stacks.
Asarco, Inc., El Paso, Texas	Blast furnace and dross furnace gases mix and then pass through a spray chamber and a fabric filter before they exit out of six stacks.

the sintering machine. Another disadvantage is the high potential for fugitive emissions. Also, the possible dilution of the strong SO_2 stream (amenable for H_2SO_4 production) by products of combustion of the hazardous wastes may result in a deterioration in the performance of the subsequent contact sulfuric acid plant. For the reasons just given, the sintering process has been eliminated from further consideration for hazardous waste disposal.

The lead refining process is essentially the final step in the formation of pure lead; hence, the introduction of extraneous materials, especially hazardous wastes, is not recommended for this process either.

Because of its high temperature [1210°C (2200°F)] and moderate residence times, the blast furnace reduction process, however, appears to be suitable for thermal disposal of hazardous wastes. The following subsections apply to the blast furnace process.

4.8.2.1 Technical Factors--

Compatibility of hazardous wastes with process fuel--Metallurgical coke is the primary fuel (as well as reducing agent) in the blast furnace reduction process. Because the main fuel is one of the reactants, the hazardous wastes cannot be substituted for any part of the fuel, but they can be introduced in addition to the coke. No compatibility problems are anticipated. The heat content of the wastes should be sufficient to sustain operating temperatures, and no additional process fuel (coke) is deemed necessary.

Graphical determination of suitable waste categories--

Process data--The typical blast furnace dimensions and operating data presented are based on blast furnace operations at the Herculaneum Lead Smelter of St. Joe Minerals Corporation, Herculaneum, Missouri.[73] The selected process conditions for determining the time-temperature relationship are as shown below:

Volume of blast furnace:	140 m^3 (4940 ft^3)
Gas volumetric flow rate:	276 Nm^3/min (10,500 scfm)
Maximum temperature:	1210°C (2200°F)
Exit gas temperature:	700°C (1300°F)

Approach--Based on computations of mean residence time (t) and fast-path residence time $\bar{t}$ for several temperatures below the maximum reaction zone temperature, process curves are plotted for the blast furnace reduction process and overlaid on the hazardous waste plots, as shown in Figure 4.8-6. Wastes in categories K015, K023, and K093 appear to be destructible.

Effects on product quality and product application--At the prevailing temperatures in the blast furnace, the hazardous constituents in the destructible wastes would undergo oxidation and/or thermal decomposition to other substances. The products of combustion would be emitted in the flue gas. No contamination of the lead bullion is expected.

Major areas of lead consumption include battery manufacturing, the production of gasoline additives, plumbing, and construction.[68,71] Any possible minor contamination is not expected to affect these applications either.

Impact on process hardware--No two blast furnaces are alike in dimensions or construction, which fact reflects a continual program to develop the most efficient design.[73] Most of the furnaces are of the Australian type, which have two rows of tuyeres. In the combustion and molten metal areas, the furnaces are generally constructed of fireclay brick about 1 foot thick. The hearth area is refractory lined. Furnace bullion is removed by refractory-lined transfer ladles. One of the destructible wastes, K015, contains halogens, and it is possible that the chlorides may affect the refractory lining.

Acceptable amounts of hazardous waste feed--Feed rates of input materials to the blast furnace vary from one plant to

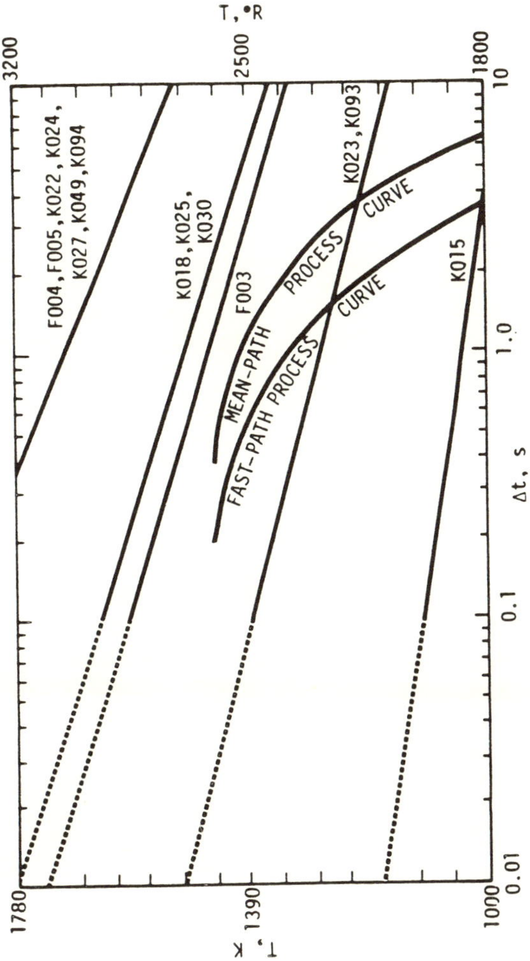

Figure 4.8-6. Hazardous waste destructibility predictions for blast furnaces in primary lead smelting.

another. Feed rates and smelting rates for a typical blast furnace are shown below:[73]

Sinter smelted	820 Mg/day (900 tons/day)
Size of charge, sinter	9.1 Mg (10 tons)
Coke in charge	9%
Air volume	276 Nm^3/min (10,500 scfm)

Assuming that the percentage of hazardous wastes in the charge is equivalent to that of coke in the charge, i.e., 9 percent by weight, then 734 Mg/day (81 tons/day) of hazardous waste can be destroyed.

Densities of the three destructible waste categories are as follows: K015, 1.64 kg/liter (13.7 lb/gal); K023, 1.37 kg/liter (11.4 lb/gal); and K093, 1.37 kg/liter (11.4 lb/gal). Assuming an average waste density of 1.44 kg/liter (12.0 lb/gal), the hazardous waste feed represents a consumption of 570 liters (150 gallons) per charge.

Provisions for waste introduction--The tuyeres available in the blast furnace for supply of combustion air could be used for introduction of the hazardous wastes. It might be preferable to introduce the atomized wastes through the lower set of tuyeres in Australian type blast furnaces. Modifications in the existing tuyere design would be necessary to atomize wastes before their introduction into the furnace. Domestic lead blast furnaces generally have 30 to 40 tuyeres at each level. These tuyeres have a diameter of 7.6 to 10 cm (3 to 4 inches).

Projected trends in process fuel--Because metallurgical coke serves the dual function of fuel and reductant in the blast furnace, no fuel substitution is possible and coke will remain the process fuel for lead blast furnaces. Additional sulfur introduced in the charge could minimize the amount of coke required.

4.8.2.2 Geographical Matching of Industry Location and Waste Generators--

Table 4.8-4 presents the geographic distribution of the destructible wastes and the SIC codes of the industries generating these wastes. The locations of the plants generating waste type K015 could not be determined. This table shows that none of the destructible waste types is generated in Missouri, where the majority of the domestic primary lead smelters are located. Significant quantities (38% of total) of waste K023 is generated in Illinois, however, which is a neighboring state to Missouri. Both K023 and K093 are also generated in Texas, where one of the primary lead smelters operated by Asarco is located.

TABLE 4.8-4. SIC CODES AND GEOGRAPHIC DISTRIBUTION OF INDUSTRIES GENERATING HAZARDOUS WASTES DESTRUCTIBLE IN PRIMARY LEAD SMELTING[19]

Hazardous waste categories	SIC codes of waste-generating industry	Geographic distribution of industries
K015	2865	Two plants (locations not known)
K023	2869	Richmond and El Sequendo, California; Kearney and Bridgeport, New Jersey; Baton Rouge, Louisiana; Bridgeville and Neville Island, Pennsylvania; Cicero and Millsdale, Illinois; Texas City, Texas
K093	2869	Texas, California, New Jersey

4.8.2.3 Environmental Factors--

Impact on air pollution control devices--Fabric filters are used to control particulates in blast furnace exhaust gases at all domestic primary lead smelters. Many of the fabrics commonly used as filter materials (e.g., Nomex, Teflon, and fiberglass) have excellent resistance to organic acids or solvents.[74] Hence, combustion products of hazardous wastes should have no harmful effect on the fabric filters.

Environmental impact--Not all sulfur in the sinter feed to the blast furnace is eliminated as SO_2. Almost half of the furnace sulfur is retained in the slag or matte.[68] Concentrations of SO_2 in the flue gas are therefore generally only a few hundred ppm. Concentrations of CO, however, are extremely high--around 25 to 50 percent.[71] Partial combustion of hazardous waste constituents could further increase CO levels. In lead smelting, afterburners are generally used to reduce CO concentrations to about 5 percent.[71] These should also control any additional CO due to waste combustion. Hydrocarbons would also be simultaneously controlled.

The fate of chlorides and fluorides resulting from the destruction of wastes in category K015 is an important consideration. Based on observed sulfur retention rates, it is expected that about 50 percent of the halogens would be retained as chlorides or fluorides in the slag and the remainder would be emitted with the flue gas. Emission measurements during test burns would help to determine accurately the fraction evolving with the flue gas.

Current disposal practices consist of conveying the slag hydraulically with a granulating water stream to a dump or tailings pond. Depending on the amount of halogens retained in the slag, the potential may exist for leaching of chlorides and fluorides as acids. This needs to be assessed in preliminary tests.

4.8.2.4 Institutional Factors--

Mine production and consumption of lead have declined since 1978, primarily because of curtailed use of lead antiknock compounds in gasoline.[71] Thus it is doubtful that any new smelters will be constructed in the United States in the near future. The industry currently employs about 2400 persons at smelters and refineries, and approximately 4700 people at mines.

Two of the primary lead smelters, St. Joe's Herculaneum smelter and Asarco's Glover smelter, are located 25 and 100 miles south of St. Louis, respectively, on the west bank of the Mississippi River. The setting of the Glover plant is rather typical of rural Missouri in the Ozarks.[75] The plant itself is surrounded by timber and cattle-grazing areas. Though easily accessible to rail and truck transportation, the plants are not close to population centers.

Many of the companies producing primary lead are vertically integrated and involved in two or more process segments. Some of these firms process other metals, such as zinc, copper, and gold, and the recovery of these metals often makes the mining and processing of lead economically feasible.[71] Because combustion of hazardous wastes is not likely to offer any cost benefits (such as fuel savings), it is doubtful that these facilities would be willing to accept wastes for destruction.

4.9 PRIMARY ZINC INDUSTRY

The major product of the primary zinc industry is metallic zinc, but zinc oxide, sulfuric acid, cadmium, and occasionally, zinc sulfate are also produced. Generally, ore is mined and concentrated at one location and then transferred to smelters for the production of zinc, zinc oxide, or both. Cadmium is normally recovered at smelters, from collected dusts and slags with sufficient cadmium content.[76]

Figure 4.9-1 shows the locations of the six existing domestic primary zinc smelters and Table 4.9-1 presents additional information regarding these smelters, including their capacities.

4.9.1 Process Descriptions

Zinc is usually found in nature as a sulfide ore called sphalerite, which also contains impurities of lead, cadmium, and

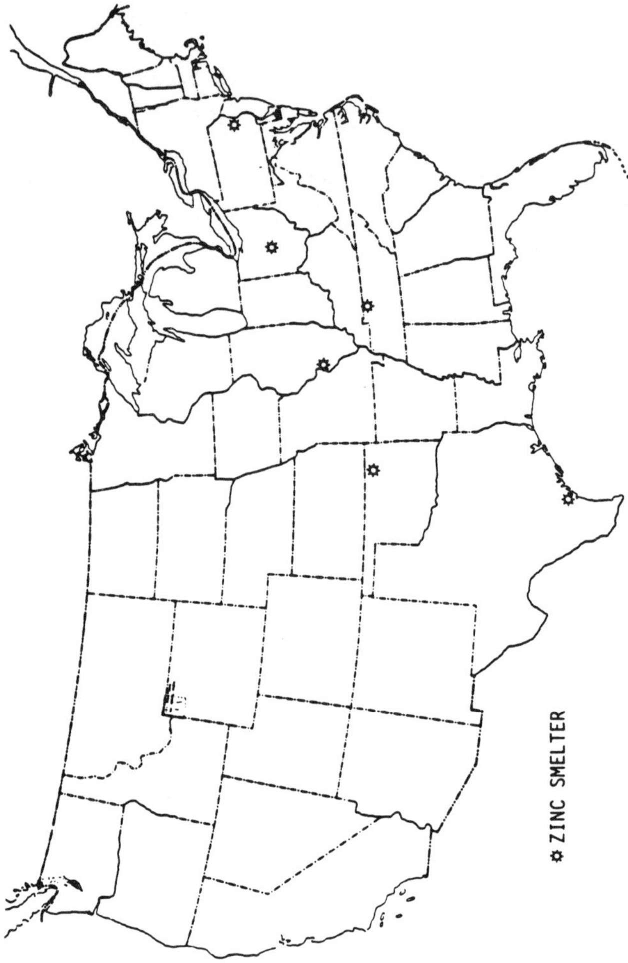

Figure 4.9-1. Distribution of primary U.S. zinc smelters.[76]

TABLE 4.9-1. PRIMARY ZINC PROCESSING PLANTS IN THE UNITED STATES[76]

Company	Location	Process type	Design capacity, Mg/yr (tons/yr)	Date plant built
Amax, Inc.	East St. Louis, Illinois	Electrolytic	76,200 (84,000)	1929
Asarco, Inc.	Corpus Christi, Texas	Electrolytic	98,000 (108,000)	1942
	Columbus, Ohio	Pyrometallurgical	20,000 (22,000)	1967
Jersey Miniere Zinc	Clarksville, Tennessee	Electrolytic	89,900 (99,000)	1978
National Zinc Co.	Bartlesville, Oklahoma	Electrolytic	45,400 (50,000)	1977
New Jersey Zinc Co.	Palmerton, Pennsylvania	Pyrometallurgical	103,000 (113,000)	1899

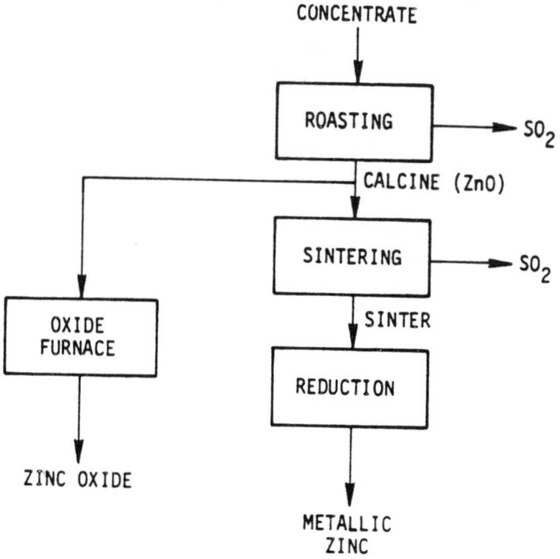

Figure 4.9-2. Flow diagram of primary pyrometallurgical zinc smelting process.[70]

minor amounts of other trace elements.[70] Generally, ore is mined and concentrated to about 62 percent zinc and 32 percent sulfur at one location, and then transferred to smelters for production of zinc and/or zinc oxide.

The smelting of zinc sulfide concentrates is carried out by either a pyrometallurgical or an electrolytic process, as illustrated in Figures 4.9-2 and 4.9-3. The initial roasting step, in which calcine is formed by removing most of the sulfur, is common to both processes. In the pyrometallurgical process, the calcine from the roasting step is sintered to form a dense, permeable feed for subsequent reduction to metallic zinc; whereas in the electrolytic process, the calcine is chemically leached and subjected to electrolytic extraction to produce 99.99+ percent pure, high-grade zinc.

The types of high-temperature furnaces employed in primary pyrometallurgical zinc manufacture are listed below:

1. Roasters: multiple-hearth, suspension (or flash), fluidized-bed

2. Sintering machines

3. Vertical/electrical retorts

4. Oxidizing furnaces: grate-type furnace, rotary (or Waelz) kiln, electrothermic furnace

Process details for these furnaces are presented in the following subsections.

4.9.1.1 Roasting Process--

Both zinc-smelting processes use one or a combination of several types of roasters to remove sulfur from the concentrate.

Multiple-hearth roasting--The multiple-hearth roaster, one of the oldest types of roasters, is currently used at only one smelter in the domestic industry. It is basically a cylindrical column 6.1 to 7.6 m (20 to 25 feet) in diameter with 7 to 16 hearths (see Figure 4.9-4). The reaction within the roaster is

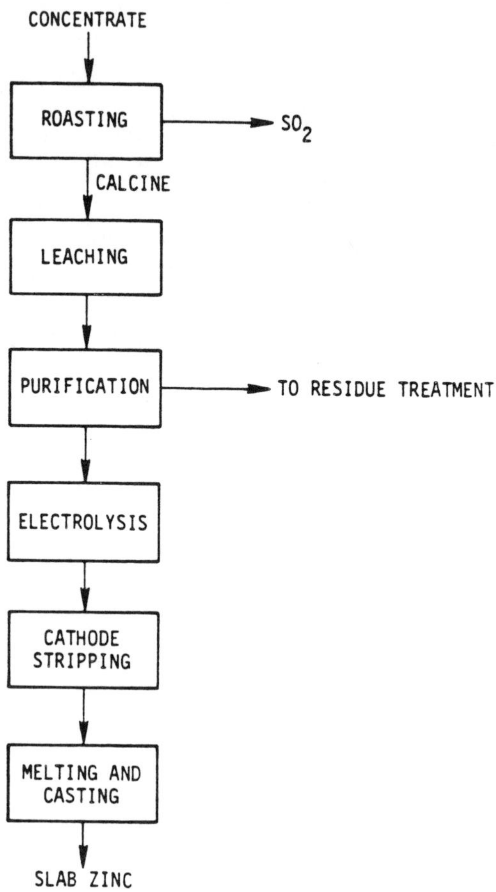

Figure 4.9-3. Flow diagram of primary electrolytic zinc smelting process.[70]

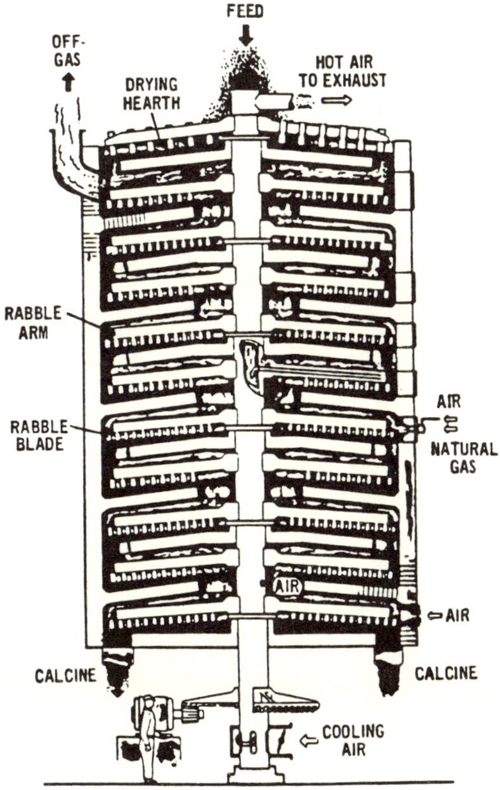

Figure 4.9-4. Multiple-hearth roasting furnace.[70]

essentially autogenous. If enough sulfur is originally present as sulfide, the following reactions occur during roasting:[76]

$$2ZnS + 3O_2 \rightarrow 2ZnO + 2SO_2$$

$$2SO_2 + O_2 \rightarrow 2SO_3$$

$$ZnO + SO_3 \rightarrow ZnSO_4$$

In pyrometallurgical reduction, only the oxide state is desired; whereas in electrolytic reduction, small amounts of the sulfate state are acceptable.

Feed capacities for such roasters average 180 Mg (200 tons) per day.[76] Average operating temperature is about 690°C (1275°F), and the lower hearths are maintained around 980°C (1800°F). Operating time depends on the type of roaster, the composition of the concentrate, and the amount of sulfur removal required. Gas, coal, or oil must be added initially to bring the charge up to reaction temperature--about 1.2 GJ/Mg of ore (1 x 10^6 Btu per ton) is required for ignition.[76]

Off-gases from the roaster contain up to 7 percent sulfur, and are routed directly to "hot" ESP's operating at temperatures around 205°C (400°F), and then to an acid plant.[76]

Suspension roasting--Suspension, or flash, roasting is a process for rapid removal of sulfur and conversion of zinc to calcine by allowing the concentrates to fall through a heated oxidizing atmosphere or blowing them into a combustion chamber.[76] Figure 4.9-5 illustrates a flash roasting furnace.

The roasting process, which is similar to the burning of pulverized coal, achieves a higher heat transfer rate than multiple-hearth roasting does. The temperature in the combustion chamber is kept around 1010°C (1840°F). Chemical reactions occurring are the same as in multiple-hearth roasting.

The SO_2 concentration in the off-gases from flash roasting averages 10 to 13 percent, which is higher than that in the off-gases from multiple-hearth furnaces. The gases are cooled to

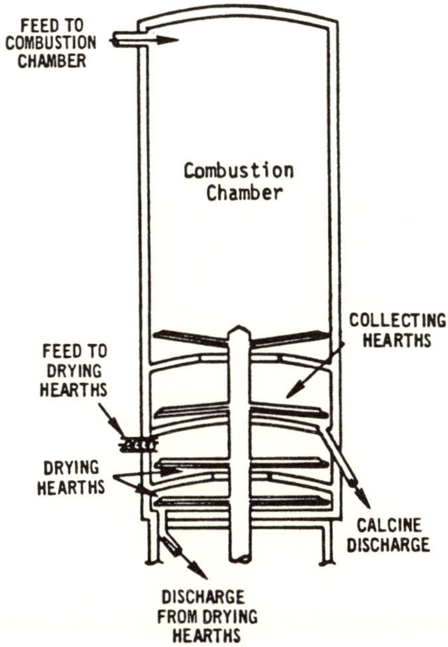

Figure 4.9-5. Suspension (flash) roasting furnace.[70]

about 400°C (750°F) in waste heat boilers, diluted with air, and humidified with water sprays before being sent to an ESP. The conditioned gases, which have an SO_2 concentration of 6 to 8 percent, are sent to an acid plant.

Fluidized-bed roasting--The fluid-bed roaster is the newest roasting system for zinc sulfide concentrates, and several domestic smelters are using it in different forms in both pyrometallurgical and electrolytic processes. The Amax and New Jersey Zinc smelters are examples.

Figure 4.9-6 illustrates a fluid-bed roaster. Roaster designs differ primarily in whether the roasters are charged with a wet slurry or a dry charge. Such roasters typically process 204 Mg (225 tons) per day of charge.[70] In the fluidized-bed process, no additional fuel is required after ignition has been achieved. Fluid-bed roasters operate at temperatures averaging 1000°C (1830°F); fluid column type roasters operate at temperatures approximately 38°C (100°F) higher. About 30 percent excess air is supplied.[77] Operating times vary, and they depend on the same factors as in other roasting processes. Natural gas, oil, or coal is used to bring the roaster feed to reaction temperatures, after which exothermic oxidation of sulfur maintains the bed temperature.

The off-gas, which contains approximately 12 percent SO_2 and exits the roaster at a temperature of about 950°C (1740°F), is cooled to 400°C (750°F) in waste heat boilers. Cyclones and ESP's are used for particulate control before the gas stream enters an acid plant.

4.9.1.2 Sintering Process--

As stated earlier, sintering of roasted charge is performed only in pyrometallurgical zinc smelting. Table 4.9-1 (presented earlier) lists only two pyrometallurgical primary zinc plants in the United States.

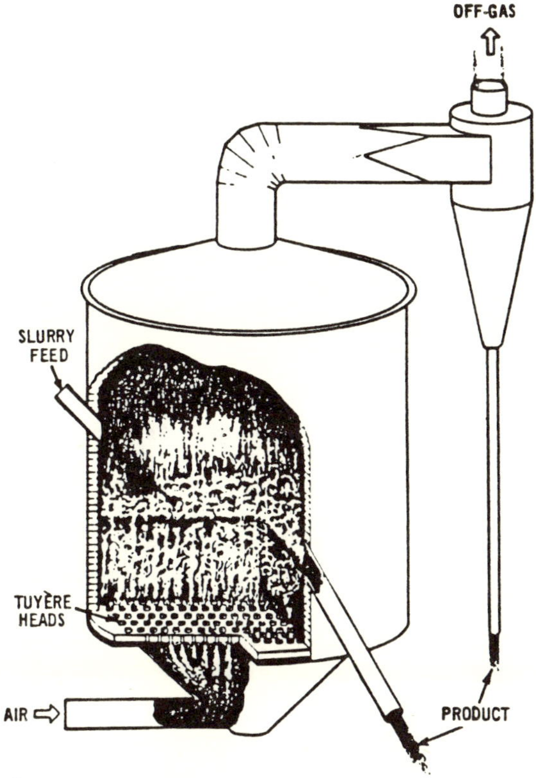

Figure 4.9-6. Fluid-bed roaster.[70]

The basic purpose of sintering in zinc smelting is the same as that for lead smelting, i.e., to produce a hard permeable mass suitable for feed to a reduction system. Sintering also volatilizes lead and cadmium impurities and discharges them into the off-gas stream, where they can be captured.

The sinter mix typically contains calcine, recycled sinter, coke or oil, sand, and other inert ingredients. Charging capacities typically range from 220 to 540 Mg (240 to 600 tons) per day. Typical operating conditions are 1040°C (1900°F) and atmospheric pressure.[76]

The sintering operation generates SO_2 and particulate matter. Because SO_2 concentrations are very low (0.1 to 2.4 percent), no SO_2 control is applied. Particulate control devices consist of various combinations of settling flues, cyclones, ESP's, and fabric filters.[76]

4.9.1.3 Retorting Process--

The retorting process is a continuous reduction/volatilization method for producing high-purity zinc from zinc oxide by reduction with carbon at elevated temperatures.[76] Depending on whether gas or electricity is used to supply the energy for reduction, the process is termed vertical retorting or electric retorting.

The primary reactions that occur in retorting consist of a reduction step and a CO regeneration step:

$$ZnO + CO \rightarrow Zn \text{ (vapor)} + CO_2$$
$$CO_2 + C \rightarrow 2CO$$

Both reactions are highly endothermic.

The feed to the retorts consists of sinter, coke, recycled zinc-bearing products, and additional binders, such as silica. No air is introduced in this reduction process. Maximum temperatures of about 1400°C (2550°F) are attained.

Emissions are minor compared with other processes in zinc smelting, such as roasting and sintering. Off-gases are cleaned by scrubbers, and because of their high CO content, serve as part of the fuel for heating the retorts.

4.9.1.4 Oxidizing Process--

This process occurs in zinc oxide production, where zinc vapor from sintering is immediately oxidized without being condensed. As with zinc metal production, zinc must first be produced in vapor form.

In the United States three types of furnaces are used for zinc oxide production: grate-type furnaces, rotary kilns, and electrothermic furnaces. Temperatures in the grate-type bed range from 980° to 1930°C (1800° to 3500°F); operating temperatures in rotary kilns are about 1320°C (2400°F). Temperatures in electrothermic furnaces vary from 900°C (1650°F) near the walls to about 1400°C (2550°F) in the main smelting zone.[76] All three furnace types emit some zinc oxide in the flue gas, and control techniques are similar to those used in other reduction operations.

4.9.2 Process Assessment for Hazardous Waste Destruction

Because roasting is the only step performed at all six domestic primary zinc smelters, it is this process that offers the greatest capacity for thermal disposal of hazardous wastes. Time-temperature requirements for all three types of roasters also appear compatible for hazardous waste destruction.

Sintering, retorting, and oxidizing processes only occur at the two domestic pyrometallurgical plants, and thus offer limited capacity for hazardous waste disposal. Possible difficulty with safely charging hazardous wastes and the inherent potential for high fugitive emissions are other disadvantages of the sintering process for hazardous waste destruction. The fact that no air is introduced in retorting excludes this process from consideration.

Rotary-kiln oxidizing furnaces would be suitable for waste destruction, but the limited number of such furnaces available does not make further consideration worthwhile.

In summary, only the roasting step in primary zinc smelting deserves further consideration with respect to hazardous waste disposal. Thus, the following subsections apply to the roasting process.

4.9.2.1 Technical Factors--

Compatibility of hazardous wastes with process fuel--Reactions occurring in all three roaster types are exothermic enough to be self-sustaining after ignition. Natural gas, oil, or coal is used to bring the charge up to reaction temperature. The lack of continuous addition of fuel in the process eliminates compatibility problems.

Graphical determination of suitable waste categories--

Process data--Multiple-hearth roasters are typically 6.1 to 7.6 m (20 to 25 ft) in diameter and 8.8 to 11 m (29 to 36 ft high).[70] The volume of off-gases leaving the multiple-hearth furnace ranges from 140 to 170 m^3/min (5000 to 6000 acfm) at 205°C (400°F).[76] The selected process conditions represent a roaster 6.1 m (20 ft) in diameter, 9.1 m (30 ft) high, with an off-gas volume of 170 m^3/min (6000 acfm). The chosen process parameters for multiple-hearth roasters can be represented as:

Working volume:	270 m^3 (9420 ft^3)
Gas volumetric flow rate:	97 Nm^3/min (3700 scfm)
Maximum reaction zone temperature:	980°C (1800°F)
Exit gas temperature:	205°C (400°F)

Suspension roasters are generally similar to multiple-hearth roasters in overall dimensions. Off-gas volumetric flow rates range from 280 to 420 m^3/min (10,000 to 15,000 acfm).[76] Based on a furnace 4.6 m (15 ft) in diameter, 6.7 m (22 ft) high, with an off-gas volumetric flow rate of 420 m^3/min (15,000 acfm), the following are the process conditions for suspension roasters:

Furnace volume:	111 m^3 (3900 ft^3)
Gas volumetric flow rate:	92 Nm^3/min (3500 scfm)
Maximum reaction zone temperature:	1010°C (1840°F)
Exit gas temperature:	927°C (1700°F) (assumed)

A fluidized-bed roaster having a diameter of 4.6 m (15 ft) and a height of 6.1 m (20 ft) is assumed based on dimensions reported in Reference 77. Typical off-gas volumetric flow rates range from 170 to 283 m^3/min (6000 to 10,000 acfm) at 950°C (1740°F).[76] The chosen conditions for determining the time-temperature relationship for fluidized-bed roasters are as follows:

Furnace volume:	99 m^3 (3500 ft^3)
Gas volumetric flow rate:	66 Nm^3/min (2500 scfm) based on 283 m^3/min at 950°C (10,000 acfm at 1740°F)
Maximum reaction zone temperature:	1000°C (1830°F)
Exit gas temperature:	950°C (1740°F)

Approach--For the assumed process conditions, computations of mean residence times (t) and fast-path residence times ($\bar{t}$) were made for each of the three roaster types for several temperatures below the maximum reaction zone temperature. These computations enabled the plotting of process curves for the three roasting processes, as shown in Figures 4.9-7, 4.9-8, and 4.9-9. Wastes in category K015 appear to be destructible in all three roasters. Wastes in categories K023 and K093 also appear destructible in suspension and fluidized-bed roasters.

Effects on product quality and product application--No harmful impact on product (zinc/zinc oxide) quality is expected

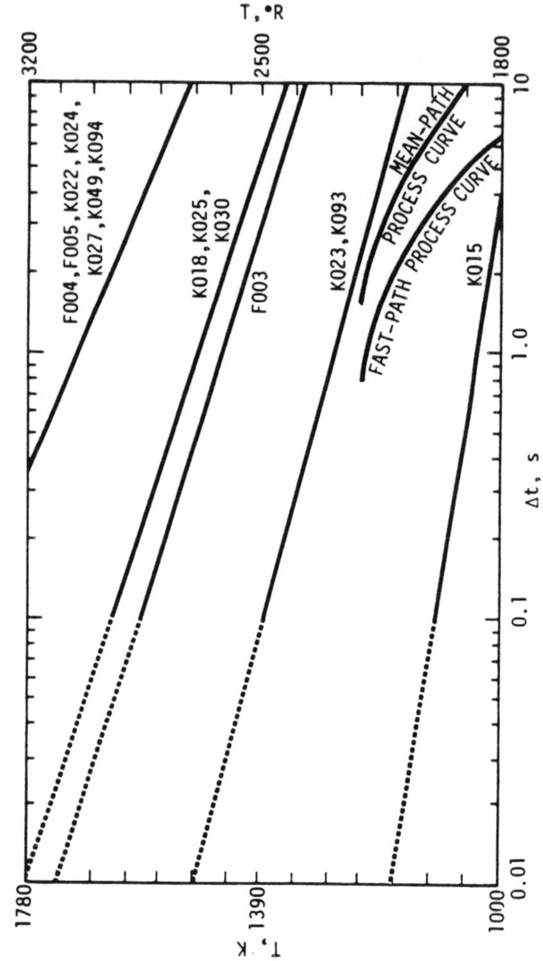

Figure 4.9-7. Hazardous waste destructibility predictions for multiple-hearth roasters in zinc smelting.

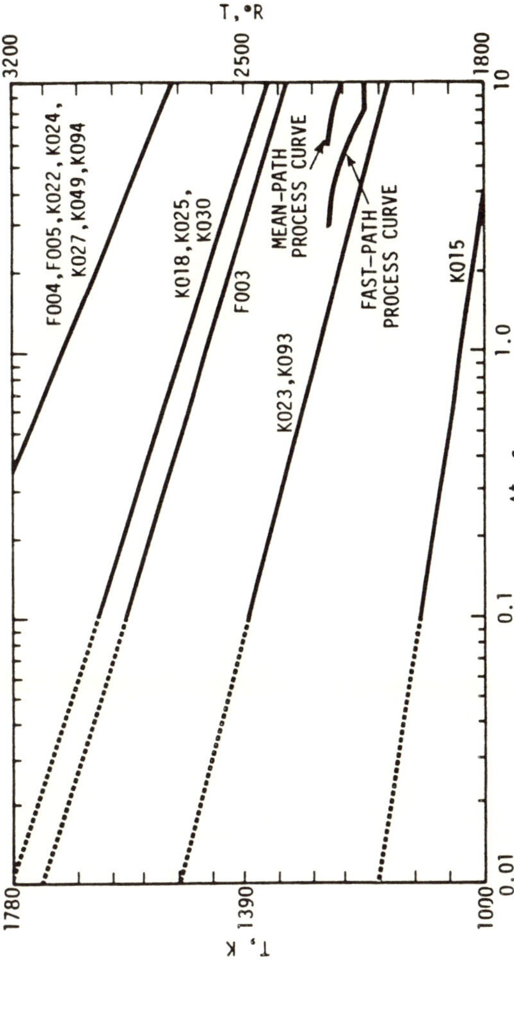

Figure 4.9-8. Hazardous waste destructibility predictions for suspension roasters in zinc smelting.

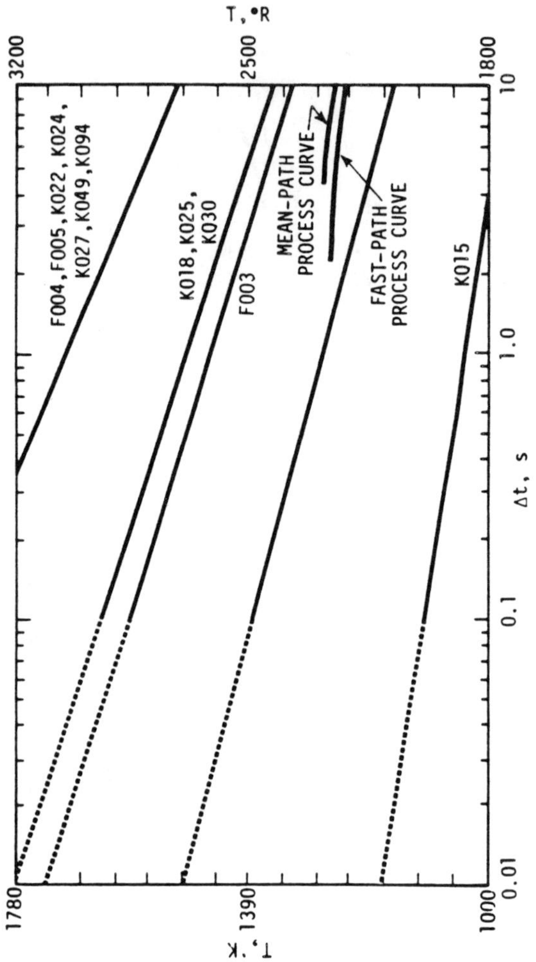

Figure 4.9-9. Hazardous waste destructibility predictions for fluidized-bed roasters in zinc smelting.

from the combustion of hazardous wastes in the roasters. It should be noted, however, that if the hazardous waste contains sufficient sulfur, additional formation of zinc sulfate ($ZnSO_4$) might occur in the roasters, which is not very desirable, especially in pyrometallurgical zinc smelting. Moreover, bed temperatures would have to be monitored closely to assure calcine quality.

Any possible minor contamination of the product, however, should not be cause for concern in the predominant zinc applications in galvanizing and alloying.

Impact on process hardware--The roasters are generally brick-lined, cylindrical steel columns covered with insulation. In suspension roasters, problems have resulted from the steel shells becoming too hot because of over-insulation.[78] Because such problems could be compounded during the firing of hazardous wastes, it is necessary that the right insulation thickness be maintained to prevent sulfation and corrosion that occur at the higher temperatures.

When hazardous wastes are fired in multiple-hearth roasters, it is recommended that the high-temperature hearths be continuously checked for plugups. In normal practice, hard accretions that form on the hearths necessitate a routine maintenance maneuver called "plowing," the frequency of which varies from 2 to 3 days for high-temperature hearths to 2 to 3 months for those operating at low temperatures.[79] The frequency of cleaning may have to be increased when firing hazardous wastes.

The destruction of hazardous wastes in the furnaces may necessitate improved maintenance procedures for screw conveyors and waste heat boilers (in the case of suspension and fluidized-bed roasters only).

Acceptable amounts of hazardous waste feed--Because fuel is generally not added to the roasters on a continuous basis, estimates of quantities of hazardous wastes destructible in roasters can be based on concentrate feed rates.

Typical concentrate feed rates to all three types of roasters average about 181 Mg (200 tons) per day. Assuming that the hazardous wastes represent 10 percent by weight of the charge, hazardous waste feed rate would be about 18.1 Mg/day (20 tons/day).

Densities of the three waste categories that are believed to be destructible in the roasters are as follows:

K015: 1.64 kg/liter (13.7 lb/gal)
K023: 1.37 kg/liter (11.4 lb/gal)
K093: 1.37 kg/liter (11.4 lb/gal)

Assuming an average density of 1.44 kg/liter (12 lb/gal), the hazardous waste feed rate represents a consumption of 12.6 kl (3330 gal) per day.

Provisions for waste introduction--In the case of the multiple-hearth roasters, the hazardous wastes should be atomized and introduced near the lower hearths of the furnace because this is where roaster temperatures are highest. "Snowballing" of concentrates should be avoided, especially when charging hazardous wastes, by maintaining temperatures in the desired range.

The suspension roaster has a large combustion space at the top and two to four hearths similar to those of a multiple-hearth furnace in the lower portion. The zinc concentrates are introduced near the top of the combustion chamber. It would be preferable to introduce the atomized hazardous wastes near this point to allow adequate residence time in the combustion space. The off-gases generally exit from a point near the top of the hearths, close to the bottom of the free combustion space.

Fluidized-bed roasters are equipped with tuyeres at the bottom to supply fluidizing air, and the hazardous wastes could be introduced through these tuyeres. Modifications in tuyere design, however, would be needed for atomizing the liquid wastes.

Projected trends in process fuel--The energy requirement for the roasting process is about 1.2 GJ/Mg (1 x 10^6 Btu/ton) of ore, which is consumed during initial ignition. Exothermic oxidation of sulfur in the ore maintains operating temperatures beyond this point. Gas, coal, and oil have been used to bring the charge to reaction temperatures; natural gas is the primary fuel.[76]

4.9.2.2 Geographical Matching of Industry Location and Waste Generators--

Table 4.9-2 presents the SIC codes and geographic distribution of the industries generating the waste categories that are destructible in the zinc ore roasting process. The geographic distribution information allows a matching of the waste generator and primary zinc smelting industry location. Waste categories K023 and K093 are both generated in Texas, where an electrolytic smelter operated by Asarco, Inc., is located. Category K023 wastes are also generated in Illinois and Pennsylvania, and each of these States has a primary zinc smelter in operation (see Table 4.9-1). Thus, the matching between industry location and generators of destructible hazardous waste categories seems to be good. Destructible wastes are also generated in significant quantities in California and New Jersey; however, no zinc smelters are located in either of these states.

TABLE 4.9-2. SIC CODES AND GEOGRAPHIC DISTRIBUTION OF INDUSTRIES GENERATING HAZARDOUS WASTES DESTRUCTIBLE IN THE ZINC ORE ROASTING PROCESS[19]

Hazardous waste category	SIC code for waste-generating industry	Geographic distribution of industries
K015	2865	Two plants (locations unknown)
K023	2869	Richmond and El Sequendo, California; Kearney and Bridgeport, New Jersey; Baton Rouge, Louisiana; Bridgeville and Neville Island, Pennsylvania; Cicero and Millsdale, Illinois; Texas City, Texas
K093	2896	Texas, California, New Jersey

4.9.2.3 Environmental Factors--

Impact on air pollution control devices--Electrostatic precipitators are used to control particulate emissions from all three types of roasters. Cyclones and waste heat boilers precede the ESP's in the case of suspension and fluidized-bed roasters. About 20 percent of the suspended dust settles out in the waste heat boilers, and more than 99.5 percent of the remaining particulates are removed in the final particulate control equipment. Any additional particulate emissions due to waste combustion should be adequately controlled by the ESP's.

Concentrations of SO_2 in roaster off-gases are sufficiently high for production of sulfuric acid; however, contact acid plants operate efficiently only when the gas stream being processed meets certain desirable characteristics. The gas stream should have a consistent flow rate, be almost completely dried, and contain no fumes or gases that may poison the catalyst in the acid plant catalytic converter.[76] Sulfur dioxide concentrations of at least 7 percent are desirable. Combustion of hazardous wastes in the roasters could have an undesirable effect on the properties of the off-gases for subsequent acid production. Final traces of objectionable materials that could result from hazardous waste combustion can be removed by special scrubbers, which are usually available at the acid plant. Off-gas analysis measurements obtained during test burns should help to identify other problem areas.

Environmental impact--Roasting processes do not generate any solid or process-water waste; however, for the suspension and fluid-bed roasters, there is a boiler blowdown from the waste heat boiler. In all the roasting processes, solids are recycled to recover byproduct metals.[76] Thus, combustion of hazardous wastes in the roasters is not expected to have any harmful impact on the environment.

4.9.2.4 Institutional Factors--

Although the primary domestic zinc smelting capacity has been declining since 1968, U.S. primary zinc demand is expected to increase at an annual rate of 2 percent through 1985. Zinc smelters employed about 4100 people in 1978, including workers involved in byproduct processes.[76] The smelters are generally located away from population centers.

The main uses of zinc are for galvanizing, alloying, and grinding into zinc dust. These applications are such that the customer would not be concerned about the impact of the waste on the zinc product. Also, zinc smelting facilities are not expected to have any major objections to accepting hazardous wastes.

4.10 SEWAGE SLUDGE INCINERATORS

4.10.1 Process Descriptions

The two major types of sewage sludge incinerators used in the United States are the multiple-hearth and the fluid-bed.[80] A third type is the flash-drying arrangement used in conjunction with a fossil-fuel or refuse-fired furnace. This type is not discussed in this report.

4.10.1.1 Multiple-Hearth Incinerator--

The multiple-hearth incinerator is currently the most widely used sewage sludge incinerator in the United States. As of 1977, approximately 340 units had been installed for wastewater sludge combustion. This incinerator is durable and relatively simple to operate, and it can handle wide fluctuations in feed quality and loading rates.

The multiple-hearth incinerator consists of vertically stacked hearths raked by rabble arms rotated from a common shaft.[81] Dewatered sludge is fed to the top hearth, and is plowed to drop holes by the rotating rabble arms. It falls to the next hearth and then the next until ash is discharged at the bottom.[82]

Because the sludge is constantly turned and broken into smaller particles as it moves across the hearths, a large surface of the sludge is exposed to the hot furnace gases. This process includes rapid and complete drying as well as combustion. As shown in Figure 4.10-1, the system has three operating zones: 1) the top hearths, where sludge feed is dried to about 48 percent moisture; 2) the incineration/deodorization zone, where the temperature is 760° to 980°C (1400° to 1800°F); and 3) the cooling zone, where the hot ash gives up heat to incoming combustion air. Exhaust gases exit at 430° to 540°C (800° to 1000°F).

Figure 4.10-2 shows an interior cut-away view of the multiple hearth incinerator. The central shaft of the furnace is a hollow iron column cast in sections; shaft speeds are adjustable from about 1/2 to 1-1/2 revolutions per minute. The hollow rabble arms are connected to machined arm sockets in the shaft. The shaft and rabble arms are air-cooled and normally insulated. A cold air tube runs up the center of the shaft. Air lances extend from the cold air tube out to the ends of each rabble arm. Ambient air of regulated pressure and volume is forced through the cold air tube and lances by means of a blower. The cold air exits from the tips of the lances and flows backward through the space between the lances and the rabble arm walls to the annular space in the central shaft (known as the hot air compartment). The flow of air cools the arms. The air is conducted through the hot air compartments to cool the shaft, and then is either discharged to the atmosphere via the exhaust gas stack or returned to the bottom hearth of the incinerator as preheated air for combustion.[80] Excess air requirements are generally 75 to 100 percent.

4.10.1.2 Fluid-Bed Incinerator--

There are approximately 60 operating fluid-bed sludge incinerators in the United States.[80] The fluid-bed incinerators are versatile devices that can be used to dispose of solid, liquid,

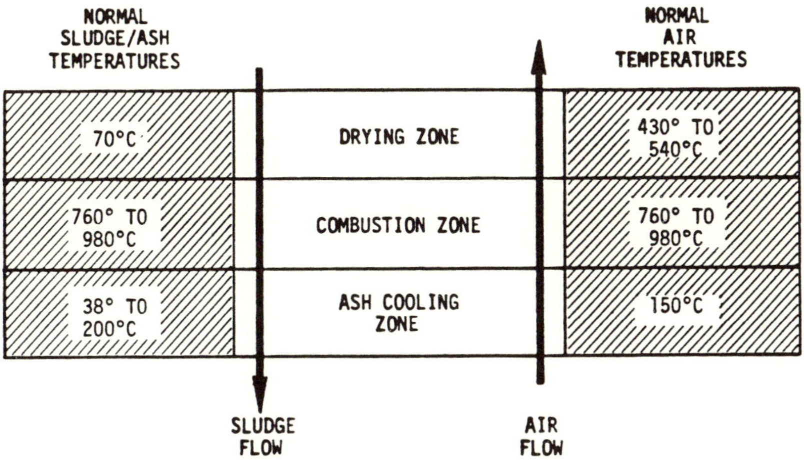

Figure 4.10-1. Process zones in multiple-hearth incinerator.

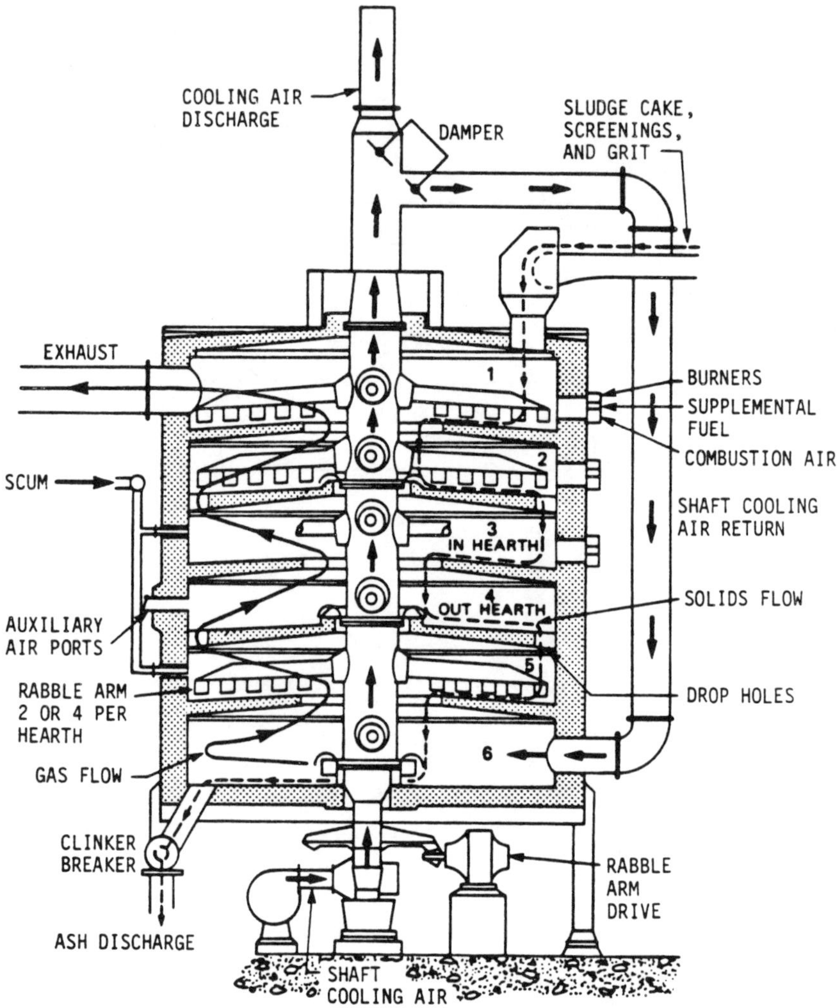

Figure 4.10-2. Cross section of a multiple-hearth incinerator.[80]

or gaseous combustible wastes. They are vertically oriented and cylindrically shaped and have a refractory-lined steel shell that contains a sand bed and fluidizing air diffusers. Air that is driven by a blower enters at the bottom of the incinerator and rises vertically through a distributor plate into a vessel containing a bed of inert granular particles or sand. The upward flow of air through the sand bed results in a dense turbulent mass that behaves much as a liquid would. Waste material to be incinerated is injected into the bed, where combustion occurs within the fluidizing media. Air passage through the bed produces strong agitation of the bed particles, which promotes rapid and relatively uniform mixing of the ignited waste material within the fluidized bed.[82] Figure 4.10-3 shows a cross section of a fluid bed incinerator.[80]

Bed temperatures are quite uniform and typically in the range of 760° to 870°C (1400° to 1600°F). The bed expands to approximately 100 percent of its rest volume. In essence, the reactor is a single chamber unit in which both drying and combustion occur in either dense or dilute phases in the sand bed. All of the combustion gases pass through the combustion zone, where the residence time is several seconds. Ash is carried out the top of the furnace and is removed by air pollution control devices, usually a venturi scrubber. Feed to the incinerator is introduced either above the bed or directly into it.[80] Excess air requirements for the fluid-bed incinerator vary from 20 to 40 percent.

4.10.2 Process Assessment for Hazardous Waste Destruction

Both the multiple-hearth and fluid-bed incinerators appear to be feasible for thermal destruction of some hazardous wastes. Both have the capability of attaining moderately high combustion temperatures and have relatively long residence times. Also, these sewage sludge incinerators are durable and designed to handle wide fluctuations in feed quality and loading rates.

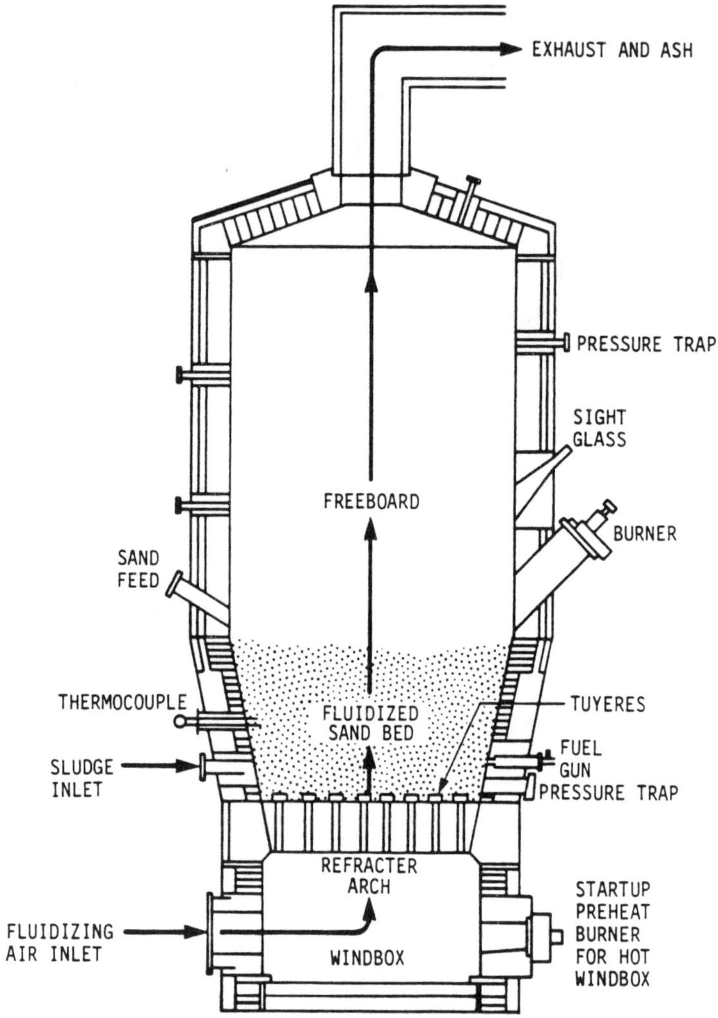

Figure 4.10-3. Cross section of a fluid-bed incinerator. [80]

4.10.2.1 Technical Factors--

Compatibility of hazardous wastes with process fuel--The compatibility of co-incinerating hazardous waste with sewage sludge is dependent on several factors: physical state and chemical composition of the hazardous waste, method of injection currently used at the incinerator, and the type of air pollution control device used. Virtually any material that can be burned can be combined with sludge in a co-combustion process.

For successful incineration, the sludge feed composition and the volatile solids content are important. Both multiple-hearth and fluid-bed incinerators provide adequate mixing of the feed with combustion gases and a residence time that insures complete combustion of sludge. For incineration of hazardous wastes, however, it may be necessary to keep temperatures higher than normal to meet the recommended destruction efficiencies. The physical state of the hazardous waste will determine which method of waste injection is most appropriate. With proper mixing, waste liquids and sludges generally can be added directly to the sewage sludge and then introduced to the incinerator. Hazardous waste can be combined with the sewage sludge in a mixing tank if proper handling practices are used. This will eliminate the need for a second feed system to handle the hazardous waste.

Proper mixing of the hazardous waste and sewage sludge is important. The feed rate of the hazardous waste should be based on 10 percent of the total heat value imput. Sludge heat value is fairly high [12 to 23 kJ/kg (5000 to 10,000 Btu/lb) dry solids]; however, the water content of most sludges requires the addition of auxiliary fuel to maintain combustion in the furnace.

Graphical determination of suitable waste categories--The criteria for determining the potential destructibility of various hazardous wastes are the time-temperature relationships. Residence time and temperatures of typical sewage sludge incinerators are as follows:

	Fluid-bed	Multiple-hearth
Maximum temperature	870°C (1600°F)	980°C (1800°F)
Exit temperature	760°C (1400°F)	430°-540°F (800°-1000°F)
Residence time	5-6 s	5 s

Figures 4.10-4 and 4.10-5 are graphic representations of the relationship of Δt and $\Delta \bar{t}$ (mean and fast-path) versus T for typical multiple-hearth and fluid-bed incinerators. The temperature-time requirements for 99.99 percent destruction of the hazardous wastes are also plotted on this graph. Because of this graph, predictions can be made regarding the probability or feasibility of meeting the 99.99 percent destruction efficiency for each hazardous waste in a sewage sludge incinerator. Results are summarized in Tables 4.10-1 and 4.10-2. It appears that none of the hazardous wastes considered in this report can be thermally destroyed in a multiple-hearth incinerator under normal operating conditions. The destruction of hazardous waste category K015 is, however, marginally feasible in a fluid-bed incinerator. Destruction of other hazardous wastes may be feasible in either furnace type if operating temperatures were increased.

Effects on product quality and product application--The incineration of sewage sludge is a waste treatment process with no end product; therefore, the question of waste impact on product quality does not arise.

Impact on process hardware--The reactors of both multiple-hearth and fluid-bed incinerators are generally refractory-lined steel, and the central shaft of the multiple-hearth incinerator is cast iron.[82] These construction materials should not be adversely affected by the inclusion of hazardous waste in the

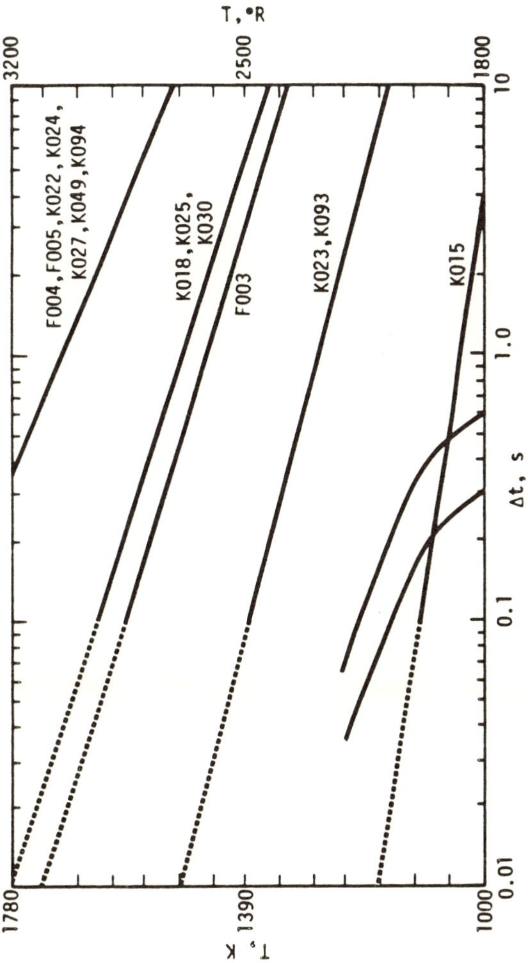

Figure 4.10-4. Hazardous waste destructibility predictions for multiple-hearth sewage sludge incinerators.

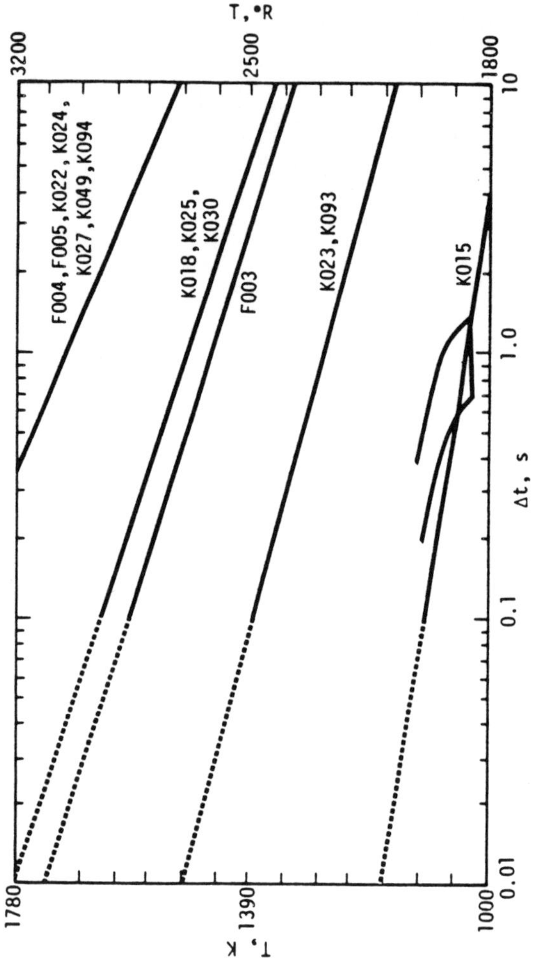

Figure 4.10-5. Hazardous waste destructibility predictions for fluid-bed sewage sludge incinerators.

TABLE 4.10-1. APPLICABILITY OF MULTIPLE-HEARTH SEWAGE SLUDGE INCINERATORS FOR THERMAL DESTRUCTION OF HAZARDOUS WASTES

K015	K023 K093	F003	K018, K025 K030	F004, F005, K022, K024 K027, K049, K094
Yes	No	No	No	No

TABLE 4.10-2. APPLICABILITY OF FLUID-BED SEWAGE SLUDGE INCINERATORS FOR THERMAL DESTRUCTION OF HAZARDOUS WASTES

K015	K023 K093	F003	K018, K025 K030	F004, F005, K022, K024 K027, K049, K094
Yes	No	No	No	No

feed. Most of the wear on process hardware is due to acid gases (HCl, H_2SO_4) and the scouring action of particulate that is already present in sewage sludge. The addition of hazardous waste in small proportions to the sludge would probably not accelerate this normal operating deterioration.

Acceptable amounts of hazardous waste feed--The net fuel value of sludge depends on the fuel value of its total combustible solids and the amount of water present. Sewage sludge generally has a high water content and, in some cases, fairly high levels of inert materials. As a result, its net fuel value is often low. Autogenous combustion of sludge is generally only possible when the solids content of the sludge is 30 to 35 percent or greater.[80] Because this is often difficult to achieve by conventional dewatering techniques, supplemental fuel is required for the combustion operation. If sludge is combined with other combustible materials in a co-combustion scheme, a furnace feed can be created that has both a low water concentration and a high enough heat value to sustain autogenous combustion, which could prove to be cost-effective.

The feed requirements will depend on the size and type of incinerator as well as the quality of the sludge. For example, a multiple-hearth incinerator processing 36 Mg/day (40 tons/day) (dry solid basis) will typically require 12×10^6 kJ/h (11×10^6 Btu/h) heat input. Based on a maximum hazardous waste input of 10 percent of the total heat input, the amount to be charged in this example should be no greater than 1×10^6 kJ/h (1×10^6 Btu/h), based on the heat content of the waste. This assumes that the waste is low in water content and inert solids.

Provisions for waste introduction--In the multiple-hearth incinerator the sludge is conveyed to the upper hearth of the incinerator, and mechanical rakes move the sludge from one hearth to the next lower hearth, where the sludge is sequentially dried and burned and the ashes cooled.[82] In the fluid-bed incinerator the sludge is injected directly into the bed, usually with screw feeders or spray nozzles. Problems with the feed system occur more often in the fluid-bed incinerator than in the multiple-hearth incinerator; the feed system will jam if the sludge is overdried.

If the waste can be added directly to the sludge before it enters the incinerator, no additional waste handling provisions are required for introducing the hazardous waste. This is especially true in the case of other waste sludges. If the hazardous waste is in a liquid state, atomization may be necessary to insure rapid and uniform vaporization and efficient combustion. The deisgn of the required additional feed and atomization system will be based on each individual furnace and waste stream. Section 2 provides a detailed discussion on hazardous waste handling and firing.

4.10.2.2 Geographic Matching--

Sewage sludge incineration has been practiced since the early part of the century and is widely used in this country. As of 1977, there were approximately 340 multiple-hearth and 60 fluid-bed incinerators in the United States; fluid-bed incinerators are a newer design and are gaining in popularity. Also, the anticipated banning of ocean disposal of sewage sludge will necessitate the use of alternative disposal methods, and the use of incinerators may increase along coastline states. Inasmuch as the use of sewage sludge incineration is widespread, there should be no problem in matching hazardous waste generation sites with locations of potential incineration processes. Hazardous waste category K015 is generated in only two plants, however; therefore the applicability of sewage sludge incinerators for the destruction of wastes in this category would be limited to incinerators near these two plant sites. If further study should prove that sewage sludge incineration was feasible for other hazardous waste categories than those indicated on Tables 4.10-1 and 4.10-2, then the widespread distribution of these incinerators would be an advantage.

4.10.2.3 Environmental Factors--

Impact on air pollution control devices--Because of the violent upward movement of combustion gases with respect to the burning sludge, particulate emissions are a major emission problem in both multiple-hearth and fluid-bed incinerators. Although dry sludge may contain from 1 to 2 percent sulfur by weight, sulfur oxides are not emitted in significant amounts in sludge burning compared with those emitted in many other combustion processes. Similarly, because incineration temperatures are lower than 980°C (1800°F) in fluidized-bed incinerators, nitrogen oxides are not formed in great amounts.[13] Nitrogen oxides may be a concern, however, if incineration temperatures are raised to destroy hazardous wastes. Table 4.10-3 provides a list of emission factors for sewage sludge incinerators based on data compiled by the EPA.[13]

TABLE 4.10-3. EMISSION FACTORS FOR SEWAGE SLUDGE INCINERATORS[a]

Pollutant	Emissions[b]			
	Uncontrolled[c]		After scrubber	
	lb/ton	kg/Mg	lb/ton	kg/Mg
Particulate	100	50	3	1.5
Sulfur dioxide	1	0.5	0.8	0.4
Carbon monoxide	Neg	Neg	Neg	Neg
Nitrogen oxide (as NO_2)	6	3	5	2.5
Hydrocarbon	1.5	0.75	1	0.5
Hydrogen chloride gas	1.5	0.75	0.3	0.15

[a] Reference 13.

[b] Unit weights in terms of dried sludge.

[c] Estimated from emission factors after scrubbers.

Whereas incineration may enable complete destruction of hazardous materials, the combustion products may not be environmentally acceptable if emitted to the atmosphere. Therefore, certain wastes should not be incinerated unless the combustion products are treated with a suitable emission control device. Wastes of concern include chlorinated hydrocarbons, materials containing heavy metals, and waste with high sulfur content.[82]

Experimental data indicate that emissions of metals increase substantially as the combustion temperature increases.[83,84] If the temperatures of these incinerators are raised above normal operating levels to insure complete combustion of the organics in the hazardous waste, this can be a significant concern. These metal emissions also tend to be enriched on smaller submicron particulate, which makes them very difficult to control with conventional air pollution control equipment.

Because of the wide variability of sludge wastes that are currently incinerated, most facilities currently depend on versatile air pollution control equipment. High-efficiency wet scrubbers usually provide the necessary versatility, as chemical solutions can be suitably varied to treat different effluent contaminants. The required addition of control equipment to maintain air quality at higher temperatures may make the burning of hazardous wastes cost-prohibitive.

Environmental impact--Fugitive emissions can occur from improper handling and transport of the hazardous waste. A spill-prevention countermeasure and control plan must be developed, along with good housekeeping practices, to minimize discharges of the hazardous waste to the environment.

4.10.2.4 Institutional Factors--

Because of the nature of the sewage sludge incineration (i.e., treatment and disposal of a waste through combustion), there is no product to be adversely affected and thus no potential exposure to a consumer.

References

1. U.S. Environmental Protection Agency. Resource Conservation and Recovery Act, Subtitle C; Background Document. Office of Solid Waste, Washington, D.C. November 14, 1980.

2. Mitre Corporation. Guidance Manual for Evaluating Permit Applications for the Operation of Hazardous Waste Incinerator Units. 1981.

3. Acurex Corporation. A Technical Overview of the Concept of Disposing of Hazardous Wastes in an Industrial Boiler. Prepared for U.S. Environmental Protection Agency, Cincinnati, Ohio. 1981.

4. Benson, S. W. The Foundations of Chemical Kinetics. McGraw-Hill. 1960. Chapters 10 and 11.

5. Weast, R. C. CRC Handbook of Chemistry and Physics. CRC Press, Inc., Boca Raton, Florida. 1980.

6. Domalski, E. S., W. H. Evans, and T. J. Jobe, Jr. Thermodynamic Data for Waste Incineration. Prepared for the American Society of Mechanical Engineers. NBSIR 78-1479. New York, New York. 1978.

7. Levenspiel, O. Chemical Reaction Engineering. Second Edition (1972). John Wiley and Sons, Inc., New York, N.Y.

8. Memorandum from L. R. Waterland and C. D. Wolbach, Acurex Corporation, Mountain View, California, to R. A. Olexsey, U.S. Environmental Protection Agency (IERL), Cincinnati, Ohio, September 21, 1981.

9. Industrial Gas Cleaning Institute. Air Pollution Control Technology and Costs in Nine Selected Areas. Stanford, Connecticut. Contract No. 68-02-0301. 1972.

10. Chews, F. H. Heavy Clay Technology. 2d Ed. Academic Press, London. 1969.

11. Kirk-Othmer. Encyclopedia of Chemical Technology. 3d Ed., Vol. 5. John Wiley & Sons, New York. 1979.

12. Personal communication from J. Blazina, Swindall Rust, to M. Taft, PEDCo Environmental. 1981.

13. U.S. Environmental Protection Agency. Compilation of Air Pollution Emission Factors. 3d Ed. Research Triangle Park, North Carolina. Publication No. AP-42, August 1977.

14. LeSourd, D. A., et al. Comprehensive Study of Air Pollution Sources to Assess the Economic Effects of Air Quality Standards. Vol. 1. U.S. Government Printing Office, Washington, D.C. December 1970.

15. Machine Design Waste Sludges Used To Make Better Insulating Bricks. 53(13), 1981.

16. Directory of Brick Manufacturers. Brick Institute of America, McLean, Virginia. 1974.

17. Product Directory of the Refractories Industry in the United States. Refractories Institute, Pittsburgh, Pennsylvania. 1972.

18. Directory, Vitrified Clay Sewer Pipe Industry. National Clay Pipe Institute, Crystal Lake, Illinois. 1972.

19. The Mitre Corp. Guidance Manual for Evaluating Permit Applications for the Operation of Hazardous Waste Incinerator Unit. EPA 68-01-6092, 1981.

20. Personal communication from G. Robinson, Clemson University, to M. Taft, PEDCo Environmental, Inc., June 1981.

21. Kirk-Othmer. Encyclopedia of Chemical Technology, 3d Ed., Vol. 4. John Wiley and Sons, New York, New York. 1978.

22. Mantell, C. L. Carbon and Graphite Handbook. John Wiley and Sons, New York, New York. 1968.

23. Gerstle, R. W., et al. Industrial Process Profiles for Environmental Use: Chapter 4. Carbon Black Industry. EPA-600/2-77-023d, February 1977.

24. PEDCo Environmental, Inc. Chemical Hazard Information Profile - Carbon Black. August 1, 1976.

25. Spinosa, E. D., et al. Summary Report on Emissions from the Glass Manufacturing Industry. EPA-600/2-79-101, April 1979.

26. Kirk-Othmer. Encyclopedia of Chemical Technology. 2d Ed., Vol. 10. John Wiley & Sons, Inc., New York, New York. 1966.

27. Stockham, J. D. The Composition of Glass Furnace Emissions. Presented at the 63rd Annual Meeting of the Air Pollution Control Association, St. Louis, Missouri, June 14-18, 1970.

28. Danielson, J. A. Air Pollution Engineering Manual. USPHS Publication No. 999-AP-40, 1967.

29. Franz, C. N., et al. Glass Furnace Particulate Emission Control Equipment. Presented at the 32nd Annual Conference on Glass Problems, University of Illinois, Urbana, Illinois, November 11, 1971.

30. Personal communication from Jerry Corbo, Johns-Manville, Parkersburg, W. Virginia, to Doug Orf, PEDCo Environmental, August 13, 1981.

31. Personal communication from Harry Grell, Libby-Owens-Ford Company, Toledo, Ohio, to Doug Orf, PEDCo Environmental, August 13, 1981.

32. Unpublished data received from Mr. R.L. Jacks of Teller Environmental Systems, Inc., New York, New York, November 7, 1972.

33. Personal communication from Hal Severin, Johns-Manville, Parkersburg, W. Virginia, to Doug Orf, PEDCo Environmental, August 13, 1981.

34. Personal communication from Ray Schick, Glass Packaging Institute, Washington, D.C., to Doug Orf, PEDCo Environmental, August 24, 1981.

35. Personal communication from Bob Laughman, Anchor Hocking Corporation, Lancaster, Ohio, to Doug Orf, PEDCo Environmental, August 14, 1981.

36. Department of Commerce. Annual Survey of Manufacturers. December 1980 (contains 1978 data).

37. Strassburger, J., et al. Blast Furnace - Theory and Practice. Vol. I. 1969.

38. Keenan, G. F., and D. M. Morrison. A Current Look at Pulverized Coal Injection for Blast Furnaces. Iron and Steel Engineer, October 1977.

39. 33 Metal Producing; World Steel Industry Data Handbook, Volume I, The United States. 1978.

40. Pietrucha, W. E., and R. L. Deily. Steel Industry in Brief: Data Book U.S.A. Institute for Iron and Steel Studies. 1979-80.

41. Sciulli, C. M., and D. G. White. Optimization of Fuel Injection into the Blast Furnace. In: Iron Making Proceedings, Volume 34, Toronto 1975. A Publication of the Iron and Steel Society of AIME. 1975.

42. Ashton, J. D., and J. E. R. Holditch. Homogenized Oil Injection at Dopasco. In: Iron Making Proceedings, Volume 34, Toronto, 1975. A publication of the Iron and Steel Society of AIME. 1975.

43. Personal communications from Robert C. McCrillis of Industrial Environmental Research Laboratories, Environmental Protection Agency, Research Triangle Park, North Carolina, September 16, 1981.

44. Mudd, S. W. Basic Open Hearth Steel Making. Physical Chemistry of Steel Making Committee, Iron & Steel Division, The Metallurgical Society of AIME. 1964.

45. AIME. Industrial Minerals and Rocks. 4th Ed. Port City Press, Baltimore, Maryland. 1975.

46. Berger, T. R. Lightweight Aggregate - Newest Concrete Material. Rock Products, 74(10), October 1971.

47. Expanded Shale Clay and Slate Institute. 1971. Lightweight Concrete - History, Applications and Economics. Washington, D.C.

48. Personal communication from L. Lawson, Solite, to M. Taft, PEDCo Environmental, Inc., September 1981.

49. Reedy, R. W. Lightweight Aggregates Part 1: Raw Materials. Pit and Quarry, 63(7), January 1971.

50. Reedy, R. W. Lightweight Aggregates Part 2: Plant Design. Pit and Quarry, 63(10), April 1971.

51. Personal Communication from Mr. J. Evans, Fuller Company, to Ms. M. Taft, PEDCo Environmental, Inc., October 1981.

52. Reedy, R. W. Lightweight Aggregates Part III: Selection of Equipment. Pit and Quarry, 64(2), August 1971.

53. Reedy, R. W. Lightweight Aggregates Part IV: Rotary Kiln Operation. Pit and Quarry, 64(5), November 1971.

54. Cohen, S. M., and N. W. Biege. Lightweight Aggregate Designing and Operating for Quality Product. Presented at the Annual Spring 1972 Meeting of the American Ceramic Society, Washington, D.C. 1972.

55. Personal communication from H. Robinson, Expanded Shale Clay and Slate Institute, to M. Taft, PEDCo Environmental, Inc., October 1981.

56. U.S. Environmental Protection Agency. Standards Support and Environmental Impact Statement. Vol. 1: Proposed Standards of Performance for Lime Manufacturing Plants. EPA-450/2-77-007a. 1977.

57. U.S. Department of the Interior. Mineral Industry Surveys, Lime in 1980. Bureau of Mines, Washington, D.C. 1980.

58. Kirk-Othmer. Encyclopedia of Chemical Technology. Vol. 14, 3d Ed. John Wiley and Sons, New York. 1978.

59. Kearney Management Consultants. Feasibility of Using Lime Kilns to Burn Hazardous Wastes. Prepared for the U.S. Environmental Protection Agency under Contract EPA 68-03-256. Alexandria, Virginia. 1981.

60. Personal Communication from Mr. F. Schwarzkopf, Kennedy Van Saun Corp., to Ms. M. Taft, PEDCo Environmental, Inc., 1981.

61. U.S. Department of the Interior. Lime, A chapter from Mineral Facts and Problems. Preprint from Bulletin 671 of the Bureau of Mines. 1980.

62. Robertson, J. L. Kiln Length, Diameter, and Speed Promote Formation of Ash Rings. Rock Products, 83(12), December 1980.

63. Personal Communication from Mr. E. Cohen, U.S. Environmental Protection Agency, Region III, to Ms. M. Taft, PEDCo Environmental, Inc., September 1981.

64. Personal Communication from Mr. Gutschich, National Lime Association, to Ms. Taft, PEDCo Environmental, Inc., July 1981.

65. U.S. Department of the Interior. Minerals Yearbook. Vol. I, Metals and Minerals. Bureau of Mines, Washington, D.C. 1978-79.

66. Hazelwood, D. L., F. J. Smith, and E. M. Gartner. Assessment of Waste Fuel Use in Cement Kilns. EPA 68-03-2586, 1981.

67. Bonner, T., et al. Draft Engineering Handbook for Hazardous Waste Incineration. EPA 68-03-2550, 1980.

68. Biswas, A. K., and W. G. Davenport. Extractive Metallurgy of Copper. 2d Ed. 1980.

69. PEDCo Environmental, Inc. Industrial Process Profiles for Environmental Use: Chapter 29, Primary Copper Industry. EPA-600/2-80-170, July 1980.

70. U.S. Environmental Protection Agency. Background Information for New Source Performance Standards: Primary Copper, Zinc and Lead Smelters. Volume I, Proposed Standards. EPA-450/2-74-002a, October 1974.

71. PEDCo Environmental, Inc. Industrial Process Profiles for Environmental Use. Chapter 27, Primary Lead Industry. EPA-600/2-80-168, July 1980.

72. Gibson, F. W. The Buick Smelter of Amax-Homestake Lead Tollers. Chapter 25, AIME World Symposium on Mining and Metallurgy of Lead and Zinc, Vol. II, The American Institute of Mining, Metallurgical, and Petroleum Engineers, Inc., New York, New York. 1970.

73. Beilstein, D. H. The Herculaneum Lead Smelter of St. Joe Minerals Corporation. Chapter 24, AIME World Symposium on Mining and Metallurgy of Lead and Zinc, Vol. II. The American Institute of Mining, Metallurgical, and Petroleum Engineers, Inc., New York, New York. 1970.

74. Cross, F. L. Baghouse Filtration of Air Pollutants. Pollution Engineering, February 1974.

75. Paul, R. B. The Glover Lead Smelter and Refinery of the American Smelting and Refining Company, Glover, Missouri. Chapter 26, AIME World Symposium on Mining and Metallurgy of Lead and Zinc, Vol. II. The American Institute of Mining, Metallurgical, and Petroleum Engineers, Inc., New York, New York. 1970.

76. PEDCo Environmental, Inc. Industrial Process Profiles for Environmental Use: Chapter 18, Primary Zinc Industry. EPA-600/2-80-169, July 1980.

77. Stankovic, G. Fluo-Solids Roaster and Acid Plant at Trepca Lead and Zinc Mines and Refineries, Mitrovica, Yugoslavia. Chapter 4, AIME World Symposium on Mining and Metallurgy of Lead and Zinc, Vol. II. The American Institute of Mining, Metallurgical, and Petroleum Engineers, Inc., New York, New York. 1970.

78. Baker, D., et al. Suspension Zinc Concentrate Roaster and Acid Plant of the Bunker Hill Company, Kellogg, Idaho. Chapter 3, AIME World Symposium on Mining and Metallurgy of

Lead and Zinc, Vol. II. The American Institute of Mining, Metallurgical, and Petroleum Engineers, Inc., New York, New York. 1970.

79. Lund, R. E., et al. Josephtown Electrothermic Zinc Smelter of St. Joe Minerals Corporation, Monaca, Pennsylvania. Chapter 20, AIME World Symposium on Mining and Metallurgy of Lead and Zinc, Vol. II. The American Institute of Mining, Metallurgical, and Petroleum Engineers, Inc., New York, New York. 1970.

80. U.S. Environmental Protection Agency. Process Design Manual for Sludge Treatment and Disposal. EPA 625/1-79-011. September 1979.

81. Conway, R. A., and R. D. Ross. Handbook of Industrial Waste Disposal. 1980.

82. Sittig, M. Incineration of Industrial Hazardous Wastes and Sludges. 1979.

83. Trichon, R. T., et al. The Fate of Trace Metals in a Fluidized Bed Sewage Sludge Incinerator. Presented at the 1981 APCA Conference.

84. Takeda, N. and M. Hiraoka. Combined Process of Pyrolysis and Combustion for Sludge Disposal. Environmental Science and Technology, 10:(12), November 1976.

Appendix A: Hazardous Wastes with Potential for Thermal Destruction[a]

Hazardous waste No.	Description of hazardous waste	Physical state	Components	Present disposal practice	Geographic distribution (% of total generation)	Annual generation, metric tons
F001	Spent halogenated solvents used in degreasing and sludges from the recovery of these solvents in degreasing operations	Organic liquid Sludges	Tetrachloroethylene Trichloroethylene Methylene chloride 1,1,1 trichloroethane Carbon tetrachloride Chlorinated fluorocarbons			99,000
F002	Spent halogenated solvents and still bottoms from the recovery of these solvents	Organic liquid Still bottoms	Tetrachloroethylene Methylene chloride Trichloroethylene 1,1,1 trichloroethane Chlorobenzene 1,1,2 trichloro-1,2,2 trifluorethane o-dichlorobenzene Trichlorofluoromethane			
F003	Spent nonhalogenated solvents and the still bottoms from the recovery of these solvents	Organic liquid Still bottoms	Xylene Acetone Ethyl acetate Ethyl benzene Ethyl ether Methyl isobutyl ketone n-butyl alcohol Cyclohexanone Methanol			
F004	Spent nonhalogenated solvents and still bottoms from recovery of these solvents	Organic liquid Still bottoms	Cresols Cresylic acid Nitrobenzene			
F005	Spent nonhalogenated solvents and the still bottoms from recovery of these solvents	Organic liquid Still bottoms	Toluene Methyl ethyl ketone Carbon disulfide Isobutanol Pyridine			
K015	Still bottoms from distillation of benzyl chloride	Organic sludge	Acetonitrile Acrylamide Benzal chloride Residuals-alkali chlorides	Incineration, landfill	(2 plants)	5,300
K016	Heavy ends or distillation residues from production of carbon tetrachloride	Organic	Heavy chlorinated hydrocarbons Hexachlorobenzene Hexachlorobutadiene Carbon tetrachloride Hexachloroethane Perchloroethylene	Landfill, incineration	W. Virginia (20%) Texas (40%) California (5%) Louisiana (14%) Alabama (13%) Kentucky (4%) Kansas (4%)	3,200
K017	Heavy ends (still bottoms) from the purification column in the production of epichlorohydrin	Organic liquid	Trichloropropane (70%) Chloroethers (14%) Dichloropropanols (10%) Epichlorohydrin (2%) Chlorinated aliphatics (4%)	Holding ponds prior to incineration	Texas Louisiana	11,000

(continued)

APPENDIX A (continued)

Hazardous waste No.	Description of hazardous waste	Physical state	Components	Present disposal practice	Geographic distribution (% of total generation)	Annual generation, Mg
K018	Heavy ends from fractionation column in ethyl chloride production	Organic liquid	Heavy chlorinated organics (43%) Trichloroethylene (32%) 1,2 dichloroethane (22%) Ethyl chloride (3%)	Land disposal, incineration	Texas Louisiana New Jersey California	35,000
K019	Heavy ends from the distillation of ethylene dichloride in ethylene dichloride production	Organic solid	Ethylene dichloride 1,1,1-trichloroethane 1,1,2-trichloroethane Tetrachloroethane Trichloroethylene Tetrachloroethylene Carbon tetrachloride Chloroform Vinyl chloride Vinylidene chloride Tars Residuals-metal chlorides (Fe, Al, Cu, Sb, Na)	Land disposal	Louisiana (49%) Texas (34%) Kentucky (7%) Puerto Rico (8%) California (2%)	6,346,000
K020	Heavy ends from the distillation of vinyl chloride in vinyl chloride monomer production	Organic solid	Ethylene dichloride 1,1,1-trichloroethane 1,1,2-trichloroethane Tetrachloroethane Trichloroethylene Tetrachloroethylene Carbon tetrachloride Chloroform Vinyl chloride Vinylidene chloride Tars Residuals-metal chlorides (Fe, Al, Cu, Sb, Na)	Incineration, landfill	Louisiana (46%) Texas (33%) Kentucky (12%) Puerto Rico (7%) California (2%)	3,776,000
K022	Distillation bottom tars from the production of phenol/acetone from cumene	Organic sludge (solid)	Tars (>90%) Phenol Acetophenone Cumy phenol Residuals-$NaSO_4$	Landfill, incineration	Widespread	45,000-100,000
K023	Distillation light ends from the production of phthalic anhydride from naphthalene	Liquid	Phthalic anhydride (90.5%) Maleic anhydride (9.5%)	Incineration, offsite disposal	Pennsylvania (48%) Illinois (38%) New Jersey (14%)	952.6
K024	Distillation bottoms from the production of phthalic anhydride from naphthalene	Organic sludge	Phthalic anhydride (<50%) 1,4 naphthoquinone (<50%) Polynuclear tarlike materials	Landfill, incineration	Pennsylvania (48%) Illinois (8%) New Jersey (14%)	2,268

(continued)

APPENDIX A (continued)

Hazardous waste No.	Description of hazardous waste	Physical state	Components	Present disposal practice	Geographic distribution (% of total generation)	Annual generation, Mg
K025	Distillation bottoms from the production of nitrobenzene by the nitration of benzene	Organic sludge	Meta-dinitrobenzene (94%) 2,4-dinitrotoluene (6%) Nitrobenzene (trace)	Landfill	New Jersey (24.8%) W. Virginia (16.9%) Texas (25.1%) Mississippi (27.1%) Louisiana (6.1%)	166-205
K026	Stripping still tails from the production of methyl ethyl pyridines	Aqueous	Paraldehyde (75%) Pyridines Picolines Sulfuric acid Soluble acetates Phenol (very low conc.)	Water treatment lagoons, sludge to landfill	Indiana[b]	720 (1973)
K027[c]	Centrifuge and distillation residues from toluene diisocyanate production	Organic sludge	Polymer and tarlike matter (90%) Ferric chloride (6%) Waste isocyanates (3%) Toluene diisocyanate Toluene Diomine	Landfilled, drums in landfill	W. Virginia Louisiana Texas Ohio	6,000
K029	Waste from the product steam stripper in the production of 1,1,1-trichloroethane	Aqueous liquid	1,2-dichloroethane (20%) 1,1,1-trichloroethane (80%) Vinyl chloride (trace) Vinylidene chloride (trace) Chloroform (trace)	Pits, landfilled	Texas (40.9%) Louisiana (59.1%)	2,350
K030	Column bottoms or heavy ends from the combined production of trichloroethylene and perchloroethylene	Organic liquid	Hexachlorobutadiene (33.8%) 1,1,2,2-tetrachloroethane (23%) Hexachlorobenzene (20%) 1,1,1,2-tetrachloroethane (6.3%) Ethylene dichloride (4%) Perchloroethylene (4.5%) beta-trichloroethane (4.5%) Pentachloroethane (3%)	Incineration, landfilled	Texas (41.8%) Louisiana (45.0%) Kentucky (6.1%) Kansas (4.4%) California (3.4%)	15,000
K032	Wastewater treatment sludge from the production of chlordane	Sludge	Hexachlorocyclopentadiene (other chlorinated cyclic organics)	Landfill	Illinois	
K033	Wastewater and scrub water from the chlorination of cyclopentadiene in the production of chlordane	Aqueous liquid	Hexachlorocyclopentadiene (other chlorinated cyclic organics)	Holding pond, filtered deep-well injection	Illinois	
K034	Filter solids from filtration of hexachlorocyclopentadiene in production of chlordane	Solids	Hexachlorocyclopentadiene (other chlorinated cyclic organics)	Landfill	Illinois	

(continued)

APPENDIX A (continued)

Hazardous waste No.	Description of hazardous waste	Physical state	Components	Present disposal practice	Geographic distribution (% of total generation)	Annual generation, Mg
K036	Still bottoms from toluene reclamation distillation in production of disulfoton	Aqueous liquid	Toluene, phosphorothioic acid esters, and other organics Residues-Na	Landfill	Missouri	120
K037	Wastewater treatment sludges from the production of disulfoton	Sludge (aqueous)	Toluene, phosphorodithioate, phosphorothioic acid esters Residues-Na	Landfill	Missouri	50
K039	Filter cake from the filtration of diethylphosphorodithioic acid in the production of phorate	Solid	Phosphorodithioic and phosphorothioic acid esters	Lagoons, holding ponds, land treatment, incineration	Missouri	
K040	Wastewater treatment sludge from the production of phorate	Sludge (aqueous)	Phorate, formaldehyde, phosphorodithioic and phosphorothioic acid esters	Lagoons, holding ponds, land treatment, incineration	Missouri	
K042	Heavy ends or distillation residues from the distillation of tetrachlorobenzene in production of 2,4,5-T	Organic sludge	Several chlorinated benzenes (hexachlorobenzene, dichlorobenzene)	Landfill	Kansas Michigan Arkansas (others not determined)	
K043	2,6-dichlorophenol waste from the production of 2,4-D	Organic liquid	2,6-dichlorophenol 2,4-dichlorophenol 2,4,6-trichlorophenol Residuals-Na Catalyst	Landfills	Arkansas	
K051	API separator sludge from the petroleum refining industry	Sludge	H_2O (53%) Oil (23%) Solids (24%) (sand, silt) Residuals-chromium, lead	Land disposal		400-800
K048	Dissolved air flotation (DAF) float from the petroleum refining industry	Sludge	H_2O (82%) Oil (12.5%) Solids (5.5%) (sand, silt) Residuals-chromium, lead	Land treatment and incineration Lagoon Landfill		4,000-29,000
K049	Stop oil emulsion solids from the petroleum refining industry	Sludge	H_2O (40%) Oil (43%) Solids (12%) (sand, silt) Residuals-chromium, lead	Land treatment and incineration Lagoon Landfill		2,000-10,000
K050	Heat exchanger bundle cleaning sludge from the petroleum refining industry	Sludge	H_2O (53%) Oil (11%) Solids (36%) (sand, silt) Residuals-chromium	Land treatment and incineration Lagoon Landfill		

(continued)

APPENDIX A (continued)

Hazardous waste No.	Description of hazardous waste	Physical state	Components	Present disposal practice	Geographic distribution (% of total generation)	Annual generation, Mg
K093[d]	Distillation light ends from the production of phthalic anhydride from ortho-xylene	Organic liquid	Phthalic anhydride (<100%) Maleic anhydride (trace)	Incineration	California (9.2%) Texas (16%) New Jersey (16%) Illinois (35%) Puerto Rico (9.3%)	2,223
K094[d]	Distillation bottoms from the production of phthalic anhydride from ortho-xylene	Organic sludge	Phthalic anhydride (<100%)	Incineration	California (9.2%) Texas (16%) New Jersey (16%) Illinois (35%) Puerto Rico (9.3%)	408
K095	Distillation bottoms from the production of 1,1,1-trichloroethane	Organic liquid	1,1,2-trichloroethane (39.7%) 1,1,1,2-tetrachloroethane (27.3%) 1,1,2,2-tetrachloroethane (31.6%) Pentachloroethane (1.4%)		Louisiana (100%)	64,420
K096	Heavy ends from the heavy ends column from the production of 1,1,1-trichloroethane	Organic liquid	1,2-dichloroethane (16.9%) 1,1,1-trichloroethane (27.3%) 1,1,2-trichloroethane (27.3%) Tetrachloroethanes and hexachloroethanes (28.3%) 1,1-dichloroethane (trace)		Louisiana (100%)	5,350
K097	Vacuum stripper discharge from the chlordane chlorinator in the production of chlordane	Organic liquid	Chlordane Heptachlor Other chlorinated wastes	Holding pond	Illinois (100%)	Not known

[a]Sources: Mite Corporation. Guidance Manual for Evaluating Permit Applications for the Operation of Hazardous Waste Incinerator Units. 1981; U.S. Environmental Protection Agency. Background Document Resource Conservation and Recovery Act. Subtitle C - Identification and Listing of Hazardous Waste. Office of Solid Waste. November 14, 1980.

[b]1980 Directory of Chemical Producers.

[c]The constituents of this waste react violently with water.

[d]Incomplete combustion can form various phthalate esters.

Appendix B: Initial Listing of High-Temperature Processes

Industry	Product	Process	Furnace type	Number of units	Process temperature °C	Process temperature °F
Organic chemicals	Acetylene	Combustion	BASF converter	3 plants	649	1200
	Biphenyl		Tublar coil reactor		600-800	1112-1472
	Carbon tetrachloride	Chlorinalysis	Reactor	11 plants	650	1202
	Carbon disulfide		Fixed bed catalytic reactor	5 plants	550-651	1022-1202
	Benzene	Hydrodealkylation	Fixed bed catalytic reactor	32 plants (4830 Mg/y)	538-649	1000-1200
	Isoprene		Catalytic reactor		600	1112
	Methacrylonitrile		Fixed bed catalytic reactor		800-900	1472-1652
	Methylcyclohexanone		Fixed bed multitublar reactor		630	1166
	Formaldehyde		Converter		816	1500
Carbon black	Carbon black	Cracking (furnace process)	Steel reactor	30 plants	760	1400
	Carbon black	Thermal cracking (thermal process)	Refractory-lined cylindrical furnace	3 plants	1320-2100	2400-3800
Petrochemicals	Olefins (ethylene)	Thermal cracking	Tubular	35 plants	816-871	1500-1600
	Synthesis gas	Catalytic steam hydro. reforming	Steam reforming		651-1050 or 1400-1800	1202-1922 or 2552-3272
Explosives	Nitric acid	Ammonia oxidation process	Catalytic oxidizer	5 plants	800-960	1472-1760
Lime	Lime	Calcination	Rotary kiln	485 kilns (90% are rotary) (21 million tons/yr)	1090-1320	2000-2400
	Lime	Calcination	Vertical kiln		1200-1300	2192-2372
					970-980	1780-1796
Clay	Bricks	Firing	Tunnel kiln or batch kiln	500 plants	900-1430	1652-2606
Phosphate rock and basic fertilizer materials	Slurried deflorinated phosphate rock	Calcination	Rotary kiln or fluid bed reactor		1400-1540	2552-2804
	Ammonia synthesis gas	Steam reforming	Tubular		550-816	1022-1499
	Nitric acid	HNO_3 formation	Catalyst bed reactor	85 plants	800-900	1472-1652
	Phosphoric acid	Electric arc process	Electric arc furnace		1320	2400
	Elemental phosphorous	Furnace charge preparation		9 plants		
		• Nodulizing	Kiln		1200-1500	2192-2732
		• Calcining	Rotary kiln		1100-1320	2012-2400
		Electric furnace process (electrothermal reduction)	Electric arc furnace		1200	2912
	Phosphorous pentoxide	Phosphorous combustion (oxidation)	Combustion chamber	4 plants	200-2000	392-3632
	Sodium polyphosphate (Na_2HPO_4)	Calcination	Rotary kiln		300-900	572-1652
	Vitreous phosphates	Fusion	Rectangular furnace		700-1000	1292-1832
Sulfur, sulfur oxides, and sulfuric acid	Spent acid combustion	Combustion	Combustion furnace	167 plants (total sulfuric acid plants)	1000-1100	1832-2012
	Sulfuric acid		Air blown concentrator		677	1250

(continued)

APPENDIX B (continued)

Industry	Product	Process	Furnace type	Number of units	Process temperature °C	Process temperature °F
Iron and steel	Agglomerated product	Sintering	Sintering plant	71 agglomerating facilities	1300-1500	2372-2732
		Pelletizing	Kiln		1200-1370	2191-2499
	Coke production	Carbonization reduction)	Coke ovens	63 (44 Mg/yr)	1150	2102
	Pig iron	Reduction smelting	Blast furnace		1540	2800
		Smelting	Electric smelter		1480-1540	2700-2800
	Steel	Melting	Electric smelter	57 furnaces at 17 mills	1570-1640	2860-3000
		Retining (oxidation)	Open hearth (reverberatory)	86 furnaces (60 Mg/yr) 33 shops	1590-1640	2900-3000
		Rolling and shaping	Soaking pit furnace		1180-1340	2155-2440
	Iron	Melting	Cupola		1480-1540	2700-2800
	Steel	Melting	Acid open hearth	8	1640	3000
		Refining	Basic oxygen converter		1640-1820	3000-3300
		Refining	Rotary oxygen converter		1640-1820	3000-3300
			Electric induction		1090-1640	2000-3000
Nonferrous metal	Titanium tetrachloride	Chlorinated	Fluidized-bed furnace	12 plants	799-899	1470-1830
		Reduction (Kroll process)	Steel pot reactor	12 plants	849-899	1560-1650
	Titanium alloy	Melting	Cylindrical copper crucible	12 plants	1800	3272
	Titanium dioxide	Oxidation	Reaction chamber	17 plants (1.2 10^6 tons/yr)	999	1830
		Calcination	Direct-fired rotary kiln	12 plants (1.2 10^6 tons/yr)	999	1830
	Alumina	Sintering	Sinter plant	8 plants	1100-1200	2012-2190
		Calcination	Rotary kiln	6,650 x 10^3 metric Mg/yr 1979	1100-1200	2012-2190
	Electrode baking	Baking	Sunken pits		1200	2190
	Aluminum	Electrolytic reduction	Aluminum pots		951-999	1742-1830
		Refining or casting	Gas-fired holding furnace		699	1290
	Aluminum hydroxide	Calcination	Rotary kiln	6 plants (total AlO_3 producers	982	1800
	Copper	Smelting	Reverberatory	11 smelters	1000-1300	1832-2372
					1480-1930	2700-3500
		Roasting	Multiple hearth roaster	4 roasters (125-650 mt/d)	200-760	392-1400
			Fluidization roaster	4 roasters (700-1500 mt/day)	649-760	1200-1400
		Smelting	Electric furnace	3 smelters	999-1300	1830-2370
			Flash smelting	1 smelter	1100	2010
		Converting	Peirce-Smith converting	15 smelters	1180-1200	2150-2190
			Hoboken converting	1 smelter	1200	2190
	Copper (blister)	Smelting	Noranda furnace	1 furnace	1200	>2190
		Electric furnace slag treatment	Electric furnace	1 furnace	1200-1300	2190-2370
		Arsenic recovery	Godfrey roaster		649-699	1200-1290
	(anode)	Fire refining	Fire refining furnace		1100	2010
		Dore furnace	Sm. reverberatory furnace		1400	2550
		Sulfation roasting	Fluidization roaster	1 roaster	374-599	705-1110

(continued)

APPENDIX B (continued)

Industry	Product	Process	Furnace type	Number of units	Process temperature °C	Process temperature °F
Nonferrous metal (continued)	Lead	Sintering	Sinter plant	6 plants	799	1470
		Smelting	Blast furnace	6 plants	215-1220	419-2230
		Metal recovery	Slag fuming furance	6 plants	999-1200	1830-2190
			Dress reverberatory furnace		999-1200	1830-2190
		Reverberatory softening	Reverberatory (1300 Mg)		<700	<1290
		Kettle softening (melting)	Kettle		700	1290
		Antimony recovery	Not named		799-899	1470-1650
		Retorting (zinc recovery)	Faber-du-Favr type furnace		1260-1320	2300-2400
		Cupelling	Cupel furnace (2,850-11,300 Kg/charge)		1150	2100
	Zinc	Ferroalloy production (ferro-mangnesa)	Waelz kilns (rotary kiln)	1 plant	1300	2370
		Multiple-hearth roasting	Multiple-hearth roaster (50-60 tons/day)	2 smelters	688 949-977	1270 avg. 1740-1790 max.
		Suspension roasting	Suspension roaster (90-320 Mg/day)	1 plant at least 5 roasters	981	1796
		Fluidized bed roaster	Fluidized bed roaster		999-1050	1830-1920
		Sintering	Dwight-Lloyd-type sinter machine (220-500 tons/day)		1040	1900
		Vertical retorting (re-duction/violatilization)	Vertical silicon-carbide retorts		1300-1400	2372-2550
		Electric retorting (re-duction/violatilization)	Resistance-type electric furnace (90 mt/day)		899-1400	1650-2550
		Oxidizing	Grate-type furnace		999-1950	1830-3540
			Rotary or Waelz kiln		1300	2370
			Electrothermic furnace		899-1400	1650-2550
		Cadmium purification and casting (melting)	Horizontal retort		788-910	1450-1670
	Magnesium	Smelting	Crucible or pot-type	4 plants	816	1500
		Calcination	Calciner		704-1650	1300-3000
	Magnesium chloride	Electrolysis	Dow electrolytic cell		710	1310
		Ferrosilicon	Not named		1170	2140
Carbon	Coke	Calcinating	Horizontal and vertical rotary calciner		1200-1400	2192-2550
		Baking	Pit furnace or ring furnace		799-999	1470-1830
	Graphite	Graphitization	Electric furnace	28 plants (total graphite)	3000	5430
	Activated carbon	Oxidation activation	Retort		802-982	1475-1800
Miscellaneous chemicals	Fused sodium chloride	Electrolysis	Electrolysis cell		804	1480
	Calcium carbide		Carbide furnace		2000-2200	3630-3990
	Sodium silicate	Melting	Regenerative tank furance	42 plants	1200-1430	2200-2600
	Mineral wool		Cupola		1650	3000

(continued)

APPENDIX B (continued)

Industry	Product	Process	Furnace type	Number of units	Process temperature °C	Process temperature °F
Agricultural	Alfalfa	Dehydrating	Dehydrator		982-1090	1800-2000
Cement	Cement	Calcining Drying	Rotary kiln Rotary kiln	158 plants w/342 kilns Avg. plant 536,000 TPY	682-1650 599	1260-3000 1110
Glass	Glass	Melting	Glass melter (50-300 T/f)	9 plants (14,060 metric T/day) (338 plants)	1370-1710 593-871 927	2500-3100 1100-1600 1700
	Ceramic Coatings	Smelting	Rotary Hearth Crucible			
Sewage sludge incinerator	Multiple hearth			19 (wastewater treatment 1977)	316-649	600-1200
	Fluid bed			2 (wastewater treatment 1977)	649-816	1200-1500
Municipal solid waste combustors	None (steam)	Incineration	Water wall		760-1320	1400-2400
Industrial sludge combustors	None	Incineration Incineration Incineration	Liquid waste combustors Fluidized bed incinerator Multiple hearth incinerator	125 installations (17,000 T/day)	649-1650 760-871 760-982	1200-3000 1400-1600 1400-1800
		Incineration	Rotary kiln incinerator		871-1650	1600-3000

Other Noyes Publications

COSTS FOR HAZARDOUS WASTE INCINERATION
Capital, Operation and Maintenance, Retrofit

by

R.J. McCormick, R.J. DeRosier, K. Lim, R. Larkin, H. Lips
Acurex Corporation

Pollution Technology Review No. 123

Relationships between capital, and operation and maintenance (O&M) costs for hazardous waste incineration and the various waste-specific, design-specific, and operational factors that affect these costs are discussed in this book. These cost relationships were to be designed so that total capital investment and unit (dollars per pound, etc.) O&M cost estimates could be calculated for a variety of waste compositions, different incineration system designs and configurations, and a wide range of system operating conditions and performance requirements.

Also covered is a methodology that can be used to estimate the costs of retrofitting/upgrading various components of existing hazardous waste incineration facilities, to comply with RCRA performance requirements.

Based on the preliminary design specifications, equipment costs are estimated using a series of empirical relationships between cost and capacity, materials of construction, or other relevant design features. Estimated costs for equipment installation, indirects, and contingency are then combined with the equipment costs to determine total capital investment.

A condensed table of contents listing **part titles, chapter titles, and selected subtitles** is given below.

ISBN 0-8155-1047-0 (1985)

274 pages